Environmental Applications of Magnetic Sorbents

Online at: https://doi.org/10.1088/978-0-7503-5909-2

Environmental Applications of Magnetic Sorbents

Edited by
Kingsley Eghonghon Ukhurebor
Department of Physics, Edo State University, Uzairue, PMB 04 Auchi, Edo State, Nigeria

Uyiosa Osagie Aigbe
Department of Mathematics and Physics, Cape Peninsula University of Technology, Cape Town, South Africa

IOP Publishing, Bristol, UK

ISBN 978-0-7503-5909-2 (ebook)
ISBN 978-0-7503-5907-8 (print)
ISBN 978-0-7503-5910-8 (myPrint)
ISBN 978-0-7503-5908-5 (mobi)

DOI 10.1088/978-0-7503-5909-2

Version: 20240501

IOP ebooks

British Library Cataloguing-in-Publication Data: A catalogue record for this book is available from the British Library.

Published by IOP Publishing, wholly owned by The Institute of Physics, London

IOP Publishing, No.2 The Distillery, Glassfields, Avon Street, Bristol, BS2 0GR, UK

US Office: IOP Publishing, Inc., 190 North Independence Mall West, Suite 601, Philadelphia, PA 19106, USA

To our families: past, present, and future.

Contents

Preface

Given the growing crisis caused by environmental contamination, primarily from industrial and domestic activities, as well as the significant amount of these contaminants released into the environment by both natural and anthropogenic activities (particularly from untreated water, which subsequently finds its way into the soil and disrupts atmospheric, aquatic, and terrestrial systems), there is an urgent need to mitigate these environmental contaminations. As a result, this book aims to provide a foundation for the development, utilization, and applications of contaminant adsorption using magnetic biosorbents for environmental sustainability and safety.

This book comprises 10 chapters that particularly focus on advanced research into the use of timely and beneficial creative techniques in the synthesis of novel magnetic biosorbents using facile synthesis methods and the characterization and the extensive applications of synthetic magnetic biosorbents in bioremediation and decontamination. The application of magnetic sorbents in seed germination and the use of magnetic sorbents to improve photovoltaic performance are included. It also discusses highlights of the challenges and perspectives of magnetic sorbent applications for environmental sustainability. Hence, this book will serve a reference book and will be of great assistance to students, professionals, practitioners, scientists, researchers, and academicians in various research domains, particularly those in the environmental and materials science domains.

Acknowledgements

First and foremost, we are most grateful to Almighty God, our creator, who kept us alive and gave us all we needed to make this book a success. We owe Him everything. We have always been and will forever be grateful to Him.

Our profound gratitude goes to our respective institutions' managements, who have always been of great assistance in our research and academic development. We are also grateful to all the chapter contributors as well as the authors and publishers whose publications were used as a basis for writing this book, without whom this book would not have been possible.

We thank our families, mentors, colleagues, friends, and all who contributed and are still contributing in one way or another to the success of our academic and research activities as well as our other endeavors. May God bless you all richly!

Editor biographies

Kingsley Eghonghon Ukhurebor

Kingsley Eghonghon Ukhurebor is a lecturer/researcher and the present acting head of the Department of Physics at Edo State University Uzairue, Nigeria, and a research fellow at the West African Science Service Centre on Climate Change and Adapted Land Use (WASCAL), Competence Centre, Ouagadougou, Burkina Faso, a Climate Institute sponsored by the Federal Ministry of Education and Research, Germany. He received his PhD in physics electronics from the University of Benin, Benin City, Nigeria. He is a member of several learned academic organizations, such as the Nigerian Young Academy (NYA), etc. His research interests are in applied physics, climate physics, environmental physics, telecommunication physics, and material science (nanotechnology). He serves as an editor and reviewer for several reputable journals and publishers, such as Springer Nature, Elsevier, the Royal Society of Chemistry (RSC), the Institute of Physics (IOP), Taylor and Francis, Wiley, the IEEE, Frontiers, Hindawi, etc. He has authored or co-authored several publications for these reputable journals and publishers. He is currently ranked among the top 50 authors in Nigeria by Scopus.

Uyiosa Osagie Aigbe

Uyiosa Osagie Aigbe is a research fellow at the Department of Mathematics and Physics, Faculty of Applied Science, Cape Peninsula University of Technology, Cape Town, South Africa. He obtained his PhD degree in physics from the prestigious University of South Africa, Pretoria, South Africa. He is currently a member of several learned academic organizations. His research interests are in applied physics, nanotechnology, fluid dynamics, water purification processes, image processing, environmental physics, and materials science. He has also served as a reviewer for numerous highly regarded journals. He has authored and co-authored several research publications.

List of contributors

Kenneth Adama
Department of Chemical Engineering, Edo State University, Uzairue, PMB 04 Auchi, Edo State, Nigeria

Uyiosa Aigbe
Department of Mathematics and Physics, Cape Peninsula University of Technology, Cape Town, South Africa

Bamikole Akinsehinde
Department of Computer Science, Aberystwyth University, Penglais Campus, West Wales, United Kingdom

Stephanie Akpeji
Department of Biological Sciences (Microbiology), Novena University, Ogume, Kwale, Delta State, Nigeria

Sefiu Olaitan Amusat
Department of Chemistry, College of Science, Engineering and Technology (CSET), University of South Africa, Florida, South Africa

Harrison Ifeanyichukwu Atagana
Department of Biotechnology and Chemistry, Vaal University of Technology, Vanderbijlpark, South Africa

Olusoji Ayeleso
Department of Environmental Resource Management, Nasarawa State University, Nasarawa State, Nigeria

Vincent Balogun
Department of Mechanical Engineering, Edo State University Uzairue, Auchi, Edo State, Nigeria

Simiso Dube
Department of Chemistry, College of Science, Engineering and Technology (CSET), University of South Africa, Florida, South Africa

Ahmed El Nemr
Environment Division, National Institute of Oceanography and Fisheries (NIOF), Kayet Bey, Elanfoushy, Alexandria, Egypt

Joseph Emegha
Department of Energy and Petroleum Studies, Novena University, Ogume, Kwale, Delta State, Nigeria

Kokolo Etiowo
Department of Chemistry, University of Cross River State, Calabar, Cross River State, Nigeria

Lucky Evbuomwan
Department of Microbiology, Edo State University, Uzairue, PMB 04 Auchi, Edo State, Nigeria

David Idisi
Fort Hare Institute of Technology, University of Fort Hare, Alice, South Africa

Udoka Bessie Igue
Department of Chemical Sciences (Biochemistry), Novena University, Ogume, Kwale, Delta State, Nigeria

Atala Jena
Department of Neurosurgery, Brigham and Women's Hospital, Harvard Medical School, Boston, MA, USA

Grace Jokthan
Africa Centre of Excellence on Technology Enhanced Learning (ACETEL), National Open University of Nigeria, Abuja, Nigeria

Temesgen Girma Kebede
Department of Chemistry, College of Science, Engineering and Technology (CSET), University of South Africa, Florida, South Africa

Rout Kerry
Department of Biotechnology, Utkal University, Vani Vihar, Bhubaneswar, Odisha, India

Kevin Frank Mearns
Department of Environmental Sciences, University of South Africa, Florida, Roodepoort, South Africa

Edson Meyer
Fort Hare Institute of Technology, University of Fort Hare, Alice, South Africa

Sekomeng Johannes Modise
Institute for Nanotechnology and Water Sustainability, College of Science, Engineering and Technology, University of South Africa, Florida Campus, Roodepoort, South Africa

Francis Montalbo
College of Informatics and Computing Sciences, Batangas State University, Batangas, Philippines

Faruk Anka Muhammad
Department of Energy and Petroleum Studies, Novena University, Ogume, Kwale, Delta State, Nigeria

Mathew Muzi Nindi
Institute for Nanotechnology and Water Sustainability (iNanoWS), College of Science, Engineering and Technology (CSET), University of South Africa, Florida Park, South Africa

Kingsley Obodo
ICTP-East African Institute for Fundamental Research, Kigali, Rwanda
and
Center for Space Research, North-West University, Potchefstroom, South Africa, 2531
and
National Institute of Theoretical and Computational Sciences, Johannesburg, 2000, South Africa

Odunayo Ojo
Department of Energy and Petroleum Studies, Novena University, Ogume, Kwale, Delta State, Nigeria

Benedict Okundaye
Centre for Urban Design, Architecture and Sustainability (CUDAS), Department of Architecture and 3D Design, School of Arts and Humanities, University of Huddersfield, United Kingdom

Timothy Imanobe Oliomogbe
Department of Energy and Petroleum Studies, Novena University, Ogume, Kwale, Delta State, Nigeria

ThankGod Onuoha
Department of Biological Sciences, Novena University, Ogume, Nigeria

Ikenna Onyeachu
Department of Chemistry, Edo State University, Uzairue, PMB 04 Auchi, Edo State, Nigeria

Adelaja Otolorin Osibote
Department of Mathematics and Physics, Faculty of Applied Sciences, Cape Peninsula University of Technology, Cape Town, South Africa

Cyril Chinedu Otali
Department of Biological Sciences, Novena University, Ogume, Kwale, Delta State, Nigeria

Oluseyi Salami
Department of Chemistry, College of Science, Engineering and Technology (CSET), University of South Africa, The Science Campus, Florida Park, Florida, South Africa

Silas Soo Tyokighir
Department of Electrical and Electronic Engineering, College of Engineering, Joseph Sarwuan Tarka University, Makurdi, Nigeria

Onyedikachi Ubani
Department of Environmental Sciences, University of South Africa, Florida, Roodepoort, South Africa

Ikechukwu Ukaga
Department of Chemistry, University of Calabar, Calabar, Cross River State, Nigeria

Kingsley Eghonghon Ukhurebor
Department of Physics, Edo State University, Uzairue, PMB 04 Auchi, Edo State, Nigeria

IOP Publishing

Environmental Applications of Magnetic Sorbents

Kingsley Eghonghon Ukhurebor and Uyiosa Osagie Aigbe

Chapter 1

An introduction to the contemporary applications of magnetic sorbents

Kingsley Eghonghon Ukhurebor, Uyiosa Osagie Aigbe, Joseph Onyeka Emegha, Kenneth Kennedy Adama, Lucky Evbuomwan, Silas Soo Tyokighir, Bamikole Olaleye Akinsehinde, Benedict Okundaye, Olusoji Anthony Ayeleso, Grace Jokthan and Ahmed El Nemr

When compared to conventional methods, the use of magnetic nanomaterials (MNMs) in pollution analysis holds great promise and presents a number of benefits. Owing to the distinctive capacity of MNMs for selective adsorption and effortless separation resulting from surface modification, stability, cost-effectiveness, availability, and biodegradability, they have significantly enhanced the extraction of various analytes from agricultural products. Conversely, conventional techniques use many organic solvents that are harmful to the environment and require longer extraction processes. Hence, the contemporary applications of MNMs and their modifications in the removal of contaminants are the main focus of this introductory chapter.

1.1 Introduction

Environmental or ecological pollutants such as mycotoxins, pesticides (agricultural materials), and pharmaceuticals (medical materials), as well as potentially detrimental materials (PDMs), may infect agricultural products, feed, and water samples, endangering the health of those who consume or come in contact with the infected items [1–4]. There are several methods by which these toxins, sometimes known as pollutants, can enter the ecosystem [5–11]. In both humans and animals, pharmaceutical substances are frequently partly digested before being eliminated through urine or feces. These substances eventually make their way to wastewater treatment facilities, where they are discharged into receiving water bodies after only partial elimination [12–18]. In addition to medicines, several other kinds of pollutants, including personal care products, chemicals or runoff from industry, micro/nano-

doi:10.1088/978-0-7503-5909-2ch1

plastics, and many more, are generally too concentrated for wastewater treatment facilities to fully remove [19–23].

Effluents and wastewater from agricultural activities [24–27], stormwater from industrial and semi-industrial regions [28], wastewater from healthcare facilities [20, 23, 29], and other indirect sources are additional routes by which pollutants and toxins can enter the environment. These contaminants can ultimately contaminate agricultural and drinking water supplies, making their way into a variety of agricultural products if they get into surface or groundwater resources. Research into detecting, monitoring, and developing technology for removing these pollutants and contaminants from agricultural products and the environment, as well as the implementation of laws to address the problem, has focused heavily on the presence of these pollutants and contaminants [20, 22, 23, 30–35].

The issues related to pollutants and contaminants, which are of growing concern, are made worse by the enormous number of substances and chemicals that are now utilized on a daily basis, both in terms of quantity and kind, as well as the annual development of new materials and chemicals. For instance, according to some statistics, the European Union (EU) has recorded more than 1.0×10^5 chemicals for use by persons, businesses, and industry, and around 400 million tons of these chemicals are generated worldwide [35]. More than 70% of the chemicals generated in the EU that pose a risk to human health and the environment are classified as compounds of major environmental concern, according to data released by EUROSTAT [36].

It is crucial to routinely check the levels of these toxic materials (pollutants) and PDMs in agricultural products and the surrounding environment due to their ubiquity and detrimental effects on humans, animals, and the ecosystem. Only then can suitable mitigation measures be taken to reduce their incidence and related negative impacts. Analysis is essential in this context, which is why there is a continuing demand for better and more effective analytical techniques. Conventional techniques for extracting and analyzing these contaminants are sometimes costly, labor-intensive, and ineffective and need many hazardous organic solvents. Thus, more affordable, speedier, more efficient, and environmentally friendly methods are required [37]. In light of their apparent promise, MNMs are considered to be a potential means of extracting these contaminants for analysis. The extraction of these contaminants utilizing MNMs is briefly discussed in this introductory chapter, which has a particular emphasis on mycotoxins, pesticides, and pharmaceuticals. This chapter also discusses the method of operation of these removal/extraction ingredients (materials) and their contemporary benefits/applications in many technological and scientific domains.

The chronic presence of contaminants in agricultural products, feed, water, and the environment—such as mycotoxins, pesticides, pharmaceuticals, and PDMs—now poses a serious threat to worldwide public health and has resulted in calls for further measures to reduce their prevalence. Regular and effective analysis to ascertain the incidence and levels of these pollutants and enable the implementation of the required proactive measures is one method of effectively managing and controlling these pollutants. Effective extraction is a crucial and frequently necessary

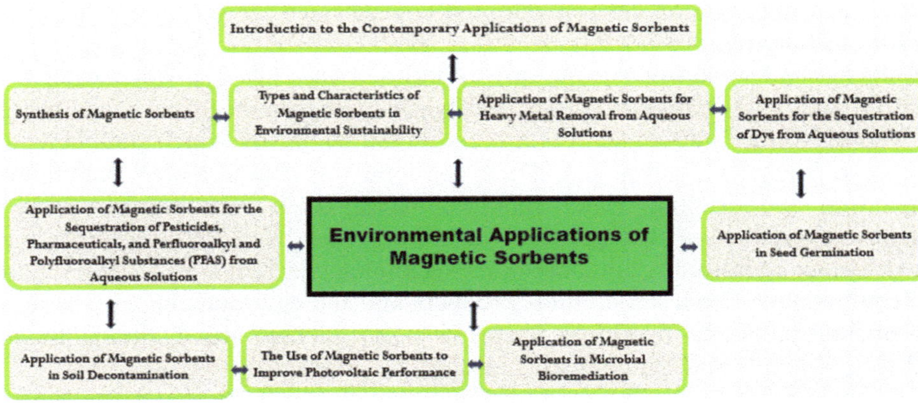

Figure 1.1. An illustration of the various chapters included in this book.

stage in the investigation of pollutants. There is an ongoing search for more efficient analytical techniques to identify and measure these contaminants because many of their traditional extraction methods have some sort of limitation. It has been determined that the use of MNMs as adsorbents to extract and remove mycotoxins, pesticides, and pharmaceuticals, and other environmental contaminants (PDMs) of concern is a promising, quick, efficient, and ecologically friendly method. With an emphasis on mycotoxins, pesticides, pharmaceuticals, and PDMs, we thoroughly examine the potential and uses of MNMs for the extraction of diverse environmental pollutants in the following parts of this chapter.

Hence, this chapter serves as the introductory chapter of this book (which is centered on the environmental applications of magnetic sorbents (MSs)) and introduces the contemporary applications of MNPs and their modifications in the removal of contaminants. Furthermore, this book attempts to discuss the synthesis of MSs in chapter 2. Chapter 3 deals with the types and characteristics of MSs in terms of environmental sustainability, while chapter 4 examines the application of MSs for heavy metal (HM) removal from aqueous solutions, and chapter 5 discusses the application of MSs for the sequestration of dyes from aqueous solutions. The application of MSs for the sequestration of pesticides, pharmaceuticals, and per- and polyfluoroalkyl substances (PFASs) from aqueous solutions is discussed in chapter 6, and the application of MSs in soil decontamination is discussed in chapter 7. Chapter 8 covers the application of MSs in seed germination, and chapter 9 examines the use of MSs to improve photovoltaic performance. Chapter 10 deals with the application of MSs in microbial bioremediation. Figure 1.1 illustratively shows the various chapters contained in this book.

1.2 Nanomaterials

Recently, nanomaterials (NMs) have been attracting a lot of interest in both research and industrial applications. They make it possible for scientists, engineers, chemists, and medical professionals to collaborate at the cellular and molecular levels to create effective advancements in the life sciences and healthcare, in addition to other technological advances. Public (community) health, environmental (ecological) safety (protection), and sustainability are achieved through the production, synthesis, and combination of nanostructure constituents (particles or materials) and NMs for the purpose of eliminating environmental or ecological toxins and PDMs. As a result of their special structure, very tiny size, useful properties, and large surface area—all of which enable the effective preconcentration and extraction of contaminants from agricultural products—they are regarded as excellent adsorbents [1, 2, 38, 39]. There are other forms of NMs as well, such as MNMs, metal NMs, carbon NMs, and quantum dots, but MNMs are the focus of this book.

1.3 Magnetic nanomaterials

MNMs are NMs that may be controlled by applying magnetic fields. Their particular advantages, including affordability, enhanced removal, precision (accuracy), selectivity, and general speed (time) of removal, together with the use of their chemical, electrical, magnetic, and thermal characteristics in a variety of investigative or analytical procedures, including removal, preconcentration, cleanup (sample treatment), detection, and chromatographic procedures, support the development of new approaches to analysis or the improvement of existing ones. Different magnetic minerals, such as iron (Fe)-based minerals (including magnetite (Fe_3O_4) and maghemite (γ-Fe_2O_3)), cobalt (Co), and nickel (Ni), together with their many derivative composites or compounds, are used in the manufacture of these MNMs [40–42]. Fe oxides, including Fe_2O_3 and Fe_3O_4, and their related ferrite derivatives, such as $CoFe_2O_4$ and $MnFe_2O_4$, are the materials most often utilized in the manufacture of MNMs. This is because, in comparison to other metals and metallic alloys such as FePt, Mn_3O_4, Ni, and Co, they are comparatively easy to prepare, have large magnetic moments, are chemically stable, and work well with a variety of biological systems [43]. It is important to remember that, despite their apparent effectiveness and efficiency, MNM treatments are often directed toward specific molecules or particles. MNMs can be functionalized or modified using various chemical groups/categories to accomplish targeted analyte extraction or selective interaction. Since their surface chemistry is often advantageous for the extraction of particular groups of chemicals, this is a significant benefit of MNMs. It is well known that Fe oxides break down organic molecules, dissolve in acidic environments, and readily react with oxygen in the atmosphere. Consequently, it is imperative to apply various protective layers, including silica, polymers, carbon NMs, and noble metals, to enhance their stability and provide novel surface characteristics and functions [44, 45]. To improve the materials' analytical applicability, the surface of the generated MNMs may be functionalized with multiple functional groups in a straightforward process that also imparts diverse physicochemical features [46]. Tables 1.1 and 1.2 list some different kinds of MNMs and magnetic alloy NMs used for the removal of contaminants.

Table 1.1. Different kinds of MNMs used for the removal of contaminants.

Types of MNMs	Descriptions and applications
γ-Fe_2O_3 and Fe_3O_4	The two main families of Fe oxide are Fe_3O_4 and γ-Fe_2O_3; these occur naturally and have appealing magnetic properties that can be useful in a variety of applications. Except for the fact that the Fe cations in Fe_3O_4 consist of the cations Fe^{2+} and Fe^{3+} and the Fe cation in γ-Fe_2O_3 is in a trivalent state, they are both soft ferrimagnetic minerals with comparable configurations [47]. Typical MNM components utilized in magnetic solid phase extraction (MSPE) are Fe_3O_4 and γ-Fe_2O_3. Numerous techniques, including the hydrothermal method, the solvothermal method, flow injection synthesis, the oxidation of MNMs, coprecipitation, flame spray pyrolysis, sol–gel synthesis, the thermal decomposition of organic precursors at high temperatures, and microemulsion, have been reported in the scientific and technological literature for the synthesis of these Fe oxides [47, 48]. As a result of its large surface area, easy synthesis, safety, efficiency, affordability, simplicity of separation and recovery, capacity to adsorb, and super-paramagnetic properties, γ-Fe_2O_3 is an excellent absorbent for the removal of HMs [49–51]. γ-Fe_2O_3 was effectively utilized by Tuutijärvi *et al* [52] to remove As (V) from water. It has also been stated that γ-Fe_2O_3 can remove Cr (VI) from tainted water [53]. Narimani-Sabegh and Noroozian [51] removed antimony (Sb) from aqueous solutions/media (such as soft drinks, bottled alcohol, water, non-alcoholic beers, and orange drinks) and synthesized γ-Fe_2O_3-based NMs from lepidocrocite by calcination. Devatha and Shivani's research [54] described a unique use of bacteria (*Bacillus subtilis*) in conjunction with γ-Fe_2O_3 NMs for the extraction or removal of cadmium (II) ions from aqueous solutions or media; their rate of recovery was about 76.40%. Rajput *et al*'s study [55] employed γ-Fe_2O_3 NMs synthesized by flame spray pyrolysis to extract Cu (II) and Pb (II) ions from liquid (water). Another potential absorbent used in a variety of scientific domains is Fe_3O_4, which is particularly useful for removing contaminants from the environment. For example, in order to extract parathion from food commodities (tomato, carrot, rice, orange, and lettuce), Piovesan *et al* [56] produced Fe_3O_4 NMs covered with chitosan (Fe_3O_4@CS). In the same way, González-Jartín *et al* [57] also produced these NMs with the intention of eliminating mycotoxins from liquid food items (beverages). Furthermore, Turan and Sahin [58] reported the use of molecularly imprinted polymer (MIP) MNMs (Fe_3O_4@EGDMA) for the removal of Ochratoxin A (OTA) from grape juice.
Neodymium (NDM) MNMs	NDM, which is part of the lanthanide group of rare earth metals, has garnered interest in a number of studies because of its potent magnetic

(*Continued*)

Table 1.1. (*Continued*)

Types of MNMs	Descriptions and applications
	properties—possibly among the strongest (most robust) known to humanity. It is now used in the production of permanent magnets used in electric motors, computer hard drive spindles, and wind turbines [59]. Several techniques, such as gel combustion, the hydrothermal process, solution coprecipitation, hydrogen plasma–metal interaction, thermal breakdown, and microemulsion, are used to create NDM-based MNMs [60]. In order to remove sunset yellow from an aqueous solution, Ahmadi *et al* (2020) produced $NdCl_3$ (NDM (III) chloride) integrated with ordered mesoporous carbon (OMC). Identical Nd_2O_3 NMs were produced and explored for acid dye extraction from an aqueous medium in another study [61]. To determine the presence of the anticancer medication raloxifene in organic samples, Chen *et al* [24] produced NDM sesquioxide covered with graphene oxide NMs and modified/improved with a glassy carbon electrode (GCE).

Table 1.2. Different kinds of magnetic alloy NMs used for the removal of contaminants.

Types of magnetic alloy NMs	Descriptions and applications
FeNi alloy MNMs	FeNi alloy MNMs are being researched extensively for a range of applications due to their appealing magnetic properties. $FeNi_3$ has good thermal stability, permeability, and substantial saturation magnetization [62]. Numerous techniques, including the hydrothermal reduction technique, the sol–gel method, spray pyrolysis, coordinated coprecipitation, and the chemical reduction technique, are used to create these alloy MNMs [63, 64]. Because of the simplicity of their synthesis and use, their ability to extract various organic composites in different extraction settings or circumstances, and their ease of separation by an external magnet, these MNMs have gained popularity among analytical scientists [65]. An $FeNi_3@SiO_2$ MNM catalyst for tetracycline degradation in a neutral environment was effectively produced by Khodadadi *et al* [66]. $FeNi_3@SiO_2$ MNMs were employed by Farooghi *et al* [67] to remove lead from an aqueous solution. A study by Nasseh *et al* [68] described the use of $FeNi_3@SiO_2$ MNMs for the extraction of metronidazole in a neutral environment.
FeCo alloy MNMs	MNMs made of FeCo alloys are soft ferromagnetic NMs that possess unique properties including high permeability, a high Curie temperature, a large saturation magnetization, low coercivity, high

	anisotropic energy, and a high anisotropic constant [68–73]. These magnetic alloy NMs find use in several technical applications, such as medication delivery, the treatment of hyperthermia, exchange-coupled NM magnets, magnetic recording media, and microwave devices [74, 75]. Some of the methods used to create FeCo magnetic alloy NMs include mechanical ball milling, interfacial diffusion, chemical vapor deposition, chemical coprecipitation, pulse laser ablation deposition (PLAD), pyrolysis, and the reductive breakdown of iron (III) acetylacetonate and cobalt (II) acetylacetonate [76–78].
FePt alloy MNMs	Researchers have become more interested in FePt hard MNMs because of their exceptional magnetic properties, which include high chemical stability, large saturation magnetization, magnetic imaging, high magneto-crystalline anisotropy, and strong magnetocaloric effects [79–81]. Biomedicine, electrocatalysis, large permanent magnets, magnetic data storage, magnetic recording media, and nanobiotechnology (NBTech) studies are among the many fields in which FePt magnetic materials are widely used [82–85]. The synthesis of these alloys involves the thermal breakdown of Fe pentacarbonyl ($Fe(Co)_5$), the reduction of Fe salts and Pt(acetylacetonate)$_2$, and the polyol process (the reduction of platinum acetylacetonate in a mixture of surfactants and polyol) [79].
FePd alloy MNMs	Due to their significant magneto-crystalline anisotropic energy, FePd alloy MNMs are hard, rigid, and strong MNMs [85, 86]. These alloys are created by a variety of techniques, including microwave irradiation, chemical synthesis based on modified FePt synthesis procedures, modified polyol procedures, and epitaxial growth electron beam deposition [87]. A Pd-rich FePd alloy material serves as a catalyst and a very effective hydrogen absorber. These alloy MNMs are used in healthcare applications [86] as well as ultrahigh-density magnetic recording media [88]. Researchers are drawn to the $Fe_{70}Pd_{30}$ material because of its martensitic conversion effect and magnetic shape memory (MSM). However, it is one of several systems. There have been recent reports of several forms of these nanosized materials, including nanotubes, nanospheres, nanohelices, and nanorods [86].

1.4 The positive and negative aspects of magnetic nanomaterials

MNMs are preferable to non-MNMs because they have a number of benefits. These include the ability to use various sorbents with varying adsorption effectiveness, a high surface area, potential functionalization or surface modification for enhanced specificity, decent dispersibility in solvents (enabled by their size), and the simplicity of using an external magnet to separate them from multifaceted matrices without the requirement for centrifugation or filtration steps [89–92]. Additionally, these materials or resources are readily recyclable and reusable—typically with the proper cleaning. Analytes adsorbed using methods such as sonication [89, 92] need fewer organic solvents and are frequently more affordable [89, 91–93].

MNMs have negative aspects despite their numerous benefits, one of which is that they tend to agglomerate and aggregate in media, which results in a decrease in their inherent magnetic (superparamagnetic or ferromagnetic) properties. This issue is resolved by functionalizing or altering the MNMs using various substances, including aptamers, metal–organic frameworks or backgrounds, silica oxide, graphene oxide, carbon nanotubes (CNTs), MIPs, covalent organic backgrounds, and immunoassay, etc [89, 91–94]. Because of their conductivity in comparison to unaltered MNMs, such treatments can accelerate electron transport rates [93]. For instance, Xu *et al* [95] used CNTs to modify MNMs in order to recover contaminants from eggs. Abdulhussein *et al* [96] reported MNMs modified with MIPs that were used to analyze thiamethoxam and thiacloprid in honey.

The thermally unstable and steady-state constituents of MNMs can either entirely or partly break down during the procedure of desorption at elevated temperatures, which reduces the accuracy and precision of the analysis. Another negative aspect of MNMs is that, when exposed to air, they can sometimes be easily oxidized or produce hydrated oxides [89]. Indeed, any Fe_3O_4 that may be present can degrade at $pH < 4$, which opens the door for chelates to form from the free ions and the target analytes. On the other hand, when $pH > 9$, the binding of -OH groups gives Fe_3O_4 particles a negative charge, which results in the creation of electrostatic repulsion between the anionic settings of the target analytes and the adsorbent forms [92].

Furthermore, compared to bigger particles, nanoadsorbents may be more challenging to extract from complex matrices because of their smaller size, especially if the sorbent is nonmagnetic. For the purpose of separating the analyte from the sample, bigger constituent parts can be readily filtered and, if they are denser compared to the samples, they can also be centrifuged. In this sense, MNMs are useful because, even at their nanosizes, they are readily removed from the material by applying an external magnet [37, 45].

1.5 The applications of magnetic sorbents for the removal of contaminants

The persistent presence of contaminants in agricultural products, drinking water, and the environment—such as mycotoxins, pesticides, and pharmaceuticals as well as PDMs—now poses a serious threat to global public health and has led to calls for further measures to reduce their prevalence [37, 45]. Regular and effective analysis to ascertain the incidence and levels of these pollutants and enable the implementation of the required proactive measures is one method of effectively managing and controlling these pollutants. Effective extraction is a crucial and frequently necessary stage in the investigation of pollutants. There is a constant search for more efficient analytical techniques to identify and measure these contaminants because many of the traditional extraction methods used have some sort of limitation. It has been determined that using NMs as adsorbents to extract and remove mycotoxins,

Figure 1.2. The main sources of ECs [97].

pesticides, pharmaceuticals, and other environmental contaminants (PDMs) of concern is a promising, quick, efficient, and ecologically friendly method [37, 45]. With an emphasis on mycotoxins, pesticides, and pharmaceuticals, this chapter attempts to thoroughly examine the potential and uses of MNMs for the extraction of diverse environmental pollutants in the following sections.

Emerging contaminants (ECs) and PDMs are becoming more dangerous [97]. The primary ECs are organic pollutants, which include those found in chemical fertilizers, medications, personal care items, hormones, polymer compounds, agricultural products, wood preservatives, cleaning products, surfactants, antiseptics, and fire retardants, among other mineral and organic compounds [98]. These compounds are often found in both the industrial sector and the natural wastewater streams that are produced by human activity. The main sources of ECs are shown in figure 1.2, which is adapted from Rathi and Kumar [97]. Most ECs are detrimental to aquatic life as well as humans, even at low doses. Cost-effective secondary (additional) methods are required, since conventional basic and additional processing facilities can scarcely remove or eradicate such harmful chemicals adequately.

Rathi and Kumar [97] diagrammatically illustrated the main EC components (figure 1.3). Advanced oxidation procedures, membrane filtration, ion exchange, adsorption, coagulation, and sedimentation are a few of the techniques used to remove ECs [97, 98]. Since it is easy to use, inexpensive to deploy, and very effective, adsorption is a great method for getting rid of ECs. Studies have shown that ECs may be removed from water and wastewater using activated carbons, modified biochars, nanoadsorbents, composite adsorbents, and other adsorbents such as MNMs [97, 98].

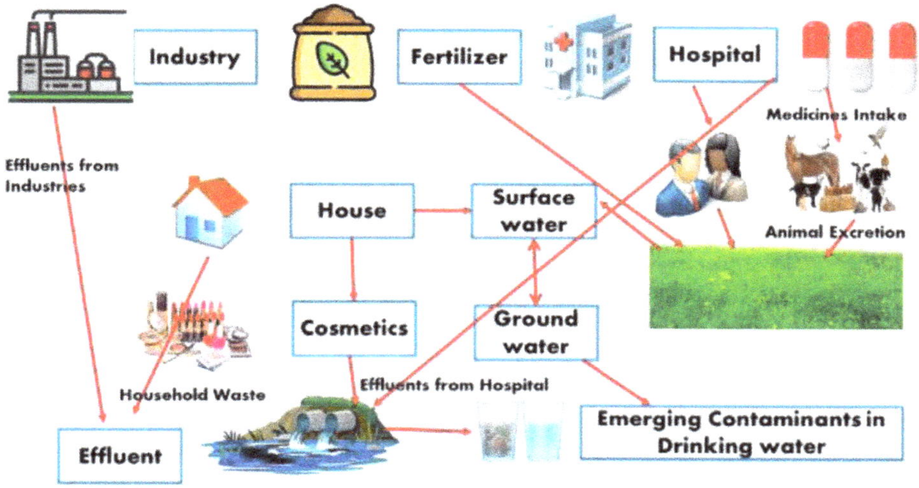

Figure 1.3. The main components of ECs. Reprinted from [97], Copyright (2021), with permission from Elsevier.

1.6 The applications of magnetic nanomaterials for the removal of pollutants from agricultural products

As a result of their special properties, which include a huge (exceptional) numbers of surface sites, a decent charge transfer capacity, and ease of separation with an external magnet, MNMs are of significant interest in agricultural product analysis [89, 91–93]. Researchers are currently beginning to take an interest in the application of MNMs in agricultural products, which is growing significantly. To facilitate the extraction of contaminants and PDMs from agricultural products, they are modified using a variety of functional substances, chemicals, and groups, such as silica oxides, graphene oxides, MIPs, ionic liquids, and graphene. Targuma *et al* [37] intensively reviewed the benefits, applications, and usefulness of several of these MNMs along with their modifications.

1.7 The applications of magnetic nanomaterials in non-target analysis

The quick identification, screening, or measurement of both recognized and unidentified analytes in a matrix is referred to as non-target analysis. There are no restrictions on the quantity of analytes that may be identified, and it is a crucial method of finding ECs by identifying, screening, or measuring matrices without the use of a reference or known standards. Such identification makes use of Orbitrap^{TM} equipment and high-resolution mass spectrometry (HRMS), in particular quadrupole HRMS with what is known as 'time-of-flight hybridization' [99]. This analysis's primary characteristic is the fingerprint, or complete ion chromatograph, that is extracted from each sample and compared to preexisting sample profiles to identify aberrations or aid in identifying or recognizing the unknowns [89, 91–93, 100]. Analytes that are unknown or of little interest can be re-examined, novel compounds can be found, and stored samples can be resampled or reanalyzed using the raw data from previous studies.

Various extraction sorbents are employed in non-target analysis [101], but MNM sorbents are the subject of this book. For instance, hierarchical micro- and meso-sphere metal–organic frameworks covered with magnetic nanospheres (H-MOF@Fe_3O_4) were used by Jin *et al* [102] to perform non-target analysis of vegetables. Using MNMs based on cellulose (Fe_3O_4@cellulose) as well as various metabolites of enniatin and beauvericin evaluated in paprika samples, non-target analysis was carried out [99]. A magnetic blade-spray tandem mass spectrometry (MBS-MS/MS) assay was developed in another investigation to perform non-target investigation in more than 204 pesticides [103]. Furthermore, it was claimed by Li *et al* [104] that a heteropore covalent organic framework covered with a magnetic nanosphere may be used to identify phytochromes and pesticides in crops. This research emphasizes how MNMs may be used to pretreat samples, including non-target analytes. In addition, MNMs can offer new and quick ways to achieve high-performance preparation of samples and identify new ECs of concern in most matrices, specifically in various agricultural products, because their surface chemistry can be tailored and modified in an exceedingly brief period of time [37, 105].

1.8 Conclusions and future directions

Some environmental pollutants, such as pesticides, medical products, mycotoxins, and PDMs, are created by humans, while other pollutants, such as insecticides, occur naturally. To maintain the safety of agricultural products and public health, there is a growing need to routinely monitor the levels of these pollutants in agricultural products and the environment, prompted by increasing numbers of studies of their detrimental impacts on the environment and human health. When compared to traditional methods, the use of MNMs in pollution analysis holds great promise and presents a number of benefits. Due to the distinct capacity of MNMs for selective adsorption and effortless separation resulting from surface modification, stability, cost-effectiveness, availability, and biodegradability, they have significantly enhanced the extraction of various analytes from agricultural products.

Conversely, conventional techniques use many organic solvents that are harmful to the environment and require longer extraction processes. The current uses of MNMs and their modifications in the removal of contaminants from agricultural products were the main focus of this introductory chapter. The most effective approach for restoring the environment and sustainability has been thought to be the development and application of magnetic nanosorbents to eliminate persistent contaminants and PDMs. By applying external magnets over several cycles, the sorbent may be recovered more rapidly, and the treatment period is shortened through the use of magnetism.

The research community is very interested in using MNMs in agricultural product analysis because of their unique chemical and physical characteristics that set them apart from other adsorbent materials and improve agricultural safety. Because they are easily separated and recovered by an external magnet and may be reused several times, they are frequently employed in the MSPE process. As discussed in this review, MNMs are easily manipulated by expanding their surface area by adding various substances, such as graphene, ionic liquids, CNTs, and polymers, among others. This improves their removal effectiveness and enriches the trace extent of the analytes targeted in multifaceted agricultural product matrices. Due to their unique extraction capacity characteristics, which also allow them to be easily modified and have a large surface area and charge transfer capacity, they are utilized in several extraction applications. These MNMs and their NMs have demonstrated a great deal of promise for the extraction of different pollutants, and they may find use in the effective analysis of other significant chemical contaminants found in agricultural products and the environment as well as in new pollutants. There are still several areas that need to be addressed in the application of some alloys (such as FeCo and FePt as well as FePd) for pollution removal, despite the fact that MNMs and their alloys have been extensively used in the fields of biomedicine, catalysis, magnetic data storage, magnetic recording media, microwave absorption, and NBTech. As a result, several aspects need to be explored in our understanding of how well MNMs may be used to extract environmental toxins, especially PDMs and ECs. Additional areas of interest may include the synthesis and/or modification of MNMs using contemporary synthetic methods, such as naturally generated plant extracts and ecologically friendly solvents like ethanol and water.

References

[1] Eleryan A, Hassaan M, Aigbe U, Ukhurebor K, Onyancha R, Kusuma H, El-Nemr M, Ragab S and El Nemr A 2023 Biochar-C-TETA as a superior adsorbent to acid yellow 17 dye from water: isothermal and kinetic studies *J. Chem. Technol. Biotechnol.* **2023** 1–38

[2] Eleryan A, Hassaan M, Aigbe U, Ukhurebor K, Onyancha R, El Nemr M, Ragab S, Hossain I and El Nemr A 2023 Kinetic and isotherm studies of acid orange 7 dye absorption using sulphonated mandarin biochar treated with TETA *Biomass Convers. Biorefin.* **2023** 1–12

[3] Anani A, Adama K, Ukhurebor K, Habib A, Abanihi V and Pal K 2004 Application of nanofibrous protein for the purification of contaminated water as a next generational sorption technology: a review *Nanotechnology* **24** 1–18

[4] Ukhurebor K, Hossain I, Pal K, Jokthan G, Osang F, Ebrima F and Katal D 2023 Applications and contemporary issues with adsorption for water monitoring and remediation: a facile review *Top. Catal.* **67** 140–55

[5] Aidonojie P, Ukhurebor K, Oaihimire I, Ngonso B, Egielewa P, Akinsehinde B, Heri S and Darmokoesoemo H 2023 Bioenergy revamping and complimenting the global environmental legal framework on the reduction of waste materials: a facile *Heliyon* **9** e12860

[6] Kusuma H, Aigbe U, Ukhurebor K, Onyancha R, Okundaye B, Ama O, Darmokoesoemo H, Widyaningrum B, Osibote O and Balogun V 2023 Biosorption of methylene blue using clove leaves waste modified with sodium hydroxide *Results Chem.* **5** 1–15

[7] Neolaka Y *et al* 2022 Synthesis of zinc (II)-natural zeolite mordenite type as a drug carrier for ibuprofen: drug release *Results Chem.* **4** 100578

[8] Neolaka Y, Riwu A, Aigbe U, Ukhurebor K, Onyancha R, Darmokoesoemo H and Kusuma H 2023 Potential of activated carbon from various sources as a low-cost adsorbent to remove heavy metals and synthetic dyes *Results Chem.* **5** 100711
[9] Onyancha R, Ukhurebor K, Aigbe U, Mogire N, Chanzu I, Kitoto V, Kusuma H and Darmokoesoemo H 2022 A review of the capabilities of carbon dots for the treatment and diagnosis of cancer-related diseases *J. Drug Deliv. Sci. Technol.* **78** 103946
[10] Onyancha K, Ukhurebor U, Aigbe O, Osibote H, Kusuma and Darmokoesoemo H A methodical review on carbon-based nanomaterials in energy-related applications *Adsorpt. Sci. Technol.* **4438286** 1–12
[11] Onyancha R, Aigbe U, Ukhurebor K and Muchiri P 2021 Facile synthesis and applications of carbon nanotubes in heavy-metal remediation and biomedical fields: a comprehensive review *J. Mol. Struct.* **1238** 130462
[12] El-Nemr M, Aigbe U, Ukhurebor K, Onyancha R, El Nemr A, Ragab S, Osibote O and Hassaan M 2022 Adsorption of Cr6+ ion using activated Pisum sativum peels-triethylenetetramine *Environ. Sci. Pollut. Res.* **29** 1–25 2022
[13] El-Nemr M, Aigbe U, Hassaan M, Ukhurebor K, Ragab S, Onyancha R, Osibote O and El Nemr A 2022 The use of biochar-NH2 produced from watermelon peels as a natural adsorbent for the removal of Cu(II) ion from water *Biomass Convers. Bior.* **2022** 1975–91
[14] Eldeeb T *et al* 2022 Biosorption of acid brown 14 dye to mandarin-CO-TETA derived from mandarin peels *Biomass Conv. Bioref.* **14** 5053–73
[15] Eleryan A *et al* 2022 Copper (II) ion removal by chemically and physically modified sawdust biochar *Biomass Convers. Bior.* **14** 9283–320
[16] Aigbe U, Ukhurebor K, Onyancha R, Okundaye B, Pal K, Osibote O, Esiekpe E, Kusuma H and Darmokoesoemo H 2022 A facile review on the sorption of heavy metals and dyes using bionanocomposites *Adsorpt. Sci. Technol.* **8030175** 1–36
[17] Aigbe U, Ukhurebor K, Onyancha R, Osibote O, Kusuma H and Darmokoeso H 2022 Measuring the velocity profile of spinning particles and its impact on Cr(VI) sequestration *Chem. Eng. Process.* **178** 1–15
[18] Eldeeb T, Aigbe U, Ukhurebor K, Onyancha R, El-Nemr M, Hassaan M, Ragab S, Osibote O and El Nemr A 2022 Adsorption of methylene blue (MB) dye on ozone, purified and sonicated sawdust biochars *Biomass Convers. Bior.* **2022** 1–23
[19] Quesada H, Baptista A, Cusioli L, Seibert D, de Oliveira Bezerra C and Bergamasco R 2019 Surface water pollution by pharmaceuticals and an alternative of removal by low-cost adsorbents: a review *Chemosphere* **222** 766–80
[20] K'oreje K, Okoth M, Van Langenhove H and Demeestere K 2020 Occurrence and treatment of contaminants of emerging concern in the African aquatic environment: literature review and a look ahead *J. Environ. Manag.* **254** 109752
[21] Rogowska J, Cieszynska-Semenowicz M, Ratajczyk W and Wolska L 2020 Micropollutants in treated wastewater *Ambio* **49** 487–503
[22] Hube S and Wu B 2021 Mitigation of emerging pollutants and pathogens in decentralized wastewater treatment processes: a review *Sci. Total Environ.* **779** 146545
[23] Yadav D, Rangabhashiyam S, Verma P, Singh P, Devi P, Kumar P, Hussain C, Gaurav G and Kumar K 2021 Environmental and health impacts of contaminants of emerging concerns: recent treatment challenges and approaches *Chemosphere* **272** 129492
[24] Chen T, Merlin J, Chen S, Anandaraj S, Elshikh M, Tseng T, Wang K, Qi D and Jiang J 2020 Sonochemical synthesis and fabrication of neodymium sesquioxide entrapped with

graphene oxide based hierarchical nanocomposite for highly sensitive electrochemical sensor of anti-cancer (raloxifene) drug *Ultrason. Sonochem.* **64** 104717

[25] Sudarni D, Aigbe U, Ukhurebor K, Onyancha R, Kusuma H, Darmokoesoemo H, Osibote O, Balogun V and Widyaningrum B 2021 Malachite green removal by activated potassium hydroxide clove leaves agro-waste biosorbent: characterization, kinetic, isotherm and thermodynamics studies *Adsorpt. Sci. Technol.* **2021** 1–15

[26] Aigbe U, Ukhurebor K, Onyancha R, Osibote O, Darmokoesoemo H and Kusuma H 2021 Fly Ash-based adsorbent for adsorption of heavy metals and dyes from aqueous solution: a review *J. Mater. Res. Technol.* **14** 2751–74

[27] Ukhurebor K, Aigbe U, Onyancha R, Nwankwo W, Osibote O, Paumo H, Ama O, Adetunji C and Siloko I 2021 Effect of hexavalent chromium on the environment and removal techniques: a review *J. Environ. Manage.* **280** 111809

[28] Spahr S, Teixidó M, Sedlak D and Luthy R 2020 Hydrophilic trace organic contaminants in urban stormwater: occurrence, toxicological relevance, and the need to enhance green stormwater infrastructure *Environ. Sci. Water Res. Technol.* **6** 15–44

[29] Khan M, Shah I, Ihsanullah I, Naushad M, Ali S, Shah S and Mohammad A 2021 Hospital wastewater as a source of environmental contamination: an overview of management practices, environmental risks, and treatment processes *J. Water Process Eng.* **41** 101990

[30] Vymazal J and Bˇrezinová T 2015 The use of constructed wetlands for removal of pesticides from agricultural runoff and drainage: a review *Environ. Int.* **75** 11–20

[31] Angeles L, Singh R, Vikesland P and Aga D 2021 Increased coverage and high confidence in suspect screening of emerging contaminants in global environmental samples *J. Hazard. Mater.* **414** 125369

[32] Slobodnik J and Dulio V 2008 NORMAN – Network of Reference Laboratories for Monitoring of Emerging Substances *The Water Framework Directive: Ecological and Chemical Status Monitoring* ed P Quevauviller *et al* (Hoboken, NJ: Wiley) 21

[33] Tijani J, Fatoba O, Babajide O and Petrik L 2016 Pharmaceuticals, endocrine disruptors, personal care products, nanomaterials and perfluorinated pollutants: a review *Environ. Chem. Lett.* **14** 27–49

[34] Cunha D, de Araujo F and Marques M 2017 Psychoactive drugs: occurrence in aquatic environment, analytical methods, and ecotoxicity—a review *Environ. Sci. Pollut. Res.* **24** 24076–91

[35] Vargas-Berrones K, Bernal-Jácome L, de León-Martínez L and Flores-Ramírez R 2020 Emerging pollutants (EPs) in Latin América: a critical review of under-studied EPs, case of study-Nonylphenol *Sci. Total Environ.* **726** 138493

[36] Gavrilescu M, Demnerová K, Aamand J, Agathos S and Fava F 2015 Emerging pollutants in the environment: present and future challenges in biomonitoring, ecological risks and bioremediation *New Biotechnol.* **32** 147–56

[37] Targuma S, Njobeh P and Ndungu P 2021 Current applications of magnetic nanomaterials for extraction of mycotoxins, pesticides, and pharmaceuticals in food commodities *Molecules* **26** 4284

[38] Eleryan A, Aigbe U, Ukhurebor K, Onyancha R, Hassaan M, Elkatory M, Ragab S, Osibote O, Kusuma H and El Nemr A 2023 Adsorption of direct blue 106 dye using zinc oxide nanoparticles prepared via green synthesis technique *Environ. Sci. Pollut. Res.* **30** 69666–82

[39] Azzouz A, Kailasa S, Lee S, Rascón A, Ballesteros E, Zhang M and Kim K 2018 Review of nanomaterials as sorbents in solid-phase extraction for environmental samples *TrAC, Trends Anal. Chem.* **108** 347–69

[40] Reyes-Gallardo E, Lasarte-Aragonés G, Lucena R, Cárdenas S and Valcárcel M 2013 Hybridization of commercial polymeric microparticles and magnetic nanoparticles for the dispersive micro-solid phase extraction of nitroaromatic hydrocarbons from water *J. Chromatogr.* A **1271** 50–5

[41] Mirzajani R and Karimi S 2019 Ultrasonic assisted synthesis of magnetic Ni–Ag bimetallic nanoparticles supported on reduced graphene oxide for sonochemical simultaneous removal of sunset yellow and tartrazine dyes by response surface optimization: application of derivative spectroph- *Ultrason. Sonochem.* **50** 239–50

[42] Onyancha R, Aigbe U, Ukhurebor K, Kusuma H, Darmokoesoemo H, Osibote O and Pal K 2022 Influence of magnetism-mediated potentialities of recyclable adsorbents of heavy metal ions from aqueous solutions—an organized review *Res. Chem.* **4** 100452

[43] Rios A and Zougagh M 2016 Recent advances in magnetic nanomaterials for improving analytical processes *TrAC, Trends Anal. Chem.* **84** 72–83

[44] De Souza K, Andrade G, Vasconcelos I, de Oliveira Viana M, Fernandes C and de Sousa E 2014 Magnetic solid-phase extraction based on mesoporous silica-coated magnetic nanoparticles for analysis of oral antidiabetic drugs in human plasma *Mater. Sci. Eng.* **40** 275–80

[45] Aigbe U, Onyancha R, Ukhurebor K and Obodo K 2020 Removal of fluoride ions using polypyrrole magnetic nanocomposite influenced by rotating magnetic field *RSC Adv.* **10** 595–609

[46] Ramadan M, Mohamed M, Almoammar H and Abd-Elsalam K 2020 Magnetic nanomaterials for purification, detection, and control of mycotoxins *Nanomycotoxicology* ed M Rai and K Abd-Elsalam (Cambridge, MA: Academic) 87–114

[47] Martinez A, Garcia-Lobato M and Perry D 2009 Study of the properties of iron oxide nanostructures *Res. Nanotechnol. Dev* **19** 184–93

[48] Yoon S 2014 Preparation and physical characterizations of superparamagnetic maghemite nanoparticles *J. Magn.* **19** 323–6

[49] Lin S, Lu D and Liu Z 2012 Removal of arsenic contaminants with magnetic γ-Fe_2O_3 nanoparticles *Chem. Eng. J.* **211** 46–52

[50] Roy A and Bhattacharya J 2012 Removal of Cu(II), Zn(II) and Pb(II) from water using microwave-assisted synthesized maghemite nanotubes *Chem. Eng. J.* **211** 493–500

[51] Narimani-Sabegh S and Noroozian E 2019 Magnetic solid-phase extraction and determination of ultra-trace amounts of antimony in aqueous solutions using maghemite nanoparticles *Food Chem.* **287** 382–9

[52] Tuutijärvi T, Lu J, Sillanpää M and Chen G 2009 As (V) adsorption on maghemite nanoparticles *J. Hazard. Mater.* **166** 1415–20

[53] Jiang W, Pelaez M, Dionysiou D, Entezari M, Tsoutsou D and O'Shea K 2013 Chromium (VI) removal by maghemite nanoparticles *Chem. Eng.* J **222** 527–33

[54] Devatha C and Shivani S 2020 Novel application of maghemite nanoparticles coated bacteria for the removal of cadmium from aqueous solution *J. Environ. Manag.* **258** 110038

[55] Rajput S, Singh L, Pittman C and Mohan D 2017 Lead (Pb^{2+}) and copper (Cu^{2+}) remediation from water using superparamagnetic maghemite (γ-Fe_2O_3) nanoparticles synthesized by flame spray pyrolysis (FSP) *J. Colloid Interface Sci.* **492** 176–90

[56] Piovesan J, Haddad V, Pereira D and Spinelli A 2018 Magnetite nanoparticles/chitosan-modified glassy carbon electrode for non-enzymatic detection of the endocrine disruptor parathion by cathodic square-wave voltammetry *J. Electroanal. Chem.* **823** 617–23
[57] González-Jartín J, de Castro Alves L, Alfonso A, Piñeiro Y, Vilar S, Gomez M, Osorio Z, Sainz M, Vieytes M, Rivas J *et al* 2019 Detoxification agents based on magnetic nanostructured particles as a novel strategy for mycotoxin mitigation in food *Food Chem.* **294** 60–6
[58] Turan E and Sahin F 2016 Molecularly imprinted biocompatible magnetic nanoparticles for specific recognition of ochratoxin a *Sensors Actuators* B **227** 668–76
[59] Riaño S and Binnemans K 2015 Extraction and separation of neodymium and dysprosium from used NdFeB magnets: an application of ionic liquids in solvent extraction towards the recycling of magnets *Green Chem.* **17** 2931–42
[60] Zinatloo-Ajabshir S, Mortazavi-Derazkola S and Salavati-Niasari M 2017 Schiff-base hydrothermal synthesis and characterization of Nd_2O_3 nanostructures for effective photocatalytic degradation of eriochrome black T dye as water contaminant *J. Mater. Sci., Mater. Electron.* **28** 17849–59
[61] Ahmadi S, Mohammadi L, Rahdar A, Rahdar S, Dehghani R, Igwegbe C and Kyzas G 2020 Acid dye removal from aqueous solution by using neodymium (iii) oxide nano-adsorbents *Nanomaterials* **10** 556
[62] Zheng J, Dong Y, Wang W, Ma Y, Hu J, Chen X and Chen X 2013 In situ loading of gold nanoparticles on Fe_3O_4@SiO_2 magnetic nanocomposites and their high catalytic activity *Nanoscale* **5** 4894–901
[63] Chen Y, Zheng F, Min Y, Wang T and Zhao Y 2012 Synthesis and properties of magnetic $FeNi_3$ alloyed microchains obtained by hydrothermal reduction *Solid State Sci.* **14** 809–13
[64] dos Santos C, Martins A, Costa B, Ribeiro T, Braga T, Soares J and Sasaki J 2016 Synthesis of FeNi alloy nanomaterials by proteic sol-gel method: crystallographic, morphological, and magnetic properties *J. Nanomater.* **2016** 1637091
[65] Mohammed A, Brouers F, Sadi S and Al-Musawi T 2018 Role of Fe3O4 magnetite nanoparticles used to coat bentonite in zinc (II) ions sequestration *Environ. Nanotechnol. Monit. Manag* **10** 17–27
[66] Khodadadi M, Panahi A, Al-Musawi T, Ehrampoush M and Mahvi A 2019 The catalytic activity of $FeNi_3$@SiO_2 magnetic nanoparticles for the degradation of tetracycline in the heterogeneous Fenton-like treatment method *J. Water Process Eng.* **32** 100943
[67] Farooghi A, Sayadi M, Rezaei M and Allahresani A 2018 An efficient removal of lead from aqueous solutions using $FeNi_3$@SiO_2 magnetic nanocomposite *Surf. Interfaces* **10** 58–64
[68] Nasseh N, Taghavi L, Barikbin B, Nasseri M and Allahresani A 2019 $FeNi_3/SiO_2$ magnetic nanocomposite as an efficient and recyclable heterogeneous Fenton-like catalyst for the oxidation of metronidazole in neutral environments: adsorption and degradation studies *Composites* B **166** 328–40
[69] Rafique M, Pan L, Iqbal M, Qiu H, Farooq M and Guo Z 2013 3-D flower like FeCo alloy nanostructures assembled with nanotriangular prism: facile synthesis, magnetic properties, and effect of NaOH on its formation *J. Alloys Compd.* **550** 423–30
[70] Gupta V, Patra M, Shukla A, Saini L, Songara S, Jani R, Vadera S and Kumar N 2014 Synthesis and investigations on microwave absorption properties of core–shell FeCo (C) alloy nanoparticles *Sci. Adv. Mater.* **6** 1196–202

[71] Braga T, Dias D, de Sousa M, Soares J and Sasaki J 2015 Synthesis of air stable FeCo alloy nanocrystallite by proteic sol–gel method using a rotary oven *J. Alloys Compd.* **622** 408–17
[72] Li X, Feng J, Du Y, Bai J, Fan H, Zhang H, Peng Y and Li F 2015 One-pot synthesis of CoFe2O4/graphene oxide hybrids and their conversion into FeCo/graphene hybrids for lightweight and highly efficient microwave absorber *J. Mater. Chem.* A **3** 5535–46
[73] Zhou J, Shu X, Wang Z, Liu Y, Wang Y, Zhou C and Kong L 2019 Hydrothermal synthesis of polyhedral FeCo alloys with enhanced electromagnetic absorption performances *J. Alloys Compd.* **794** 68–75
[74] Abbas M, Islam M, Rao B, Ogawa T, Takahashi M and Kim C 2013 One-pot synthesis of high magnetization air-stable FeCo nanoparticles by modified polyol method *Mater. Lett.* **91** 326–9
[75] Çelik Ö and Fırat T 2018 Synthesis of FeCo magnetic nanoalloys and investigation of heating properties for magnetic fluid hyperthermia *J. Magn. Magn. Mater.* **456** 11–6
[76] Wang W, Jing Y, He S, Wang J and Zhai J 2014 Surface modification and bioconjugation of FeCo magnetic nanoparticles with proteins *Colloids Surf. B Biointerfaces* **117** 449–56
[77] Galiote N, Oliveira F and Lima F 2019 FeCo-NC oxygen reduction electrocatalysts: activity of the different compounds produced during the synthesis via pyrolysis *Appl. Catal. B Environ.* **253** 300–8
[78] Kishimoto M, Latiff H, Kita E and Yanagihara H 2019 Morphology and magnetic properties of FeCo particles synthesized with different compositions of Co and Fe through co-precipitation, flux treatment, and reduction *J. Magn. Magn. Mater.* **476** 229–33
[79] Bai L, Wan H and Street S 2009 Preparation of ultrafine FePt nanoparticles by chemical reduction in PAMAM-OH template *Colloids Surf. A Physicochem. Eng. Asp.* **349** 23–8
[80] Ovejero J, Velasco V, Abel F, Crespo P, Herrasti P, Hernando A and Hadjipanayis G 2017 Colloidal nanoparticle clusters to produce large FePt nanocrystals *Mater. Des.* **113** 391–6
[81] Duan X, Wu C, Wang X, Tian X, Pei W, Wang K and Wang Q 2019 Evolutions of microstructure and magnetic property of wet-chemical synthesized FePt nanoparticles assisted by high magnetic field *J. Alloys Compd.* **797** 1372–7
[82] Liu Y, Yang K, Cheng L, Zhu J, Ma X, Xu H, Li Y, Guo L, Gu H and Liu Z 2013 PEGylated FePt@Fe2O3 core–shell magnetic nanoparticles: potential theranostic applications and *in vivo* toxicity studies *Nanomed. Nanotechnol. Biol. Med.* **9** 1077–88
[83] Shi Y, Lin M, Jiang X and Liang S 2015 Recent advances in FePt nanoparticles for biomedicine *J. Nanomater.* **2015** 467873
[84] Akdeniz M and Mekhrabov A 2019 Size dependent stability and surface energy of amorphous FePt nanoalloy *J. Alloys Compd.* **788** 787–98
[85] Meng Z, Li G, Zhu N, Ho C, Leung C and Wong W 2017 One-pot synthesis of ferromagnetic FePd nanoparticles from single-source organometallic precursors and size effect of metal fraction in polymer chain *J. Organomet. Chem.* **849** 10–6
[86] Yamamoto S, Takao S, Muraishi S, Xu C and Taya M 2019 Synthesis of Fe70Pd30 nanoparticles and their surface modification by zwitterionic linker *Mater. Chem. Phys.* **234** 237–44
[87] Van N, Trung T, Nam N, Phu N, Hai N and Luong N 2013 Hard magnetic properties of FePd nanoparticles *Eur. Phys. J. Appl. Phys.* **64** 10403
[88] Luong N, Trung T, Loan T, Kien L, Hong T and Nam N 2016 Magnetic properties of FePd nanoparticles prepared by sonoelectrodeposition *J. Electron. Mater.* **45** 4309–13

[89] Khan F, Mubarak N, Khalid M, Walvekar R, Abdullah E, Mazari S, Nizamuddin S and Karri R 2020 Magnetic nanoadsorbents' potential route for heavy metals removal—a review *Environ. Sci. Pollut. Res.* **27** 24342–56
[90] Faraji M and Yamini Y 2021 Application of magnetic nanomaterials in food analysis *Magnetic Nanomaterials in Analytical Chemistry* ed M Ahmadi, A Afkhami and T Madrakian (Amsterdam: Elsevier) 87–120 94
[91] Faraji M, Shirani M and Rashidi-Nodeh H 2021 The recent advances in magnetic sorbents and their applications *TrAC, Trends Anal. Chem.* **141** 116302
[92] Andrade-Eiroa A, Canle M, Leroy-Cancellieri V and Cerdà V 2016 Solid-phase extraction of organic compounds: a critical review (Part I) *TrAC, Trends Anal. Chem.* **80** 641–54
[93] Ahmadi M, Ghoorchian A, Dashtian K, Kamalabadi M, Madrakian T and Afkhami A 2021 Application of magnetic nanomaterials in electroanalytical methods: a review *Talanta* **225** 121974
[94] Mehdinia A, Khodaee N and Jabbari A 2015 Fabrication of graphene/Fe_3O_4@polythiophene nanocomposite and its application in the magnetic solid-phase extraction of polycyclic aromatic hydrocarbons from environmental water samples *Anal. Chim. Acta* **868** 1–9
[95] Xu X, Xu X, Han M, Qiu S and Hou X 2019 Development of a modified QuEChERS method based on magnetic multiwalled carbon nanotubes for the simultaneous determination of veterinary drugs, pesticides and mycotoxins in eggs by UPLC-MS/MS *Food Chem.* **276** 419–26
[96] Abdulhussein A, Jamil A and Bakar N 2021 Magnetic molecularly imprinted polymer nanoparticles for the extraction and clean-up of thiamethoxam and thiacloprid in light and dark honey *Food Chem.* **359** 129936
[97] Rathi B and Kumar P 2021 Application of adsorption process for effective removal of emerging contaminants from water and wastewater *Environ. Pollut.* **280** 116995
[98] 2023 *Adsorption Applications for Environmental Sustainability* ed K Ukhurebor, O Aigbe and B Onyancha (Bristol: IOP Publishing)
[99] García-Nicolás M, Arroyo-Manzanares N, Campillo N and Viñas P 2021 Cellulose-ferrite nanocomposite for monitoring enniatins and beauvericins in paprika by liquid chromatography and high-resolution mass spectrometry *Talanta* **226** 122144
[100] Ballin N and Laursen K 2019 To target or not to target? Definitions and nomenclature for targeted versus non-targeted analytical food authentication *Trends Food Sci. Technol.* **86** 537–43
[101] Milman B and Zhurkovich I 2017 The chemical space for non-target analysis *TrAC, Trends Anal. Chem.* **97** 179–87
[102] Jin Y, Qi Y, Tang C and Shao B 2021 Hierarchical micro- and mesoporous metal–organic framework–based magnetic nanospheres for the nontargeted analysis of chemical hazards in vegetables *J. Mater. Chem.* A **9** 9056–65
[103] Rickert D, Singh V, Thirukumaran M, Grandy J, Belinato J, Lashgari M and Pawliszyn J 2020 Comprehensive analysis of multiresidue pesticides from process water obtained from wastewater treatment facilities using solid-phase microextraction *Environ. Sci. Technol.* **54** 15789–99
[104] Li W, Jiang H-X, Geng Y, Wang X-H, Gao R-Z, Tang A-N and Kong D-M 2020 Facile removal of phytochromes and efficient recovery of pesticides using heteropore covalent

organic framework–based magnetic nanospheres and electrospun film *ACS Appl. Mater. Interfaces* **12** 20922–32

[105] Cheng D, Ngo H, Guo W, Chang S, Nguyen D, Liu Y, Wei Q and Wei D 2020 A critical review on antibiotics and hormones in swine wastewater: water pollution problems and control approaches *J. Hazard. Mater.* **387** 121682

IOP Publishing

Environmental Applications of Magnetic Sorbents

Kingsley Eghonghon Ukhurebor and Uyiosa Osagie Aigbe

Chapter 2

The synthesis of magnetic sorbents

Kokolo M Etiowo, Ikenna B Onyeachu and Ikechukwu C Ukaga

The conventional methods used to synthesize magnetic adsorbents (MAs) and magnetic sorbents (MSs) are comprehensively summarized in this chapter of this book (titled: 'Environmental Applications of Magnetic Sorbents'). It includes various types of synthesis, such as single methods and combinations of two or more methods (hybrid methods). The MA and MS synthesis methods described in this chapter are seen to be responsible for the shape and size (dimensions) as well as the distribution and surface chemistry of the resulting adsorbents or sorbents.

2.1 Introduction

The method of synthesis used to produce a magnetic adsorbent (MAs) or a magnetic sorbent (MSs) is responsible for its dimensions (i.e. the shape and size), the distribution and size of the particles, and the surface chemistry of the adsorbents or sorbents thus produced [1]. It is therefore responsible for the magnetic property of the adsorbent (such as paramagnetic, superparamagnetic, ferromagnetic, etc). It is worth noting that the method of preparation determines to a large extent the degree of structural defects (impurities), which also influences the distribution of such impurities or defects within the structure of the particle. This gives rise to the magnetic property or behavior of the adsorbent [1]. The stability of the adsorbent, its compatibility, and its absorptive performance are also determined by the method of synthesis [2–4]. The properties of most MAs or MSs strongly rely on their size and size distribution. Thus, their method of synthesis is of great importance.

2.2 The coprecipitation method

Due to its low cost, simple procedure, and rapid reaction (RXN), the coprecipitation method (C-PCTM) is known to be one of the most common methods or techniques of synthesizing or producing magnetic nanoparticles (MNPs) employed as adsorbents or sorbents. The standard C-PCTM procedure involves the mixture of two

doi:10.1088/978-0-7503-5909-2ch2

solutions, precipitation based on acid/alkali (pH) settings, accompanied by filtration, washing, and drying.

The C-PCTM has repeatedly been shown to be a convenient method for the synthesis of ferrite (iron (Fe) oxide) nanoparticles (NPs) of controlled size and magnetic properties. MAs and MSs have been synthesized or produced from negatively valued Fe mud, which is a waste product (effluent) produced by water treatment plants (especially those that purify or decontaminate groundwater), by Fe^{2+}/Fe^{3+} coprecipitation [5]. This was done utilizing ascorbic acid (AC) as a reducing agent and nitric, nitrous, or azotic acid (acid wastewater) as the solution for leaching. The synthesized MAs or MSs were used to adsorb methyl blue from wastewater. The result revealed a high adsorption of around 87.30 mg g^{-1} and a decent magnetic response, which proves that these materials have excellent potential for use in the purification, treatment, and management of dye wastewater [5].

Ahribesh *et al* [6] noted that different synthesis conditions, such as the Fe salt concentration (CCT), the nature or condition of the base utilized for the C-PCTM, the ratio of Fe^{2+} (the Fe^{3+} ratio of $OH^-/(Fe^{2+}+Fe^{3+})$), the addition rate of the base solution, the temperature, and method of drying, greatly influence the properties of MAs or MSs synthesized by C-PCTM. Meng *et al* [7] examined the influence of the synthesis parameters on the properties of sepiolite-based MAs or MSs; their results indicated that sepiolite-based MAs or MSs prepared, arranged, and organized employing sodium hydroxide (NaOH) had a less magnetic composition (low magnetization) but higher or better adsorption capacity (ASC) when compared to those produced or synthesized utilizing ammonia (NH_3). Also, the order in which the various reagents were mixed or combined had very little effect, influence, or impact on the features or properties of these adsorbents or sorbents, specifically those produced using NaOH. $FeCl_3.6H_2O$ and $FeCl_2.4H_2O$ in deionized water were used to prepare, arrange, and organize Fe salt solutions along with a solution of ferrous salt [7]. These solutions were thereafter mixed or combined as well as heated to prepare, arrange, and organize nano-Fe_3O_4 with NH_3, in which water served as a precipitant in the combined solution (mixture). Also, their findings indicated that the CCT of Fe^{2+}/Fe^{3+} in the combined or mixed solution had the greatest impact, influence, or effect on the yield (product) of nano-Fe_3O_4. The temperature also had an impact, influence, or effect on the size (dimension) of the particles. In this case, the temperature was proportional to the size (dimension) of the nano-Fe_3O_4 particle before it was reduced. Figure 2.1 by Meng *et al* [7] shows a schematic illustration of the production or synthesis of Fe_3O_4 NPs using the C-PCTM.

2.3 The solvothermal method

The solvothermal method (SVTM) is a process or procedure that occurs in a closed (isolated) RXN container (vessel). It involves a chemical RXN, namely the decomposition, breakdown, or disintegration of precursors as in the presence of a solvent at a temperature greater than that of the boiling point of the solvent [8]. It is seen to be a suitable low-cost, eco-friendly synthesis method for the production of microcarbons or nanocarbons such as carbon dots, carbon nanotubes, carbon

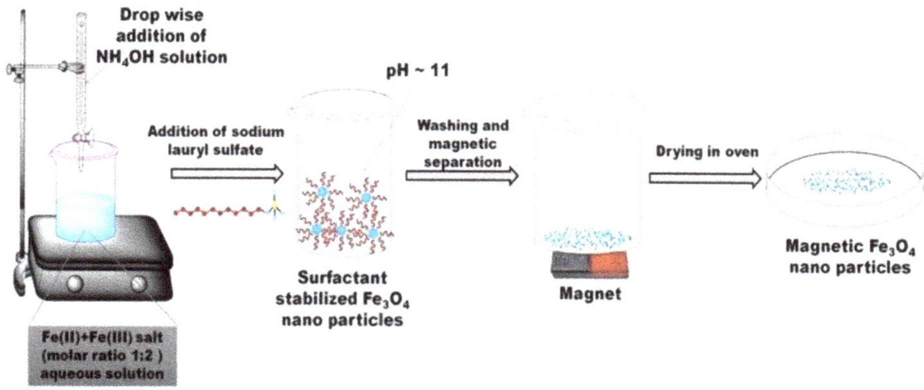

Figure 2.1. A schematic illustration of the production or synthesis of Fe_3O_4 NPs using C-PCTM [7].

spheres, and many more. This is a result of its low-temperature RXN, its environmental or ecological friendliness, and the high yields of this method or technique of synthesis [9].

The SVTM used to synthesize MAs and MSs improves the structural or compositional defects/limitations as well as the chemical defects/limitations of the crystal development process (growth). It does this by adjusting, regulating, or altering the factors or parameters which impact, influence, or affect the rate of the crystal development process (growth) to accomplish the purpose or objective of adjusting, regulating, or altering the morphology (structure) of the crystal and size (dimension) as well as the degree or extent of functionalization [9]. The factors or parameters that influence or affect the rate of the crystal development process (growth) are: the temperature, the pressure, the solvent type, the pH of the RXN medium, the nature of any additives, and the chemical composition, structure, and properties of the precursors [8, 9]. SVTM was used to synthesize four different magnetic Fe_3O_4 NPs using different sodium (Na) salts [10]. NaOAc and Na_2CO_3, together with a mixture of NaOAc and Na_3Cit as well as a mixture of NaOAc and $Na_2C_2O_4$ were the Na salts used for this synthesis. Findings obtained using x-ray diffraction (XRD) as well as scanning electron microscopy (SEM) images showed that the second sample (Na_2CO_3) had the lowest average NP and crystalline sizes (dimensions) of 29 and 48 mm, respectively. The adsorption results also indicated that the second sample (Na_2CO_3) had the highest UV adsorption in solutions whose pH was about ten.

Chen *et al* [11] used the SVTM to synthesize size-controlled superparamagnetic Fe oxide NPs in an ethylene glycol/diethylene glycol (EG/DE4) binary solvent system. When the $V_{\mathrm{Eqn}}/V_{\mathrm{DEG}}$ values were varied from 100/0 to 80/20, 60/40 and 40/60, the Fe_3O_4 nanospheres produced had average diameters of about 700, 500, 300, and 100 mm, respectively. The saturation magnetization (SM) of Fe_3O_4 NPs at these particle sizes were 85.41, 80.28, 75.94, and 72.14 emu g^{-1}, respectively.

Pachfule *et al* [12] reported the structure (composition) and structural properties of metal–organic framework (MOF) isomers, which were obtained from a partially fluorinated link using the SVTM of synthesis (production). The solvothermal RXN of $Cu(NO_3)_2{\bullet}3H_2O$ with 4,4-(hexamfluoroisoproylidene) bi 5(benzone acid) ($C_{17}H_{10}F_6O_4$, H_2hfbba) and terminal monoclentate ligand 3-methylpyridine (3-picoline/3-mepy) in the presence of N,N-dimethylformamide (DMF) and N,N-diethylformamide (DEF) solvents resulted in an increase of two structurally or mechanically different two-dimensional (2D) fluorinated MOFs (F-MOFs). The effects of the chosen solvent were clearly reflected or revealed in the obtained structures.

2.4 The thermal decomposition method

Generally, the thermal decomposition method (TDM), also known as thermolysis, is a process in which heat is added to a chemical RXN, causing the reactants to be broken down into two or more smaller products. In the synthesis of MAs or MSs, the TDM is seen as one of the most vital techniques for the synthesis of MAs or MSs in which the arrangement is highly regulated (controlled) [13]. Here, the dimensions (sizes) of the synthesized particles yield a standard deviation ranging from 0.1 to 0.15 when fitted into a log-normal distribution, which is much smaller than the standard deviation of the particle dimensions synthesized by the C-PCTM (0.2–0.4) [13].

According to published reports, the TDM is used for the synthesis of MNPs (especially Fe oxide NPs). Coordination or organometallic compounds are used as the precursors [14–16]. The decomposition of these organometallic precursors has been shown by recent studies to modify magnetic samples and exhibits advantages over the conventional methods used to synthesize samples such as size control, a narrow (thin) size distribution, excellent crystalline structure, and the ability to be used for mass production [14–16].

Although the highlighted benefits are some of the advantages of the TDM, the high costs (huge amounts) of the precursors/surfactants and the likely toxicity (noxiousness) are some of the obvious general disadvantages of the TDM. Maity *et al* [15] synthesized MNPs (Fe_3O_4) using a solvent-free TDM. The NP size was controlled (managed) by adjusting, regulating, or altering the RXN time or temperature. For an increased RXN temperature of 330 °C and a reaction time of four hours, the average particle size was seen to be approximately 9 nm and the magnetism of the MSs was approximately 76 emu g^{-1}. A simple TDM was used by Daengsakul *et al* [17] to study the magnetic features as well as the cytotoxicity of $La_{0.7}$ $Sr_{0.3}$ MnO_3 NPs. The experiment was carried out using a RXN time of six hours and RXN temperatures of 600 °C, 700 °C, 800 °C, 900 °C, and 1000 °C. The results indicated that all the samples prepared had a perovskite structure which changed from cubic to rhombohedral with an increase in RXN temperature. The saturated magnetization and coercive field (Hc) which was evaluated by simple vibrating magnetometry showed that the samples had soft ferromagnetic components and an SM of approximately 9–55 emu g^{-1} and Hc values of approximately −8 to 37 Oe.

2.5 The sol–gel method

The sol–gel method (S-GM) is a chemical technique used for the synthesis or production of various nanostructures, specifically metal-oxide NPs. In this technique, the molecular precursor (predominantly an alkali metal) is dissolved or liquified in water or alcohol and converted by heating and stirring via the process of hydrolysis or/and alcoholysis [3, 18]. The basis of the S-GM is the synthesis or production of a homogenous sol (solution) from the precursors and its modification, conversion, or transformation into a gel. Since the gel obtained from the hydrolysis/alcoholysis process is wet, an appropriate drying method is applied to obtain the required or anticipated features, properties, benefits, and application of the gel. The S-GM is seen to be cost-effective (cheap), and as a result of its low RXN temperature, it is applicable for the synthesis or production of MAs or MSs of controlled chemical composition, structure, and polarity [3, 18, 19]. There are different factors or parameters that affect the kinetics (KT), growth, hydrolysis, and condensation RXNs of a S-GM. They are: the solvent class or group, temperature, precursors, catalysts, pH, the additives employed, and the mechanical agitation used [3].

Due to the uniqueness of the properties and characteristics of the materials formed using the S-GM, it has found application in many fields, such as surface engineering, biosensors, optics, electronics, energy, and separation technology. It has also found application in photocatalysis and desorption sciences [18–20]. The synthesis, production, and characterization of magnetic $CoFe_{1.9}Cr_{0.1}O_4$ NPs by the S-GM and their application as an adsorbent or sorbent for water management and purification (treatment) was investigated by Amar *et al* [20]. Their study reported the adsorptive confiscation or removal of a very noxious cationic dye, namely methylene blue (MB) from aqueous solution (QS) utilizing spinel ferrite, $CoFe_{1.9}Cr_{0.1}O_4$ (CFC) MAs or MSs. The CFC provider was synthesized by the S-GM and characterized using XRD, Fourier transform infrared spectroscopy (FTIR), and SEM. The impact and influence of the various experimental and analytic variables or parameters on the confiscation or removal of MB, such as contact time, initial dye CCT, adsorbent dosage, solution pH, and temperature were all considered and investigated. The findings of this study showed that approximately 94.00% of the MB was confiscated or removed under the optimal operational and analytic settings. The adsorption KT revealed that the adsorption parameters and data were better defined and described by the pseudo-second-order model (PSO). The adsorption isotherm (AIT) followed the Langmuir isotherm model (LIM) and the highest monolayer ASC was found to be around 11.41 mg g^{-1}.

In a study by Li *et al* [21], the preparation, adsorption parameters, and properties of composites of magnetic chitosan sorbents or adsorbents (CsFeAC) were modified through the use of magnetic macroparticles in addition to an extremely permeable activated carbon carrier utilizing the S-GM. SEM, the Brunauer–Emmett–Teller (BET) theory, FTIR, XRD, thermal gravimetric analysis (TGA), and ultrasonic machining (USM) methods were employed for the characterization of the sorbent or adsorbent. A batch test was carried out to investigate the Cu^{2+} adsorption features,

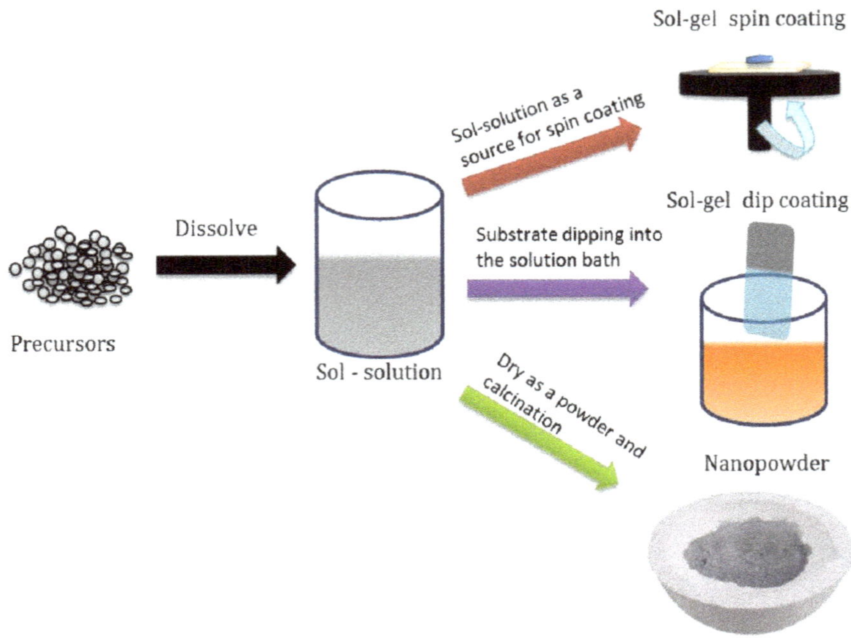

Figure 2.2. A schematic representation of the S-GM [23].

parameters, and properties of CsFeAC at diverse pH values and contact times as well as the effects of the initial Cu^{2+} CCT and temperature. The adsorption was a good fit for the LIM and followed the PSO model, signifying that monolayer adsorption may have taken place. Also, the rate-limiting phase/step was found to be the chemical chelation RXN. The saturated ASC was obtained and found to be around 216.60 mg g^{-1}. BET and XRD investigations revealed that the higher (advanced) specific surface area and lower crystallinity of CsFeAC usefully enhanced the ASC and rate.

Thiagarajan *et al* [22] schematically illustrated the S-GM as shown in figure 2.2.

2.6 The hydrothermal reaction method

The hydrothermal reaction method (HTRM) is similar to the SVTM. However, in this method, the solvent is water. It is the synthesis that occurs through a chemical RXN in a QS heated to a temperature greater than the boiling point of water [23]. The HTRM has been widely reported to be utilized for the preparation and production of NPs, specifically metal oxides [4]. This method is one of the most efficacious means of growing crystals from several diverse constituents [3]. The particle size and distribution area are a function of the precursor CCT and residence time (RXN time) [2, 3]. As the precursor CCT increases, the particle size and size distribution increase, while at a short RXN time, a monodispersed particle is produced. One of the foremost benefits or advantages of HTRM is the sluggish or moderate RXN KT at a given or known temperature. However, the RXN KT of crystallization can be increased by using microwave heat [24].

Akbarzadeh *et al* [2] studied the use of the HTRM for the preparation of magnetic Fe_3O_4@C NPs for dye adsorption (e.g. MB adsorption). Fe_3O_4@C NPs were organized, prepared, and produced from $FeCl_2$ and $FeCl_3$ together with glucose. Following the HTRM, they were treated at about 160 °C for approximately six hours. Using a characterization process that included TEM, XDS, XPS, and an infrared spectrometer, the adsorption parameters, behaviors, and activities of MB on Fe_3O_4@C NPs were examined, studied, investigated, and reported. The findings and results showed that Fe_3O_4@C NPs effectively adsorbed MB with a huge ASC of about 117 mg g^{-1}. The AIT was a good match for the Temkins model. The adsorption attained an equilibrium (balance position) within three hours and the KT followed the PSO model. The upsurge in the strength of the ions (ionic strength) inhibited and demonstrated the adsorption, and the reduction in the pH resulted in a drastic reduction in the ASC. Exceedingly environmentally friendly and biodegradable (recyclable) MAs or MSs based on montmorillonite ($CoFe_2O_4$/MNT) were produced (fabricated) by means of a simplistic HTRM to harvest tetracycline (TC) and ciprofloxacin (CIP) from effluent (toxic water). The prepared sorbent or adsorbent was characterized via the XRD, FTIR-SEM, and vibrating sample magnetometry (VSM) techniques to comprehend as well as investigate its structure and composition as well as its morphology and magnetic process. The influence and impacts of the experimental parameters or variables such as the pH of the solution, time of adsorption, initial CCT, and ionic strength (ion strength) were all observed, examined, studied, and investigated in detail. The experimental as well as the analytical adsorption process and data obtained for TC and CIP were respectively found to be good fits for the PSO KT model and the LIM. The highest adsorption of TC and CIP possibly ranges between 240.90 and 224.00 mg g^{-1}. A thermodynamic (temperature-related) investigation showed that the adsorption parameters or data were spontaneous. The additives as well as the antibiotics were broken down further under visible light and the MAs or MSs could correspondingly be thermally recycled, redeveloped, or regenerated.

Another study was conducted using the HTRM, in which MAs or MSs produced from waste Fe mud were used to confiscate heavy metals (HMs) effectively and optimally from smelting effluents (wastewater, to be specific) [25]. The findings revealed that the Fe content of the MAs or MSs was 41.80 wt%, 0.025% times more than that of the Fe mud, and this was as a result of the hydrothermal reaction of the dissolved nonferrous contaminants (such as quartz and albite) under alkaline conditions, regardless of the inclusion or exclusion of AC. The percentage of ferrihydrite was 92.70% in dry Fe mud before the addition of AC, and this progressively diminished to 58.10% after increasing the molar ratio (MR) of AC to Fe in the HTRM treatment process. The strongest SM of 16.29 emu g^{-1} was detected in the produced NA-4 when the AC-to-Fe MR was one. The maximum surface site CCT of 1.31 mmel g^{-1} was detected in NA-2 when the MR was 2.0×10^{-2}. The effect of the HTRM when used to modify the waste Fe mud to an MA was the reductive dissolution of ferrihydrite to form siderite, which was thereafter re-oxidized to maghemite. When 12.50 g l^{-1} of MA-2 was utilized for the treatment of smelting effluent (i.e. wastewater), very close to 100% (99% to be specific) removal of the HMs (Cu^{2+}, Zn^{2+}, Pb^{2+} and Cl^{2+}) was achieved. The primary mechanism of the Cu^{2+} and Zn^{2+} adsorption by the sorbent or adsorbent was cationic exchange.

2.7 The direct precipitation method

The direct precipitation method (DPM), also known as chemical precipitation, is the formation of a separable solid from a solution. It is used to remove metal ions from QS. Thus, it is a good method for producing MAs or MSs to be applied for wastewater cleaning. DPM is one of the oldest methods of synthesizing MNPs, MSs, or MAs and offers rigorous control over their size and shape (dimension) in a simple manner [26]. In this method, two processes are involved: nucleation and the growth of nuclei. However, to achieve monodispersed magnetic particles, the two stages are separated and nucleation is avoided during the growth process.

The DPM and the characterization of ZnO NPs were studied by Moharram *et al* [27]. The NPs were prepared by the hydrolysis and condensation of zinc acetate dehydrated by potassium hydroxide in an alcoholic solution at a very low temperature. A thermal gravimetric analysis (TGA) of the precursor was performed with the aim of identifying the range of temperatures over which the weight loss and the influence of heat and temperature (thermal effect) are crucial or significant. XRD of the prepared and produced specimens showed that the main crystallographic structure was hexagonal. According to Scherer's formula, the average size (dimension) of the NPs was found to be 22.4±0.60 nm. The adsorption peak of the prepared and produced sample was about 298 mm, which was categorically blue-shifted when compared to that of the bulk material (360 mm). Magnetic ferrites (MFs) were produced at ambient temperature via the DPM in QS at changing pH values and were utilized as innovative adsorbents or sorbents for HM-containing effluent (wastewater) treatment [28]. The MFs were applied for the confiscation of Cd^{2+} ions from effluent (wastewater). They were characterized using their settling velocity (SV), XRD and SEM analysis, and a vibrating sample magnetometer. The influence of the pH value and contact time on the adsorption parameters was studied, examined, and explored. The MFs had an SM value of 82.30 emu g^{-1} and an SV of 2.00%, signifying that the separation process in a QS under a magnetic field was not a difficult one. The adsorption of Cd^{2+} MFs followed the PSO KT and the LIM. The most appropriate pH setting for the production of MFs with optimum Cd^{2+} ASC was 9.0, and the highest ASC of the ferrite was 160.91 mg g^{-1}. A cost analysis was also conducted as part of this study, which revealed that the MFs could be produced as a cost-effective (cheap or affordable) adsorbent or sorbent for the management and treatment of wastewater containing Cd.

A nanosized metal oxide, namely copper (Cu) oxide, was synthesized using the DPM and was characterized utilizing XRD, TEM, and magnetic measurement techniques [29]. The XRD results showed that the Cu oxide formed was CuO and had a monoclinic structure. The magnetic measurements revealed that CuO had one unpaired electron and was naturally paramagnetic. The TEM results indicated that the size (dimension) of the CuO particles varied from 12 to 35 nm. The study therefore concluded that DPM is a convenient, easy, and effective method of producing magnetic nanosized Cu oxide compared to other conventional means (such as TDM).

2.8 Recent advancements in the synthesis of magnetic adsorbents and magnetic sorbents

The above conventional methods of synthesizing MAs have been extensively studied. However, advanced techniques are being developed based on conventional techniques and principles; these advanced or improved techniques are characterized by energy-saving ability, greater efficiency, simplicity, and an increased adsorption performance in most cases [4].

According to Osman *et al* [30], figure 2.3 illustratively summarizes most of the recent techniques/methods used for producing or synthesizing MSs, which include

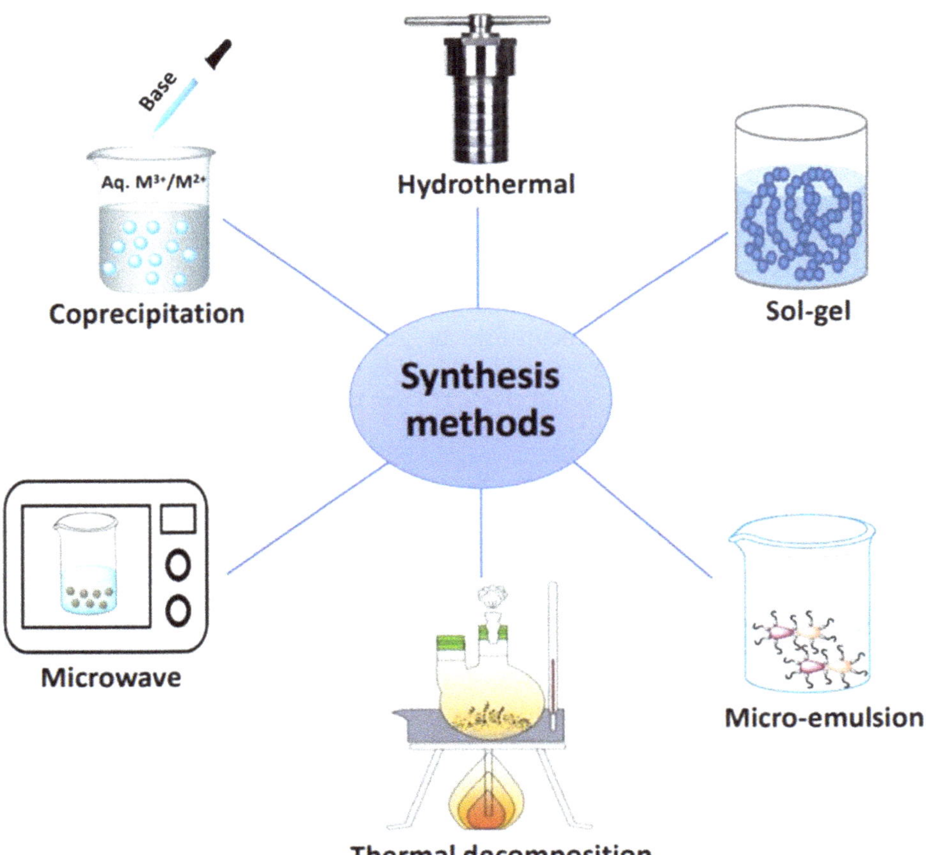

Figure 2.3. Techniques for the synthesis or production of MSs. Reproduced from [31]. CC BY 4.0.

coprecipitation, the hydrothermal method, thermal decomposition, and polyol, microwave, sol–gel, and micro-emulsion methods. More details of these techniques/methods can be found in [30].

Some of these advancements include bioderived MAs or MSs (produced by the conversion of biomass to MAs or MSs) and the green synthesis (GS) of MAs or MSs (achieved through the use of plant extracts and solvent-free precursors) for MA or MS production [30–34].

2.8.1 Green synthesis

Many natural biomasses and plants have abundant (readily available) functional groups that can promote or encourage the formation and ASC of some MAs or MSs [30–35]. MNPs were synthesized utilizing a cheaper (low cost) and greener means in an open-air setting or environment using crude latex of *Jatropha curcas* (JC) and leaf extract of *Cinnamomum tamala* (CT) [36]. The characterization of the MNPs was performed using dynamic light scattering (DLS) ultraviolet–visible (UV–vis) spectroscopy, FTIR, powder XRD, and field emission SEM (FE-SEM). The size (dimension) range of the produced MNPs was found to be 20.00–42.00 mm for JC-Fe_3O_4 and 26.00–35.00 nm for CT-Fe_3O_4 via the FE-SEM images. The impacts of the produced MNPs in effluent (wastewater) treatment and management revealed that both synthesized MNPs were effective in confiscating effluent materials such as organic dyes, toxic metal ions, and HMs from water.

Finally, Bassim *et al* [37] reported the GS of Fe_3O_4 NPs and their application and benefits in effluent (wastewater) treatment and management. In this report, an extract of *Citrus aurantium* (CA) was utilized in a GS procedure to formulate and produce Fe_3O_4 NPs. The green synthesized Fe_3O_4 NPs were employed to reduce the quantity of MB dye in effluent (wastewater); the results revealed that Fe_3O_4/CA is an active absorbent or sorbent for the removal of MB from QS: 93.14% confiscation was attained at a solution pH of 8.98, an adsorbent or sorbent dose of around 997.99 mg l^{-1}, an initial dye CCT of 10.22 mg l^{-1}, and a contact time (duration) of 43.70 min.

2.9 Conclusions/future views

The investigation of novel means or approaches for the synthesis or production of MAs is continuously developing. The major or conventional methods of synthesis of MAs are comprehensively summarized in this chapter. In most of the research literature, the synthesis or production of MAs is grouped into the direct utilization of MNPs, the attachment of prearranged adsorbents or sorbents and prearranged MNPs, the synthesis of MNPs and the co-synthesis of adsorbents or sorbents, and the synthesis of adsorbents or sorbents and MNPs. The methods of synthesis represented in this chapter include every type of synthesis or production, whether as a single method of synthesis or as a combination of two or more methods.

Although great work has been done on the synthesis of MAs or MSs, some challenges remain. One such challenge is the production of structure-controlled MAs or MSs to enhance the adsorption site coverage of the adsorbent material. However, to date, the methods of synthesis have achieved the production of size-controlled MAs or MSs. Another key challenge that should be looked into is the development of simulations that will provide or offer more insights into the design and fabrication of ideal MAs or MSs. This would enable the synthesis of efficient, economical, and eco-friendly MAs or MSs for use in different adsorption processes and applications (wastewater, HMs, dyes, and other contaminants).

References and further reading

[1] Akbarzadeh A, Samiel M and Davaran S 2012 Magnetic nanoparticles: preparation, physical properties and application in biomedicine *Nanoscale Res. Lett.* **7** 144 https://nanscalereslett.com/content/7/1/144

[2] Wu R, Liu J, Zhao L, Zhong X, Xie J, Yu B, Ma X, Yang S, Wang H and Liu Y 2014 Hydrothermal preparation of magnetic Fe_3O_4@ C nano particles for dye adsorption *J. Environ. Chem. Eng.* **2** 907–13

[3] Majidi S, Sehrig F, Farkhani S, Golovjeh M and Akbarzadeh A 2014 Current methods for the synthesis of magnetic nanoparticles *Artif. Cells, Nanomed. Biotechnol.*

[4] Phouthavong V, Yan R, Nijpanich S, Hagio T, Ichino R, Kong L and Li L 2022 Magnetic adsorbents for wastewater treatment; advancements in their synthesis methods *Nanomaterials* **15** 1053

[5] Liu J, Yu Y, Zhu S, Yang J, Song J, Fan W, Yu H, Bian D and Huo M 2018 Synthesis and characterisation of magnetic adsorbent from negatively-valued iron mud for methylene blue adsorption *PLoS One* **1392** E0191229

[6] Ahribesh A, Lazarevic S, Jankovic-Castan I, Jokic B, Spasojevic V, Radetic T, Janackovic D and Petrovic R 2017 Influence of the synthesis parameters on the properties of the sepolite-based magnetic adsorbents *Powder Technol.* **305** 260–9

[7] Meng H, Zhang Z, Zhao F, Qiu T and Yang J 2013 Orthogonal optimisation design for preparation of fe_3o_4 nanoparticles via coprecipitation *Appl. Surf. Sci.* **280** 679–85

[8] Demazeau G 2010 Review. Solvothermal Processes: Definition, Key Factors Governing the Involved Chemical Reactions and New Trends *Z. für Naturforschung* **B*65b*** *999–1006*

[9] Huo Y, Xiu S, Meng L and Quan B 2022 Solvothermal synthesis and applications of micro/nano carbons: a review *Chem. Eng. J.* **451** 138572

[10] Jamshidiyan M, Shirani A and Alahyarizadeh G 2017 Solvothermal synthesis and characterisation of magnetic Fe_3O_4 nanoparticle by different sodium salt sources *Mater. Sci.-Pol.* **35** 50–7

[11] Chen Y, Zhang J, Wang Z and Zhou Z 2019 Solvothermal synthesis of size-controlled monodispersed superparamagnetic iron oxide nanoparticles *Appl. Sci.* **9** 5157

[12] Pachfule P, Das R, Poddar P and Banerjee R 2011 Solvothermal synthesis, structure and properties of metal organic framework isomers derived from a partially fluorinated link *Cryst. Growth Des.* **11** 1215–22

[13] Tarlaj P, Morules M, Veintemillas-Verdaguer S, Gonzalez-Carreno T and Serna C 2003 The preparation of magnetic nanoparticles for applications in biomedicine *J. Phys. D: Appl. Phys.* **36** 182–97 https://iopscience.iop.org(0022–3727136/13/202)

[14] Cotin G *et al* 2018 Unravelling the thermal decomposition parameters for the synthesis of anisotropic iron oxide nanoparticles *Nanomaterials* **8** 881

[15] Maity D, Choo S, Yi J, Ding J and Xue J 2009 Synthesis of magnetite nanoparticles via a solvent free thermal decomposition route *J. Magn. Magn. Mater.* **321** 1256–9

[16] Bakr E, El-Nahass M, Hamada W and Fayed T 2021 Facile synthesis of superparamagnetic Fe_3O_4@Noble metal core–shell nanoparticles by thermal decomposition and hydrothermal methods; comparative study and catalytic applications *RSC Adv.* **11** 781
[17] Daengsakul S, Mongkolkachit C, Thomas C, Siri S, Thomas I, Amornkitbanrung V and Maensiri S 2009 A simple thermal decomposition synthesis, magnetic properties and cytotoxicity of $La_{0.7}Sr_{0.3}Mno_3$ nanoparticles *Appl. Phys.; Mater. Sci. Process.* **96** 691–9
[18] Bokov D, Jalil A, Chupradil S, Suksatan W, Ansari M, Shewael I, Valieu G and Kianfar E 2021 Nanomaterials by sol–gel method: synthesis and applications *Adv. Mater. Sci. Eng.*
[19] Dippong T, Levei E, Petean I, Borodi G and Cadar O 2021 Sol–gel synthesis structure, morphology and magnetic properties of $Ni_{0.6}Mn_{0.4}Fe_2O_4$ nanoparticles embedded in SiO_2 matrix *Nanomaterials* **11** 3455
[20] Amar I, Sharif A, Omer N, Akale N, Altohami F and AbdulQadir M 2018 Synthesis and characterisation of magnetic $CoFe_{1.9}Cr_{0.1}O_4$ nanoparticles by sol–gel method and their applications as an adsorbent for water treatment *Proc. First Conf. Eng. Sci. Technol.*
[21] Li J, Jiang B, Liu Y, Qiu C, Hu J, Qian G, Wenshan G and Ngo H 2017 Preparation and adsorption properties of magnetic chitosan composite adsorbent for Cu^{2+} removal *J. Clean. Prod.*
[22] Thiagarajan S, Sanmugam A and Vikraman D 2017 Facile methodology of sol–gel synthesis for metal oxide nanostructures *Recent Applications in Sol–Gel Synthesis* (London: IntechOpen)
[23] Feng S and Li G 2017 Hydrothermal and solvothermal synthesis *Modern Inorganic Synthetic Chemistry* (Amsterdam: Elsevier) 73–104
[24] Wang P, Sun Q, Zhang Y and Cao J 2020 Hydrothermal synthesis of magnetic zeolite p from fly ash and its properties *Mater. Res. Express* **7** 016104
[25] Zhu S *et al* 2018 Hydrothermal synthesis of a magnetic adsorbent from wasted iron mud for effective removal of heavy metals from smelting wastewater *Environ. Sci. Pollut. Res.* **25** 22710–24
[26] Cregan V, Myers T G, Mitchell S L, Ribera H and Schwarzwälder M C 2016 Nanoparticle growth via the precipitation method *Pro. Ind. Math.* **26** 357–64
[27] Moharram A, Mansour S, Hussein M and Rasheard M 2014 Direct precipitation and characterisation of ZnO nanoparticles *J. Nanomater.* **2014** 716210
[28] Liu F, Zhou K, Chen Q, Wang A and Chen W 2018 Comparative study on the synthesis of magnetic ferrite adsorbent for the removal of Cd(11) from wastewater *Adsorpt. Sci. Technol.* **36** 1456–69
[29] Dahiya R 2018 One pot facile synthesis of nanosized copper oxide by direct precipitation method *Int. J. Res. Anal. Rev.* **5** 478–82
[30] Osman A *et al* 2023 Methods to prepare biosorbents and magnetic sorbents for water treat: a review *Environ. Chem. Lett.* **21** 2337–98
[31] Eleryan A, Aigbe U O, Ukhurebor K E, Onyancha R B, Hassaan M A, Elkatory M R, Ragab S, Osibote O A, Kusuma H S and El Nemr A 2023 Adsorption of direct blue 106 dye using zinc oxide nanoparticles prepared via green synthesis technique *Env. Sci. Pollut. Res.* **30** 69666–82
[32] Emegha J O, Oliomogbe T I, Okpoghono J, Babalola A V, Ejelonu C A, Elete D E and Ukhurebor K E 2023 Green biosorbents for the degradation of petroleum contaminants ed K E Ukhurebor, U O Aigbe and R B Onyancha *Adsorption Applications for Environmental Sustainability* (Bristol: Institute of Physics Publishing) pp 12-1–12-20

[33] Aigbe U O, Ukhurebor K E, Onyancha R B, Okundaye B and Osibote O A 2023 Green nanomaterials in wastewater treatment ed K Pal *Green Nanoarchitectonics* (Jenny Stanford Publishing, CRC Press)
[34] Adetunji C O *et al* 2021 Bionanomaterials for green bionanotechnology ed R P Singh and K R B Singh *Bionanomaterials: Fundamentals and Biomedical Applications* (Bristol: Institute of Physics Publishing) pp 10-1–10-24
[35] Zhou F, Repo E, Sillanpaa M, Meng Y, Yin D and Tang W 2014 Green synthesis of magnetic EDTA-and/or DTPA—cross-Linked chitosan adsorbent for highly efficient removal of metals *Ind. Eng. Chem. Res.* **54** 1271–81
[36] Das C, Sen S, Singh T, Ghosh T, Paul S, Kim T, Jeon S, Matti D, Kim J and Biswas G 2020 Green synthesis, characterization and application of natural product coated magnetite nanoparticles for wastewater treatment *Nanomaterials* **10** 1615
[37] Bassim S, Mageed A and Abdulrazak A 2022 Green synthesis of Fe_3O_4 nanoparticles and its applications in wastewater treatment *Inorganics* **2022** 260
[38] Demazaeu G 2008 Solvothermal processes; new trends in material chemistry *J. Phys.; Conf. Ser.* **121** 082003

Chapter 3

The types and characteristics of magnetic sorbents used for environmental sustainability

Joseph Onyeka Emegha, Timothy Imanobe Oliomogbe, Odunayo Tope Ojo, Faruk Anka Muhammad, Stephanie Clara Akpeji and Kingsley Eghonghon Ukhurebor

In recent times, the utilization of different biomaterials for sorption purposes has presented promising opportunities to alleviate environmental pollution caused by both inorganic and organic substances. Extensive research has been conducted to investigate the potential use of magnetic derivatives in improving the manipulation abilities of biosorbents. Various biomaterials of diverse origins can undergo magnetic modification to acquire intelligent properties, enabling them to respond specifically to external magnetic fields. This chapter provides a concise overview of the different kinds and characteristics of magnetically modified sorbents in terms of their role in promoting environmental sustainability.

3.1 Introduction

The rapid process of urbanization, coupled with the industrial revolution and the increasing pressure from population growth, has significantly impacted the global environment. The various unsustainable practices associated with modern civilization have resulted in substantial waste generation and consequent pollution [1–3]. Furthermore, the heightened pace at which raw materials are being consumed by major industries has led to the disposal of significant amounts of chemical pollutants and radioactive waste in our surroundings. This detrimental practice poses a severe threat to the Earth's biosphere, causing irreversible harm [4, 5]. The production of waste is the primary contributor to the depletion of resources and energy, resulting in negative environmental impacts. The issue of hazardous waste is commonly associated with urban growth and varies significantly between different cities. It is believed that with an increase in industrialization, there will also be a corresponding rise in waste production unless proper scientific waste management practices are implemented [6]. There is a stark contrast between the states of waste management

doi:10.1088/978-0-7503-5909-2ch3

in developed and developing nations; developing nations face challenges in implementing proper mechanisms for waste collection and disposal [4]. Improper disposal practices, such as unregulated dumping on the outskirts of settlements, have resulted in overburdened landfills that cause detrimental effects on the environment. These include contaminating soil and groundwater, compromising air quality, and contributing to global warming. In the majority of developing countries, an estimated 90% of municipal solid waste is improperly managed through open dumps and unregulated landfills [4, 5]. This unscientific method not only leads to severe air, water, and soil pollution but also prevents the proper utilization or the realization of the extraction value of the waste materials [4, 7–9].

Currently, there is a growing trend to use metals in various process industries. Consequently, the waste generated from these industries often contains metal ions that pose significant and long-lasting environmental contamination because of their nonbiodegradable properties [10]. Heavy metal (HM) water pollutants such as chromium, zinc, lead, mercury, and cadmium have been receiving increasing attention. This can be mainly attributed to the extensive utilization of HMs in various industrial sectors such as chrome plating, dye production, and battery manufacture [11, 12]. Today, numerous researchers are focusing their attention on the elimination of harmful substances from soil and effluents [13]. Accordingly, extensive research has been conducted by scientists to explore a variety of innovative processes aimed at efficiently eliminating pollutants from contaminated water and soil.

In addition to the three primary categories of remediation methods, namely chemical processes, physical techniques, and biological approaches [2, 11], a variety of mechanisms are commonly employed to efficiently reduce or eliminate pollutants. These include electron-beam irradiation, incineration, extraction, air sparging, biodegradation, and adsorption. These methods have been extensively employed in environmental remediation efforts [14]. The removal of pollutants through adsorption is commonly employed in this field due to its straightforward application, cost-effectiveness, and environmental friendliness [15]. However, the effectiveness of adsorption as a method for removing pollutants is dependent on the characteristics of the different materials employed as adsorbents. These materials include activated carbon, zeolites, iron oxides, and silica [2, 11, 15].

The utilization of biological resources such as bacteria, algae, and fungi has proven to be an ecologically sustainable and economically efficient way to effectively remove and retrieve contaminants through various mechanisms [4, 11, 16–18]. Biosorbents are readily available at low cost and are effective for adsorption purposes. There have been numerous recent reports of the development of heavy metal remediation using immobilized microorganisms [15, 19]. Furthermore, the application of magnetic separation has emerged as a prospective technique for addressing environmental remediation concerns, primarily due to its efficacy in eliminating contaminants without giving rise to supplementary pollutants such as flocculants. Moreover, this method enables the prompt and effective elimination of substantial amounts of waste within a brief timeframe [20]. Magnetic particles have demonstrated significant promise as effective sorbents for the elimination of heavy metal ions by means of adsorption. The use of an external magnetic field addresses the issue of separation, guaranteeing optimal and proficient outcomes [11]. The utilization of both bioadsorption and magnetic separation techniques presents various benefits, including improved ecological compatibility, economic feasibility, and versatile implementation. This chapter describes the characteristics of magnetic nanoparticles and proposes their use as green sorbents for environmental cleanup solutions.

3.2 The properties of magnetic sorbents

Magnetic nanoparticles possess a range of fascinating characteristics, including their size-dependent behavior, interaction capabilities, magnetic separation abilities, and their specificity for target applications with a particular surface chemistry. Magnetic nanoparticles, such as magnetite, nano zero-valent iron, and maghemite, are being increasingly applied in the domains of medicine, molecular biology, and environmental remediation [21]. These versatile materials offer promising solutions with which to tackle a range of issues including environmental pollution [21]. Typically, magnetite or maghemite is frequently chosen as the primary material for magnetic iron oxide nanoparticles (MIONs).

MIONs can be categorized into three main classes based on their behavior: paramagnetic, ferromagnetic, and superparamagnetic [22, 23]. The paramagnetic nature is characterized by random orientations of magnetic dipoles at typical temperatures, caused by unpaired electrons. This leads to a small positive susceptibility and the occurrence of weak interactions when subjected to a magnetic field. Ferromagnetic materials maintain their magnetization due to the structure of their domains, even without an external magnetic field. However, when these materials undergo a significant change, such as a decrease in particle size to a size less than the domain size, their ferromagnetic properties are affected [22]. Superparamagnetic materials typically exhibit higher magnetic susceptibility compared to paramagnetic materials due to the alignment of the entire nanoparticle's magnetic moment with the direction of the applied magnetic field. The magnetic properties of nanoparticles play a crucial role in their various applications [22]. Figure 3.1, as modified and redrawn from [22], shows the hysteresis loops demonstrating the magnetization behavior of superparamagnetic nanoparticles.

Generally, the utilization of functionalized magnetic particles (FMP) as adsorbent materials in magnetic solid-phase extraction (MSPE) has become a leading approach within the analytical field [24]. To achieve this objective, the target material is isolated from the liquid solution through the utilization of an external magnetic field. This is done by introducing functionalized magnetic particles into the sample and allowing these particles to absorb the analyte onto their surface. The FMPs possess the ability to disperse again in a solution when the external magnetic field is no longer present. This characteristic enhances the ease of washing and desorption processes [11, 22], as illustrated in figure 3.2. Moreover, FMPs have the potential to enhance mass transfer through the expansion of the contact surface area between the analyte and solution. [25].

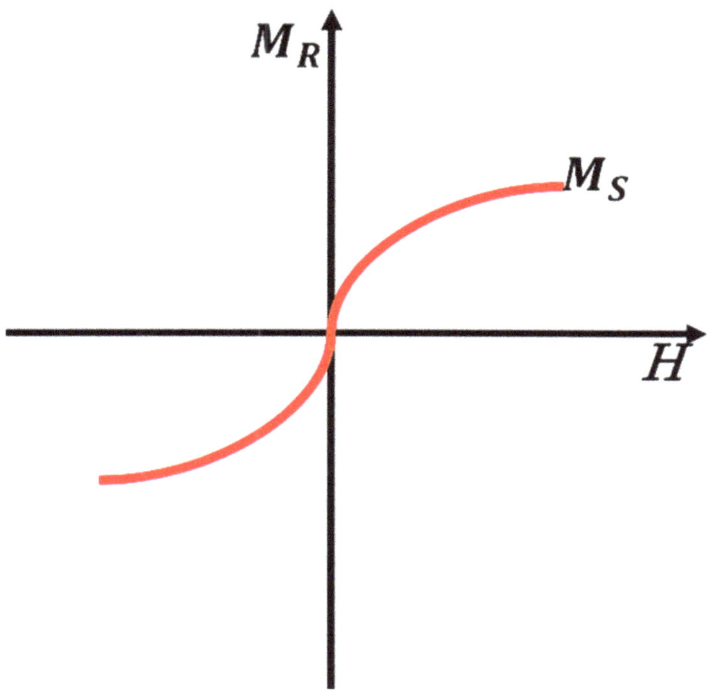

Figure 3.1. Hysteresis loop demonstrating the magnetization behavior of superparamagnetic nanoparticles.

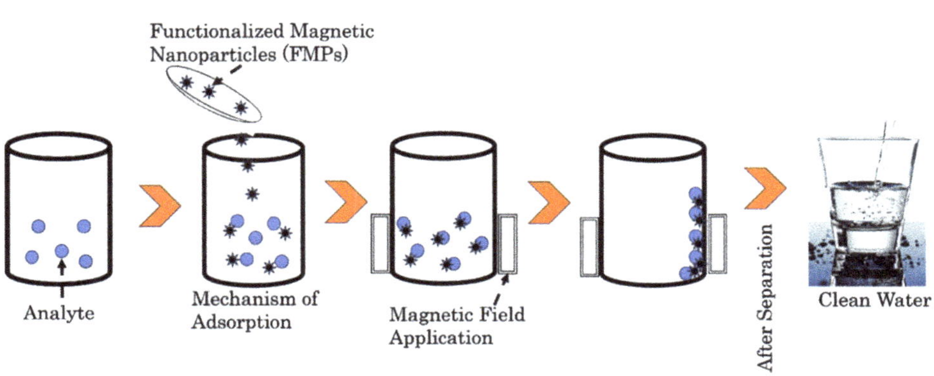

Figure 3.2. A diagram illustrating the use of functionalized particles in MSPE.

Iron oxides (Fe_3O_4 and Fe_2O_3) are frequently chosen from the various FMPs due to their favorable attributes. These include the ease of preparation on a large scale, due to a straightforward and efficient process. Additionally, iron oxides boast numerous functional groups on their surface, which allows for convenient modifications. Their superparamagnetic properties eliminate the need for centrifugation or filtration during operation, making them highly practical. Moreover, these materials can be easily recovered and reused while also exhibiting excellent dispersibility in aqueous solutions that facilitates the rapid attainment of extraction equilibrium [25]. In contrast to traditional solid-phase extraction methods, MSPE offers the advantage of convenient sample pretreatment. This achievement is facilitated by the efficient and seamless separation of phases accomplished through the utilization of an externally applied magnetic field [26]. The use of magnetic sorbents has become increasingly popular due to their effectiveness in various areas, including science, technology, and engineering. Moreover, they are being increasingly employed for the analysis of trace metals. In recent times, there has been a notable increase in the implementation of advanced methods for assessing and extracting heavy metal ions from contaminated samples using micro- or nano-magnetic particles [11, 24].

3.3 Magnetic sorbent materials

Different types of sorbent materials, like ion exchangers and polymers such as C18 or octadecyl, carbon nanotubes, activated carbon, and magnetic particles have effectively been utilized in the process of solid-phase extraction. Magnetic materials such as nickel, iron, and cobalt, as well as their corresponding oxides, are commonly used in the composition of magnetic sorbents [1, 9, 11, 22]. These materials can also be combined with superparamagnetic or ferromagnetic components to enhance their magnetic properties. When magnetized, the magnetic nanoparticles act as small permanent magnets [11]. As a result of their magnetic properties, they can form aggregates or structures resembling lattices due to the interactions between them [24]. Moreover, ferromagnetic particles exhibit a continuous magnetic field that enables them to easily adopt a lattice structure after the external magnetic field is no longer present. On the other hand, superparamagnetic particles do not retain any residual magnetism once the magnetic field has been eliminated [24]. Permanent magnetic particles are available in various sizes ranging from the microscale to the nanoscale (1–100 nm). The exceptional ability of these particles to disperse, combined with their significant surface area and high surface-to-volume ratio, has ignited considerable interest in technological research. As a result of these properties, they possess an effective adsorption capacity [27, 28]. Nanosized ferromagnetic particles possess immense potential in the fields of environmental and biological analysis. This is due to their ease of surface modification, simplified large-scale production methods, and effective recyclability [29]. Inorganic magnetic materials, such as Fe_3O_4, have a natural inclination to arrange in lattice formations that impact their magnetic characteristics. However, their limited specificity makes them unsuitable for samples containing multiple components. To overcome this constraint, it is imperative to alter the magnetic coating of nanoparticles by affixing distinct active groups to them [28]. This process, known as 'shell' coating, involves applying coatings onto the surface of FMPs [11, 24, 30]. The coating process can add inorganic substances such as silica and alumina or organic compounds like surfactants or polymers. This is done to enhance their chemical stability, resistance to oxidation, and ability to selectively absorb certain ions [30–32]. These FMP-attached shells serve various analytical and environmental pollutant treatment purposes [24]. A possible approach for advanced environmental remediation involves the use of a magnetically stabilized fluidized bed. In this method, fine magnetic particles can be made to fluidize within a filter column that is held in place by magnets. The waste effluent can then be pumped through the column for treatment [33].

To enhance the analytical capabilities and environmental remediation processes, the use of 'shells' attached to FMPs is common practice. These shells serve various purposes, including the treatment of environmental pollutants [33]. One method for advanced environmental remediation is the utilization of a magnetically stabilized fluidized bed [33]. FMPs can be efficiently immobilized and confined to a magnetic filter column during this procedure, allowing the waste effluent to flow through the column with ease [34]. This enables the efficient separation and treatment of pollutants. Complex FMPs, such as titano-magnetite, have emerged as a viable alternative [34]. Certain specific microorganisms possess the unique ability to produce inorganic FMPs (such as ferrous sulfate) through metabolic processes [34]. These FMPs have the potential used in the efficient and environmentally friendly removal of HMs from various sources [35]. Certain types of magnetotactic bacteria found in nature can orient themselves parallel to the direction of a magnetic field. Enzymatic processes have proven to be highly effective in efficiently eliminating organic pollutants from wastewater [36]. The use of magnetite particles coated with surfactants has gained attention among researchers for removing organic pollutants such as 2-hydroxyphenol [37]. Polymer-coated magnetic sorbent materials with a composite coating of vermiculite and iron oxide have been utilized to effectively and rapidly separate oil pollutants from water spills [38].

3.4 The characterization of magnetic sorbents

To fully harness the potential of FMPs as sorbent materials, it is essential to establish comprehensive guidelines for characterizing engineered nanomaterials [39, 40]. As in the case of larger materials, nanomaterials also necessitate the assessment of various attributes including molecular structure and chemical composition [39]. In addition, nanomaterial characterization is highly concerned with various aspects including dimensions and uniformity, porosity or pore size, surface area measurement, shape and structural analysis, wetting properties evaluation, zeta potential determination, adsorption behavior assessment (in terms of capacity and propensity for aggregation), as well as the distribution of bonded constituents and impurities [1, 11, 12, 38–41]. The inclusion of magnetic nanomaterials in environments or their incorporation into materials can give rise to diverse undesirable outcomes, such as clustering, clumping, and indiscriminate uptake by other substances. These interactions may originate from various intermolecular associations occurring at the interfaces between nanomaterials and molecules, as well as systems that enable such interactions [39–41].

Magnetic materials' surface characteristics in a given medium depend on their physical and chemical properties. In addition, several factors, such as the presence of organic macromolecules, temperature, ionic strength, and pH, have a significant impact on additional characteristics, including surface charge, dissolution rate, hydration behavior, and stability in size distribution in dispersion. The likelihood of nanoparticle combination or aggregation also has an effect [39, 40]. By carefully investigating these features, we can gain a more thorough understanding of the behaviors displayed by nanomaterials when they interact with live creatures. The following are brief explanations of numerous methodologies used to investigate the unique physical and chemical features of magnetic nanoparticles, as well as their primary strengths and drawbacks for technological investigations [39–41].

3.4.1 Scanning electron microscopy

Electron microscopy (EM) employs accelerated electron beams and electromagnetic or electrostatic lenses to generate high-resolution images. This is achieved utilizing the shorter wavelengths of electrons, which are smaller than those of visible light photons. Scanning electron microscopy (SEM), specifically used for surface imaging, entails scanning an electron beam across a sample's surface. Through interaction with the specimen, this process generates signals that offer insight into its atomic composition and topographic characteristics [42, 43]. The incident electrons result in the emission of elastic scattering, specifically backscattered electrons. Moreover, the phenomenon of low-energy secondary electrons can emerge because of inelastic scattering and cathodoluminescence. This latter term denotes the generation of distinct x-ray radiation from atoms situated either on or adjacent to the surface material within the examined sample [43]. In SEM, the detection of secondary electrons is a widely used method for detecting emissions. It has the capability to achieve resolutions smaller than 1 nm [43].

The dimensions, size distributions, and morphologies of nanomaterials can be directly measured using SEM. It is important to note that the drying and contrast enhancement procedures may result in specimen shrinkage and modification of the properties inherent to the nanomaterials [44]. Moreover, when subjected to an electron-beam scan, biomolecular samples that lack conductivity often accumulate charge and fail to properly redirect the electron beam. As a result, this can cause imaging defects or distortions in the resulting images. To prepare the sample, it is often necessary to apply a thin layer of electrically conductive material onto the biomolecules [44]. Due to the necessity for cryogenic freezing in electron microscopy to image the surface groups attached to nanoparticles, it is not possible to study the size of nanomaterials under physiological conditions [44, 45].

One notable example is environmental SEM (ESEM), which allows for the imaging of samples in their native condition without any alterations or prior preparations [46, 47]. The ESEM sample chamber operates within a controlled atmospheric environment, specifically in a low-pressure gas range of 10–50 Torr. Furthermore, it maintains optimal humidity levels to effectively eliminate any charging artefacts that may occur. As a result, it is no longer necessary to coat samples with conductive materials for enhanced conductivity [47]. One limitation that is common among several EM techniques, such as SEM, is the requirement for sample preparation methods that can be destructive. This limitation prevents analysis using alternative methods [47]. Furthermore, the occurrence of biased statistics for the size distribution of diverse samples is inevitable in SEM because of the limited number of particles analyzed within the scanning region [39]. Through SEM analysis, Adlnasab *et al* [48] conducted a study of the physical characteristics of a novel magnetic biosorbent designed for removing arsenic from polluted water sources. It was found that the sorbent exhibited a layered structure with agglomerated magnetic nanoparticles. Another study by Abdul Rahman *et al* [49] also reported similar findings of agglomerated, spherical, and rough images measuring 3.51–3.95 mm in diameter for an inorganic sorbent intended for heavy metal removal from industrial wastewater.

3.4.2 Transmission electron microscopy

Transmission electron microscopy (TEM) is widely utilized in the field of nanomaterial characterization. It offers high-resolution images and chemical analysis, enabling researchers to observe nanomaterials at atomic scales [50]. In the traditional TEM mode, a thin foil specimen is permeated by an incident electron beam. Throughout this process, incident electrons interact with the specimen, resulting in their transformation into unscattered electrons and various forms of scattered electrons, including elastically and inelastically scattered electrons [51]. The primary factor determining the magnification of TEM is the positioning of the objective lens in relation to both the specimen and its image plane. The distance separating these elements plays a critical role in achieving the desired magnification levels [51]. Electrons are steered via a series of electromagnetic lenses before being projected onto a screen, whether they have been scattered or remain unaltered. This device generates diverse types of images, including patterns from electron diffraction, images with variations in amplitude contrast, phase-contrast images, and representations resembling shadows. The intensity levels within these images are determined by the concentration of unscattered electrons [51].

The high spatial resolution of TEM allows for detailed examination and analysis of the morphology and structure of nanomaterials. Furthermore, TEM can be combined with various analytical techniques to support a range of applications. For example, electron energy loss spectroscopy allows for accurate chemical analyses, while energy dispersive x-ray spectroscopy provides a quantitative exploration of the chemical composition and electronic structure of nanomaterials [46]. Both TEM and SEM can analyze the size, shape, aggregation levels, and dispersion characteristics of nanomaterials. However, TEM offers superior spatial resolution compared to SEM and allows for additional analytical measurements [44].

Despite the benefits of TEM, it is important to acknowledge that there are also certain disadvantages associated with its use [51]. One limitation to consider is the requirement for a strong vacuum and a reduced sample thickness when performing TEM measurements to facilitate penetration by the electron beam [44].

Typically, when utilizing high-resolution electron microscopy for imaging purposes, it is possible to examine only a small portion of the specimen within a given timeframe. As a result, there may be inadequate statistical sampling. Furthermore, the use of 2D TEM for examining 3D specimens can lead to numerous artefacts and a lack of depth sensitivity in individual TEM images. A further constraint pertains to the requirement of thin specimens that are capable of effectively transmitting electrons for image generation. Specifically, when performing high-resolution TEM or electron spectroscopy, specimen thicknesses below 50 nm are necessary. Conducting thorough preparations of thin specimens significantly raises the chance of modifying the structure of the sample, resulting in a time-consuming process for TEM analysis. A significant concern with TEM technology is the risk of specimens being harmed or destroyed by powerful electron beams. However, there is an intriguing approach known as wet TEM that enables the analysis of nanomaterials in a liquid environment. This technique allows for the determination of parameters such as particle size, dispersion, displacement (dynamic), and agglomeration [52, 53]. Furthermore, a newly developed wet-scanning TEM imaging system has been designed to facilitate the observation of samples fully immersed in a liquid phase. This innovative system overcomes the challenges related to low contrast and potential shifts of objects that often arise when ESEM is used to capture images of a liquid's surface [54]. Wet-mode scanning TEM allows for the observation of nanoscale details with high contrast, even when water is present at thicknesses of up to a few micrometers. This can be achieved without the need for additional contrast agents or stains. [39].

A study conducted by He *et al* [55] examined TEM images and observed the presence of spherical nanoparticles with an average diameter of 10 ± 2.5nm. In contrast, Khodosova *et al* [56] reported that cobalt ferrite ($CoFe_2O_4$) particles used as magnetic nanosorbents exhibit irregular shapes when combined with bentonite and spinel, ranging in size from 20 to 120 nm. The findings from TEM analysis showed that the distribution of spinel nanoparticles within the aluminosilicate phase is statistically spread throughout its volume. This spatial arrangement leads to a homogenizing effect on micropore contributions and significantly decreases both the mesopore volume and the specific surface area of the sorbent by approximately 29%–35% [56].

3.4.3 Atomic force microscopy

Atomic force microscopy (AFM) is a valuable technique that, like SEM and TEM, provides insights into numerous properties of nanomaterials. These encompass dimensions, morphology, composition, adsorption characteristics, dispersibility, and aggregation tendencies. AFM employs distinct modes for scanning purposes, such as noncontact mode, contact mode, and intermittent sample contact mode [45, 57]. AFM enables the analysis of nanomaterials' size and shape in a physiological setting and provides insights into the interactions between these materials when introduced to biological systems. For example, it enables real-time observation of how nanomaterials interact with supported lipid bilayers, an aspect that cannot be achieved using traditional electron microscopy techniques [58]. AFM is becoming increasingly valuable in the field of biomaterial imaging, as it enables visualization without significant harm to a wide range of natural surfaces. The primary advantage of AFM lies in its ability to capture detailed images of different types of materials at an incredibly precise subnanometer scale, even when immersed in aqueous solutions. [59]. One significant limitation, however, is that the dimensions of the cantilever tip often exceed those of the nanomaterials being studied. As a result, there is an undesired overestimate in lateral size measurements for the samples [46, 47]. Unlike fluorescence techniques, AFM does not have the inherent capability to precisely identify or locate individual molecules. Nevertheless, recent progress in single-molecule force spectroscopy has overcome this limitation by incorporating ligands, cell adhesion molecules, or chemical groups onto AFM cantilever tips. These modified probes now enable the detection of individual functional molecules on cell surfaces [60].

3.4.4 Raman scattering

Raman scattering (RS) is a commonly utilized technique in the field of nanomaterials and nanostructure analysis. This method offers high-resolution imaging capabilities for light-transparent materials without the need for sample preparation, allowing for convenient *in situ* experimentation [61]. RS is a technique used to investigate the occurrence of inelastic photon scattering. During this process, the photons of incident light with varying frequencies interact with the electric dipoles found in molecules [62].

RS is a phenomenon that occurs when incident photons interact with molecules, resulting in the generation of scattered photons at different frequencies. These frequency shifts are associated with the vibrational states of the molecules. In the Raman spectrum, photons that are scattered at lower frequencies are referred to as Stokes lines, whereas those scattered at higher frequencies are called anti-Stokes lines [62]. In the realm of spectroscopy, Raman and infrared (IR) techniques are often regarded as complementary to each other. It is commonly observed that vibrational modes which exhibit activity in Raman spectroscopy tend to be inactive in IR spectroscopy, and vice versa. This trend holds particularly true for small symmetrical molecules, where Raman transitions occur due to the modulation of molecular polarizability by nuclear motion rather than through a significant change in dipole moment [62]. An important benefit of RS is its applicability in analyzing biological samples immersed in aqueous solutions. This suitability stems from the weak scattering of water molecules, making them suitable for RS measurements. Moreover, the comprehensive molecular data provided by RS can be utilized to examine the structures and levels of tissue components. This highlights the potential of RS in identifying irregularities within tissues [63]. On the other hand, while conventional RS techniques are capable of indirectly characterizing nanomaterials by examining changes in spectral line broadening and shift to determine average size and distribution, they lack the spatial resolution necessary for distinguishing between various domains. This limitation hinders their application in nanotechnology [61].

Other limitations associated with conventional RS methods include susceptibility to external factors such as fluorescent interference and a limited cross-sectional area. This necessitates the use of high-intensity laser excitation and substantial quantities of sample material to generate sufficiently strong RS signals [64]. On the other hand, the use of surface-enhanced RS (SERS) has proven to be highly effective in intensifying RS signals and improving spatial resolution. This technique involves immobilizing molecules on metallic structures [65, 66]. SERS has various applications in metallic nanoparticle research. First, it allows for the analysis of surface functionalization on these particles. SERS can also be used to monitor changes in the conformation of proteins that are bound to metallic nanoparticles [39].

Tip-enhanced RS (TERS) has emerged as an innovative technique through the application of a novel approach inspired by Raman near-field scanning optical microscopy. This method involves confining the light field to surpass diffraction-limited resolution, allowing for enhanced results in analysis. Unlike traditional methods that utilize an optical fiber, TERS exploits the use of an apertureless metallic tip to enhance surface signals in Raman spectra (known as the SERS effect) [67, 68]. In comparison to conventional RS methods, SERS and TERS techniques not only reveal insights into the structural, chemical, and electronic characteristics of nanomaterials but also offer valuable topographical information. Traditional RS is limited to providing structural, chemical, and electronic information only. The enhanced functionality of SERS and TERS sets them apart from conventional RS techniques. However, achieving consistent measurements in SERS can pose challenges due to the inherent variability in particle shapes and undesired particle aggregation. This poses a difficulty when using SERS for imaging studies conducted in living organisms or laboratory environments [61, 65].

3.4.5 Infrared spectroscopy

In general, a molecule can absorb IR radiation when it experiences variations in its dipole moment over time and these oscillations align with the frequency of the incident IR light [42, 69]. The material absorbs IR radiation, leading to the transfer of energy and subsequent changes in covalent bond stretching, bending, or twisting. These alterations can be described as stationary states in the molecular vibrational Hamiltonian that correspond to normal modes [62]. Materials that lack dipole moments, such as N_2 and O_2 diatomic molecules, are unable to absorb IR radiation [69]. Typically, the vibrations observed within a molecule are associated with pairs of interconnected atoms or covalent bonds. To gain a comprehensive understanding of their characteristics, it is necessary to examine these components as combinations of normal modes. As a result, the IR spectrum can provide valuable insights into the molecular structure by demonstrating absorption or transmission based on the incident IR frequency [62].

Fourier-transform IR (FTIR) spectroscopy is an extensively used technique in the field of nanomaterial applications. It utilizes specific spectral bands to detect and analyze the conjugation between nanomaterials and molecules, such as proteins binding to nanoparticle surfaces. In addition, FTIR spectroscopy can provide valuable insights into the conformational states of bound proteins. Furthermore, this technique has demonstrated effective results when studying materials at the nanoscale, such as validating the presence of functional molecules that are covalently bonded to carbon nanoparticles [70, 71].

Attenuated total reflection Fourier-transform IR (ATR-FTIR) spectroscopy is an innovative method that combines the use of total internal reflection and IR spectroscopy to examine the molecular composition of materials found on solid–air or solid–liquid interfaces. This technique offers advantages such as the avoidance of complex sample preparation procedures and improved reproducibility compared to traditional IR spectroscopy methods [42]. ATR-FTIR systems produce evanescent waves inside an internal reflection element (a crystal) through total internal reflectance. This allows them to propagate into the sample material itself, reaching depths ranging from 0.5 to 5 μm. It is worth noting that these waves rapidly decrease in intensity as they move away from the interface [69].

The ATR-FTIR technique is commonly used to analyze changes in surface properties and identify the chemical characteristics of polymer surfaces. This technique entails the absorption of IR waves that are nearly disappearing, operating at frequencies aligned with the vibrational modes of the sample present at the interface between the internal reflection element and the sample. ATR-FTIR spectroscopy proves useful in examining the surface properties of nanomaterials [39, 69]. Nevertheless, this technique has certain constraints in terms of sensitivity when analyzing nanometer-scale surfaces, as its penetrative ability is restricted by the incident IR wavelength [39]. Khodosova *et al* [56] conducted a study of the FTIR spectra of nanocomponent sorbents created by doping natural bentonite with nanosized cobalt ferrite spinel. Their analysis revealed that the prominent component in the spectra was dioctahedral smectite, accompanied by impurities such as quartz and kaolinite. Notably, there were two distinct ranges of vibrations specific to dioctahedral smectite which allowed for the identification of the band corresponding to its stretching vibrations. Furthermore, another study also highlighted that FTIR analyses performed on magnetic sorbents developed for arsenate removal exhibited spectral characteristics falling within certain ranges (400 – 4000 cm^{-1}) [48].

3.4.6 Nuclear magnetic resonance

Nuclear magnetic resonance (NMR) is a highly effective analytical technique that offers insights into the local environment of materials, particularly amorphous substances, polymers, and molecules [39]. Unlike imaging and diffraction methods that offer insight into the overall structure of crystals, NMR delves into specific characteristics at a more localized scale. This ability makes it highly advantageous for studying materials that lack long-range order or crystalline properties [72]. Moreover, NMR spectroscopy provides valuable methodologies for investigating the dynamic interactions among different substances under varying conditions, in addition to analyzing their structures and compositions [72]. A range of different measurements can be used to evaluate relaxation, molecular structure, and molecular movement. These measurements rely on the use of specially created sequences involving radiofrequency and/or gradient pulses [72].

The application of NMR spectroscopy has proven valuable in studying and characterizing different physiochemical properties of nanomaterials. This method enables the analysis of structural composition, level of purity, and functional attributes in diverse substances such as dendrimers, polymers, and fullerene derivatives. Furthermore, NMR can effectively examine alterations in molecular conformations that arise during interactions between ligands and nanomaterials [39, 58]. Pulsed-field gradient NMR has been employed to evaluate the diffusive properties of nanomaterials. This technique allows for calculations regarding the sizes and interactions between different species under investigation. Unlike optical methods, NMR is nondestructive and does not necessitate complex sample preparation procedures. However, it should be noted that the relatively lower detection sensitivity of NMR necessitates a larger quantity of the sample to obtain accurate measurements [45]. The utilization of high-resolution magic angle spinning NMR has gained significant popularity in the fields of biology and biomedicine due to its ability to generate spectra that are on par with those produced by high-resolution NMR techniques. This technique is particularly valuable for examining tissues and cells with diverse properties, although it may be time-intensive when a specific signal-to-noise ratio needs to be attained for spectral analysis [73]. The utilization of high-resolution magic angle spinning NMR has proven to be highly effective in accurately analyzing ligands affixed to surfaces and altered surfaces. This technique has been successfully employed in the examination of the synthetic processes involved in immobilizing cyclo-peptides on nanoparticles based on poly(vinylidene fluoride), as well as investigating thiol-derivatized silver clusters produced through thermolysis [39, 73].

3.4.7 X-ray diffraction

X-ray diffraction (XRD) is a crucial tool in several x-ray spectroscopy techniques because it accurately identifies the atomic-scale structures of crystalline materials [45, 62]. XRD is a scientific method in which an intense beam of x-rays is projected onto the surfaces of a crystalline sample to gather valuable insights about its atomic structure. This technique operates based on the fundamental principles described by Bragg's law [62].

XRD utilizes x-ray scattering to analyze and determine the dimensions, morphology, and lattice distortion of crystalline substances. It might not be as useful for research into disorganized materials [45]. Despite being a widely used technique for studying material structures at the atomic level, the XRD technique's practical application is hampered by issues such as crystal growth challenges, and one of its drawbacks is that it cannot provide results for multiple conformations or binding states of the sample [45]. Another restriction of the technique is the limited intensity of the diffracted x-rays, which is especially noticeable in materials with low atomic numbers. When compared to electron diffractions, which yield higher intensities in such instances, these limits become more obvious [74, 75]. A recent study used XRD to demonstrate a groundbreaking technique for interpreting the structure of big molecules that used ultrafast pulses from a hard x-ray free-electron laser. This approach shows promise for situations in which conventional radiation sources are

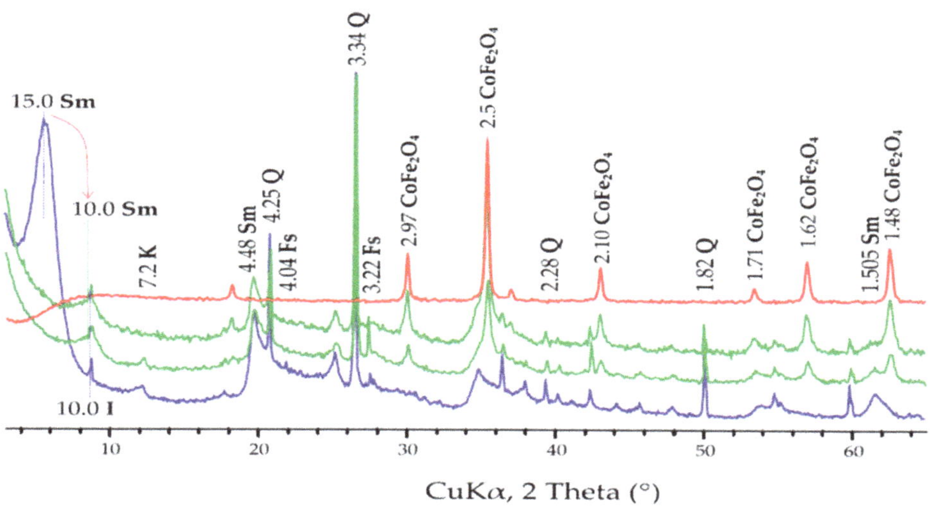

Figure 3.3. XRD patterns of investigated samples of $CoFe_2O_4$ and composites. Reproduced from [56]. CC BY 4.0.

unable to provide a result due to the lack of a crystal of sufficient size or there is a concern about a lack of radiation damage resistance in the sample [39, 76].

The XRD pattern shown in figure 3.3 depicts the structures and compositions of bentonite and cobalt ferrite ($CoFe_2O_4$) spinels, which are magnetic nanosorbents [56]. This XRD analysis provides valuable information about the chemical composition of the fabricated cobalt ferrite, indicating its constituent elements: Co 14.44, Fe 28.73, and O (56.83). Moreover, it acts as a catalyst on the surface of the sorbent, leading to new properties [56]. Previous studies have documented comparable diffraction patterns in both unmodified and modified biochar materials while investigating their effectiveness in removing Cr and acid orange 7 dye from aqueous solutions [77], as well as the utilization of Fe_2O_3fly-ash-derived magnetic adsorbents sourced from Pulang Pisau's power plant [78].

3.4.8 Small-angle x-ray scattering

Small-angle x-ray scattering (SAXS) can offer useful information about a variety of substances, including both crystalline and amorphous materials such as polymers, proteins, and nanomaterials, unlike XRD, which is exclusively applicable to crystalline materials. SAXS is therefore a flexible method for examining a range of properties of these compounds [45]. The SAXS technique examines the elastic scattering of a fraction of an x-ray beam produced by a sample, which generates a two-dimensional scattering pattern on an x-ray detector situated at a right angle to the direction of the incoming beam. SAXS is instrumental in analyzing polymer and nanomaterial-bioconjugate systems that are dissolved in solution, allowing their characterization. By analyzing scattered x-rays within a specific range of angles

(0.1°–3°), valuable information about their orientation, size distribution, shape, and overall structure can be obtained through SAXS analysis [45].

SAXS uses small-angle scattering to analyze non-repetitive structures, eliminating the need for samples to be precisely crystallized. This simplifies the preparation process and makes SAXS a noninvasive technique [76]. Alternatively, SAXS measurements can offer comprehensive insights into the structure of a material, capturing averaged characteristics instead of focusing on specific grain-level observations [76]. However, the use of this feature might be unfavorable when a high level of resolution is needed. In contrast, recent developments in SAXS have facilitated improved measurement accuracy. This has been achieved through the integration of synchrotrons as a high-energy x-ray source, yielding higher-resolution results [76].

X-ray spectroscopic techniques, such as x-ray absorption spectroscopy, provide valuable insights into the chemical state and symmetries of the x-ray absorption site. These techniques offer a range of structural information, including coordination numbers and the interatomic distances of ligands and nearby atoms connected to the absorbing element. By analyzing the spectra of x-ray absorption near edge structures, it is possible to extract details about their structural characteristics [79].

Modifications in the porous composition of biocarbons derived from pulp feedstocks have been investigated through an analysis of SAXS spectra. The observations indicated a rise in scattering intensity and pore radii within the range of 4.5–35 nm [80]. Through an examination of SAXS data, it has been determined that non-magnetically activated carbon exhibits a smaller specific area compared to carbon synthesized through one-stage and two-stage processes [80]. In another study, the scattering data from hypercrosslinked polystyrenes and the corresponding nanocomposites was collected using a SAXS diffractometer. The results indicated that the average size and distribution values aligned commendably with the findings of prior research [81].

3.4.9 Scanning tunneling microscopy

The scanning tunneling microscopy (STM) method creates high-resolution pictures of the electron density on conductive or semiconductive surfaces using a quantum tunneling current. Using this technique, it is possible to see atomic-scale features and determine whether molecules are affixed to conductive substrates [39]. A typical step in scanning probe microscopy involves placing a sensitive probe close to an object's surface to see and study its reactions. The basic parts of an STM system include a highly accurate scanning tip, an *x*–*y*–*z* piezo scanner for controlling the tip's horizontal and vertical movements, a coarse control unit to place the tip close enough to the sample to achieve the desired tunneling range, a vibration isolation stage, and feedback regulation electronics [39, 74]. Applying a low voltage causes electron tunneling by maintaining some gaps between the tip and the sample. As the scanning tip moves across the sample in the *x*–*y* plane, changes in the current response are produced. Ultimately, a charge density map is produced [82]. In an alternative approach, the use of feedback electronics allows for the adjustment of tip height to accurately capture and image the topography of the sample [82]. When it comes to characterizing molecules using STM or electron microscopy techniques, a common approach is to embed the samples in a matrix to maintain their natural structures. Subsequently, a thin metallic layer, such as gold, is applied to the sample surfaces prior to image capture [83]. Conventional electron microscopy techniques are unable to visualize some molecules in their natural state without the use of a time-consuming sample preparation process. However, STM offers several advantages over EM techniques by not only addressing these limitations but also providing high-resolution imaging at the atomic scale utilizing specialized tips made of a platinum-iridium (Pt-Ir) alloy with extremely sharp ends [83]. Despite the numerous potential benefits offered by high-resolution STM in analyzing nanoscale materials, such as providing insights into their dimensions, morphology, architecture, and level of aggregation and dispersion, a scarcity of studies utilizing gold or carbon substrates has been observed [84]. One of the primary challenges of this technique lies in ensuring that the sample's conductive surface meets the requirements and effectively detecting its electronic structure on the surface [84]. Regrettably, most materials are non-conductive, and it is not always straightforward to establish a direct correlation between the electronic properties of their surfaces and their surface topography. However, STM continues to be a highly valuable method for analyzing the atomic structures of conductive materials such as graphene, fullerenes, and carbon nanotubes [84].

3.4.10 Mass spectrometry

Mass spectrometry (MS) is a widely utilized analytical method in scientific investigations that enables the analysis of particles or molecules by examining their mass, elemental composition, and chemical structure [85]. The fundamental concept behind this technique involves the separation of charged particles according to their unique mass-to-charge ratios [39].

Mass spectrometry offers a significant level of accuracy and precision in determining molecular weight. In addition, it provides high detection sensitivity, requiring only very small amounts of samples ranging from 10^{-9} to 10^{-21} mol. Different mass spectrometry techniques are frequently employed to accurately depict various physicochemical properties of nanomaterials, such as their mass, composition, and structure. These methods differ in terms of the ion sources used, separation techniques applied, and detector systems utilized [47, 85]. Two commonly employed methods for ionization in combination with mass spectrometry analyzers are matrix-assisted laser desorption/ionization (MALDI) and electrospray ionization. These techniques efficiently convert molecular derivatives into ions by evaporating them at lower temperatures to minimize fragmentation or degradation. On the other hand, inductively coupled plasma ionization is predominantly used for analyzing nanomaterials that contain metal elements [46, 47]. Various mass spectrometry methods have significant utilizations in the field of nanomaterial research. For instance, time-of-flight MS (TOF-MS) is employed to ascertain the size and dispersion patterns within these materials. Another technique, known as MALDI-TOF-MS, facilitates the determination of molecular weights in macromolecules, polymers, and dendrimers. In addition, this method can be effectively utilized for examining protein nanoparticle interactions [54, 85]. To verify the successful binding of a modified contrast agent to a functionalized nanoparticle, researchers employ inductively coupled plasma mass spectrometry. In addition, secondary ion mass spectrometry is utilized to analyze and evaluate the elemental and molecular characteristics of the outer layer of these nanoparticles. Although mass spectrometry techniques are widely used for analyzing the physicochemical properties of various molecules, accurately identifying molecular species still presents challenges due to limitations in existing MS spectral databases [85, 86]. This is particularly evident in cases such as analyzing MALDI-TOF-MS outcomes. Moreover, the utilization of MS techniques in studying nanomaterials has primarily been limited to the characterization of bioconjugates. This constraint arises from factors such as the high costs associated with acquiring instrumentation, potential sample damage during analysis, and a scarcity of specialized instruments typically used for other research purposes [45, 86].

3.5 Conclusions

Magnetic materials are highly accessible and cost-effective sorbents that offer a diverse range of applications for environmental remediation. These sorbents have undergone further advancements to become 'smart sorbents;' such enhancements enhance their efficiency by causing them to specifically respond to strong magnetic fields. As a result, they can be separated from waste effluents with accuracy and effectiveness during the remediation process, making significant contributions to sustainability initiatives. The revolutionary potential of these magnetic nanomaterials in the fields of engineering technology, biosciences, and environmental studies is significant, leading to a vast array of practical uses. To prioritize ecological preservation and uphold sustainability principles, it is becoming crucial to employ eco-friendly magnetic sorbents that do not produce any detrimental byproducts or toxic waste. The advancement of magnetic nanomaterial production for sustainable environmental applications has necessitated an examination of their manufacture and implementation in sorption systems. Accurate characterization methods are vital for evaluating these measures, especially within the realm of environmental remediation. A summary detailing each technique's merits and drawbacks assists in selecting suitable approaches to analyze potential nanomaterials intended for utilization as magnetic sorbents.

References

[1] Khan Y, Sadia H, Ali Shah S Z, Khan M N, Shah A A, Ullah N, Ullah M F *et al* 2022 Classification, synthetic, and characterization approaches to nanoparticles, and their applications in various fields of nanotechnology: a review *Catalysts* **12** 1386

[2] Oliomogbe T I, Emegha J O and Ukhurebor K E 2023 Microorganism derived biosorbent in the sequestration of contaminants from the soil *Adsorption Applications for Environmental Sustainability* (Bristol: IOP Publishing)

[3] Raj A, Jhariya M K and Harne S S 2018 Threats to biodiversity and conservation strategies *Forests, Climate Change and Biodiversity* ed K K Sood (New Delhi: Kalyani Publishers) 304–20 ISBN 9789327289947

[4] Mondal S and Palit D 2019 Effective role of microorganism in waste management and environmental sustainability *Sustainable Agriculture, Forest and Environmental Management* (Singapore: Springer Nature)

[5] Ukhurebor K E, Aigbe U O, Onyanche R B *et al* 2023 Introduction to the state of the art and relevant aspects of the applications of adsorption for environmental safety and sustainability *Adsorption Applications for Environmental Sustainability* (Bristol: IOP Publishing)

[6] Edema-Sillo T and Emegha J O 2022 Effect of Environmental Degradation on the Inhabitants of Obodo Community in Warri South Local Government Area of Delta State, Nigeria *J. Appl. Sci. Environ. Manage.* **26** 1951–5

[7] Ukhurebor K E, Aigbe U O, Onyanche R B *et al* 2022 An overview of the emergence and challenges of land reclamation: issues and prospect *Appl. Environ. Soil. Sci.* **5889823** 1–14

[8] Sameera V, Naga Deepthi C H, Srinu Babu G and Ravi Teja Y 2011 Role of biosorption in environmental cleanup *J. Microbial. Biochem. Technol.* **R1** 001

[9] Rasheed T 2022 Magnetic nanomaterials: greener and sustainable alternatives for the adsorption of hazardous environmental contaminants *J. Clean. Prod.* **362** 132338

[10] Li H, Li Z, Liu T, Xiao X, Peng Z and Deng L 2008 A novel technology for biosorption and recovery hexavalent chromium in wastewater by bio-functional magnetic beads *Bioresour. Technol.* **99** 6271–9

[11] Kanjilal T and Bhattacharjee C 2018 Green applications of magnetic sorbents for environmental remediation *Organic Pollutants in Wastewater I: Methods of Analysis, Removal and Treatment* (Millersville, PA: Materials Research Forum)

[12] Zhu J, Wei S, Chen M, Gu H, Rapole S B, Pallavkar S, Ho T C, Happer J and Guo Z 2013 Magnetic nanocomposites for environmental remediation *Adv. Pow. Technol.* **24** 459–67

[13] Emegha J O, Oliomogbe T I, Okpoghono J, Babalola A V, Ejelonu C A, Dennis Eyetan Elete D E and Ukhurebor K E 2023 Green biosorbents for the degradation of petroleum contaminants *Adsorption Applications for Environmental Sustainability* (Bristol: IOP Publishing)

[14] Vahabisani A 2020 Exploring the effects of microalgal biomass on the oil behavior in a sand-water system *Master of Applied Science (Civil Engineering) Thesis* Concordia University, Montréal https://spectrum.library.concordia.ca/id/eprint/987883/

[15] Ukhurebor K E, Aigbe U O, Onyanche R B *et al* 2021 Developments, utilization and applications of nanobiosensors for environmental sustainability and safety *Bionanomaterials for Environmental and Agricultural Applications* (Bristol: IOP Publishing)

[16] Bai R S and Abraham T E 2003 Studies on chromium (VI) adsorption-desorption using immobilized fungal biomass *Bioresour. Technol.* **87** 17–26

[17] Chen J Z, Tao X C, Xu J, Zhang T and Liu Z L 2005 Biosorption of lead, chromium and mercury by immobilized microcystis aeruginosa in a column *Proc. Biochem.* **40** 3675–9

[18] Nur K K, Jeppe N L, Meral Y and Gonul D 2007 Characterization of a bacterial consortium for effective treatment of wastewaters with reactive dyes and Cr (VI) *Chemosphere* **67** 826–31

[19] Lim M W, Von Lau E and Poh P E 2016 A comprehensive guide of remediation technologies for oil contaminated soil—present works and future directions *Mar. Pollut. Bull.* **109** 14–45

[20] Elehinafe F B, Agboola O, Vershima A D and Bamigboye G O 2022 Insights on the advanced separation processes in water pollution analyses and wastewater treatment—a review *S. Afr. J. Chem. Eng.* **42** 188–200
[21] Tang S C N and Lo I M C 2013 Magnetic nanoparticles: essential factors for sustainable environmental applications *Water. Res.* **47** 2613–32
[22] Lakshmanan R 2013 Application of magnetic nanoparticles and reactive filter materials for wastewater treatment *PhD Thesis* Royal Institute of Technology School of Biotechnology Stockholm https://www.diva-portal.org/smash/get/diva2:665773/FULLTEXT01.pdf
[23] Gun'ko Y K and Brougham D F 2009 Magnetic nanomaterials as MRI contrast agents *Magnetic Nanomaterials* ed S R Kumar (New York: Wiley)
[24] Cavallini M, Bystrenova E, Timko M, Koneracka M, Zavisova V and Kopcansky P 2008 Multiple-length-scale patterning of magnetic nanoparticles by stamp assisted deposition *J. Phys. Condens. Matter.* **21** 204144
[25] Li X S, Zhu G T, Luo Y B, Yuan B F and Feng Y Q 2013 Synthesis and applications of functionalized magnetic materials in sample preparation *Trends Anal. Chem.* **45** 233–47
[26] Aguilar-Arteaga K, Rodriguez J A and Barrado E 2010 Magnetic solids in analytical chemistry: a review *Anal. Chim. Acta* **674** 157–65
[27] Rangreez T A, Inamuddin A M, Asiri B G and Alhogbi M 2017 Synthesis and ion-exchange properties of graphene th(IV) phosphate composite cation exchanger: its applications in the selective separation of lead metal ions *Int. J. Environ. Res. Public Health* **14** 828
[28] Kumar A, Naushad M, Rana A, InamuddinPreeti G, Sharma A A *et al* 2017 ZnSe–WO_3 nano-hetero-assembly stacked on Gum ghatti for photo-degradative removal of bisphenol a: symbiose of adsorption and photocatalysis *Int. J. Biol. Macromol.* **104** 1172–84
[29] Zhal Y H, Duan S, He Q, Yang X H and Han Q 2010 Solid phase extraction and preconcentration of trace mercury(II) from aqueous solution using magnetic nanoparticles doped with 1,5-diphenylcarbazide *Microchim. Acta* **169** 353–60
[30] Mei-Jiang H, Peng-Yan Z, Zhao Y, Hu X and Zhen-Lian H 2012 Zincon immobilized silica coated magnetic Fe_3O_4 nanoparticles for solid phase extraction and determination of trace lead in natural and drinking waters by graphite furnace atomic absorption spectrometry *Talanta* **94** 251–6
[31] Karatapanis A E, Fiamegos Y and Stalikas C D 2011 Silica modified magnetic nanoparticles functionalized with cetylpyridinium bromide for the pre concentration of metals after complexation with 8-hydroxyquinoline *Talanta* **84** 834–9
[32] Faraji M, Yamini Y, Saleh A, Rezace M, Ghambarian M and Hassani R 2010 A nanoparticle based solid phase extraction procedure followed by flow injection inductively coupled plasma-optical emission spectrometry to determine some heavy metal ions in water samples *Anal. Chim. Acta* **659** 172–7
[33] Ambashta R D and Sillanpaa M 2010 Water purification using magnetic assistance: a review *J. Hazard. Mater.* **180** 38–49
[34] Yang Z and Langdon A G 2004 The use of magnetic bed conditioning and pH control to enhance filtration by natural titanomagnetite *Water Res.* **38** 3304–12
[35] Watson J H P and Ellwood D C 1994 Biomagnetic separation and extraction process for heavy metals from solution *Miner. Eng.* **7** 1017–28
[36] Bahaj A S, James P A B and Moeschler F D 2002 Efficiency enhancements through the use of magnetic field gradient in orientation magnetic separation for the removal of pollutants by magnetotactic bacteria *Sep. Sci. Technol.* **37** 3661–71

[37] Peng Z G, Hidayat K and Uddin M S 2003 Extraction of 2-hydroxyphenol by surfactant coated nano-sized magnetic particles *Korean J. Chem. Eng* **20** 896–901
[38] Burbano A A, Gascó G, Horst F, Lassalle V and Méndez A 2023 Production, characteristics and use of magnetic biochar nanocomposites as sorbents *Biomass Bioenergy* **172** 106772
[39] Lin P-C, Lin S, Wang P C and Sridhar R 2014 Techniques for physicochemical characterization of nanomaterials *Biotechnol. Adv.* **32** 711–26
[40] Aragaw T A, Bogale F M and Aragaw B A 2021 Iron-based nanoparticles in wastewater treatment: a review on synthesis methods, applications, and removal mechanisms *J. Saudi Chem. Soc.* **25** 101280
[41] French R A, Jacobson A R, Kim B, Isley S L, Penn R L and Baveye P C 2009 Influence of ionic strength, pH, and cation valence on aggregation kinetics of titanium dioxide nanoparticles *Environ. Sci. Technol.* **43** 1354–9
[42] Johal M S and Johnson L E 2018 *Understanding Nanomaterials* (Boca Raton, FL: CRC Press) 2nd edn
[43] Samrot A, Sahithya C S, Selvarani J, Purayil S K and Ponnaiah P 2021 A review on synthesis, characterization and potential biological applications of superparamagnetic iron oxide nanoparticles *Curr. Res. Green Sustain. Chem.* **4** 100042
[44] Hall J B, Dobrovolskaia M A, Patri A K and McNeil S E 2007 Characterization of nanoparticles for therapeutics *Nanomedicine (London)* **2** 789–803
[45] Sapsford K E, Tyner K M, Dair B J, Deschamps J R and Medintz I L 2011 Analyzing nanomaterial bioconjugates: a review of current and emerging purification and characterization techniques *Anal. Chem.* **83** 4453–88
[46] Tiede K, Boxall A B A, Tear S P, Lewis J, David H and Hassellöv M 2008 Detection and characterization of engineered nanoparticles in food and the environment *Food Addit. Contam. Part* A **25** 795–821
[47] Gmoshinski I V, Khotimchenko S A, Popov V O, Dzantiev B B, Zherdev A V, Demin V F *et al* 2013 Nanomaterials and nanotechnologies: methods of analysis and control *Russ. Chem. Rev.* **82** 48
[48] Adlnasab L, Djafarzadeh N and Maghsodi A 2020 A new magnetic bio-sorbent for arsenate removal from the contaminated water: characterization, isotherms, and kinetics *Environ. Health Eng. Manag.* **7** 49–58
[49] Abdul Rahman A S, Fizal A N S, Khalil N A, Ahmad Yahaya A N, Hossain M S and Zulkifli M 2023 Fabrication and characterization of magnetic cellulose–chitosan–alginate composite hydrogel bead bio-sorbent *Polymers* **15** 2494
[50] Goodhew P 2011 *General Introduction to Transmission Electron Microscopy (TEM)* (New York: Wiley)
[51] Williams D B and Carter C B 2009 *Transmission Electron Microscopy: A Textbook for Materials* (Berlin: Springer)
[52] Hondow N, Wang P, Brydson R, Holton M, Rees P and Summers H 2020 TEM analysis of nanoparticle dispersions with application towards the quantification of *in vitro* cellular uptake *J. Phys.: Conf. Ser.* **371** 012020
[53] Carlton C E and Ferreira P J 2012 In situ TEM nanoindentation of nanoparticles *Micron* **43** 1134–9
[54] Ponce A, Mejía-Rosales S and José-Yacamán M 2012 Scanning transmission electron microscopy methods for the analysis of nanoparticles *Nanoparticles in Biology and Medicine* (Totowa, NJ: Humana Press) Methods in Molecular Biology 906 453–71

[55] He S, Zhong L, Duan J, Feng Y, Yang B and Yang L 2017 Bioremediation of wastewater by iron oxide-biochar nanocomposites loaded with photosynthetic bacteria *Front. Microbiol.* **8** 823
[56] Khodosova N, Novikova L, Tomina E, Belchinskaya L, Zhabin A *et al* 2022 Magnetic nanosorbents based on bentonite and CoFe2O4 spinel *Minerals* **12** 1474
[57] Hinterdorfer P, Garcia-Parajo M F and Dufrene Y F 2012 Single-molecule imaging of cell surfaces using near-field nanoscopy *Acc. Chem. Res.* **45** 327–36
[58] Patri A K, Dobrovolskaia M A, Stern S T and McNeil S E 2006 Preclinical characterization of engineered nanoparticles intended for cancer therapeutics *Nanotechnology for Cancer Therapy* (Boca Raton, FL: CRC Press) 7
[59] Parot P, Dufrêne Y F, Hinterdorfer P, Le Grimellec C, Navajas D, Pellequer J-L *et al* 2007 Past, present and future of atomic force microscopy in life sciences and medicine *J. Mol. Recognit.* **20** 418–31
[60] Dufrêne Y F and Garcia-Parajo M F 2012 Recent progress in cell surface nanoscopy: light and force in the near-field *Nano. Today* **7** 390–403
[61] Popovic Z V, Dohcevic-Mitrovic Z, Scepanovic M, Grujic-Brojcin M and Askrabic S 2011 Raman scattering on nanomaterials and nanostructures *Ann. Phys.* **523** 62–74
[62] Smith W E and Rodger C 2000 Surface-Enhanced Raman Scattering (SERS), Applications *Encyclopedia of Spectroscopy and Spectrometry* ed J C Lindon (Cambridge, MA: Academic Press) 2nd edn 2329–34
[63] 2012 *Raman Spectroscopy for Nanomaterials Characterization* ed C S Kumar (Berlin: Springer)
[64] Chang H W, Hsu P C and Tsai Y C 2012 Ag/carbon nanotubes for surface-enhanced Raman scattering *Raman Spectroscopy For Nanomaterials Characterization* ed C S R Kumar (Berlin: Springer) 119–35
[65] Lee H M, Jin S M, Kim H M and Suh Y D 2013 Single-molecule surface-enhanced Raman spectroscopy: a perspective on the current status *Phys. Chem. Chem. Phys.* **15** 5276–87
[66] Wilson A J and Willets K A 2013 Surface-enhanced Raman scattering imaging using noble metal nanoparticles *Wiley Interdiscip. Rev. Nanomed. Nanobiotechnol.* **5** 180–9
[67] Ando J, Yano T A, Fujita K and Kawata S 2013 Metal nanoparticles for nano-imaging and nano-analysis *Phys. Chem. Chem. Phys.* **15** 13713–22
[68] Wang Y and Irudayaraj J 2013 Surface-enhanced Raman spectroscopy at single-molecule scale and its implications in biology *Philos. Trans. R. Soc. Lond B: Biol. Sci.* **368** 20120026
[69] Johal M S 2011 *Understanding Nanomaterials* (Boca Raton, FL: CRC Press)
[70] Baudot C, Tan C M and Kong J C 2010 FTIR spectroscopy as a tool for nano-material characterization *Infrared. Phys. Technol.* **53** 434–8
[71] Staurt B H 2004 *Infrared Spectroscopy: Fundamentals and Applications* (Hoboken, NJ: Wiley)
[72] André J S, Myrna J S and Ronald S 2018 Environmental nuclear magnetic resonance spectroscopy: an overview and a primer *Anal. Chem.* **90** 628–39
[73] Alam T and Jenkins J H R-M A S 2012 NMR spectroscopy in material science *Advanced Aspects of Spectroscopy* ed M Farrukh (London: IntechOpen)
[74] Cullity B D 1956 *Elements of X-ray Diffraction* (Reading, MA: Addison-Wesley)
[75] Emegha J O, Ukhurebor K E, Aigbe U, Olofinjana B, Azi S O and Eleruja M A 2022 Effect of deposition temperature on the properties of copper–zinc sulphide thin films using mixed copper and zinc dithiocarbamate precursors *Gazi Univ. J. Sci.* **35** 1556–70

[76] Rao C N R and Biswas K 2009 Characterization of nanomaterials by physical methods *Annu. Rev. Anal. Chem.* **2** 435–62

[77] Chella S, Ehsan D, Kumud M T, Pranas B, TaeYoung K, Edita B and Amit B 2020 Synthesis and characterization of magnetic biochar adsorbents for the removal of Cr(VI) and acid orange 7 dye from aqueous solution *Environ. Sci. Pollut. Res.* **27** 32874–87

[78] Wardani D A P, Rosmainar L, Iqbal R M and Simarmata S N 2020 Synthesis and characterization of magnetic adsorbent based on Fe_2O_3-fly ash from Pulang Pisau's power plant of Central Kalimantan *IOP Conf. Ser.: Mater. Sci. Eng.* **980** 012014

[79] Zanchet D, Hall B D and Ugarte D 1999 X-ray Characterization of Nanoparticles *Characterization of Nanophase Materials* ed Z L Wang (Weinheim: Wiley-VCH Verlag GmbH) 13–36

[80] Soloviy C, Malovanyy M, Bordun I, Ivashchyshyn F, Borysiuk A and Kulyk Y 2020 Structural, magnetic and adsorption characteristics of magnetically susceptible carbon sorbents based on natural raw materials *J. Water Land Dev.* **47** 160–8

[81] Alexander V P, Vadim A D, Vladimir V V, Sergei V A and Kseniya I L 2014 Structure and sorption properties of hypercrosslinked polystyrenes and magnetic nanocomposite materials based on them *J. Polym. Res.* **21** 406

[82] Bonnell D 2001 *Scanning Probe Microscopy and Spectroscopy: Theory, Techniques, and Applications* (Weinheim: Wiley-VCH)

[83] Kocum C, Cimen E K and Piskin E 2004 Imaging of poly(NIPA-co-MAH)-HIgG conjugate with scanning tunneling microscopy *J. Biomater. Sci. Polym. Ed.* **15** 1513–20

[84] Wang H and Chu P K 2013 Surface characterization of biomaterials *Characterization of Biomaterials* ed A Bandyopadhyay and S Bose (Cambridge, MA: Academic Press) 105–74

[85] Barker J 1999 *Mass Spectrometry, Analytical Chemistry by Open Learning* (New York: Wiley) 2nd edn

[86] Schlunegger U P 1980 *Advanced Mass Spectrometry: Applications in Organic and Analytical Chemistry* (Oxford: Pergamon)

Chapter 4

The application of magnetic sorbents for heavy metal removal from aqueous solutions

Uyiosa Osagie Aigbe, Kingsley Eghonghon Ukhurebor, Efosa Aigbe, Benedict Okudaye, Adelaja Otolorin Osibote and Ahmed El Nemr

Magnetic nanoparticles (MNPs) have garnered a lot of attention lately due to their potential use as nanosorbent materials with which to address environmental issues, thanks to their special physical and chemical characteristics that set them apart from conventional nanosorbents. Many tailored MNP types for the confiscation of numerous pollutants can be created using the technique of functionalization, which is achieved by anchoring certain functional groups (FGs) on their surface. This chapter presents a comprehensive review of recent advances in the synthesis of various MNPs and functionalization approaches as well as the various applications of MNPs in resolving environmental issues such as contaminant removal and detection.

4.1 Introduction

The dreadful water quality conditions caused by natural contamination have gained international attention and led to calls for a workable solution. Only about forty percent of superficial water (lakes, rivers, transitional, and coastal waters) is in a reasonable biologic position, and 38% has a good chemical status, according to a report by the European Environmental Agency. Both natural and inert contaminants can enter superficial waters and groundwater via agricultural runoff, sewage plants, industrial effluents, and other human actions. Natural contaminants, such as heavy metals (HMs), aromatic chemicals, antibiotics, dyes, and insecticides are generally unsafe and resistant to microbial breakdown. Water becomes hazardous to drink once these contaminants enter water bodies, and it can sometimes take a very long time to eliminate these contaminants from the water [1].

The legislation related to drinking water and wastewater has grown stricter because of the health risks and environmental concerns. This has encouraged the

doi:10.1088/978-0-7503-5909-2ch4

development of effective techniques that can meet the concentration values that have been set. Many technologies, including coagulation/precipitation, solvent extraction, electrodeposition, ion exchange, membrane filtration, electrodialysis, and advanced oxidation processes, have been established for HM confiscation from manufacturing seepage [2, 3]. Yet, most of those systems have several constraints that must be addressed before these technologies can be applied extensively, such as cost, complexity, efficiency, and sludge creation. Because of its low cost, convenience of use, regeneration potential, sludge-free operation, and extraordinary retention efficiency when used with the right sorbent, sorption is considered one of the most favorable technologies. Due to their superior characteristics compared to those of conventional sorbents (huge surface area (SA), extraordinary quantity of functional sites on their surface, negligible intraparticle diffusion rate (INDR), and extraordinary sorption capacities), MNPs have been studied as nanosorbent materials for HM sequestration. Their characteristics make their use very promising for contaminated water remediation, as they reduce costs and create only limited pollution [2, 4].

In recent years, many concerns have been raised about the continuing degradation of the environment brought about by anthropogenic activities, especially industrial ones. Modern civilization expects the creation of innovative technology solutions capable of fostering a sustainable and more positive industrial sector. With fewer limits and a cleaner approach, nanotechnology (NT) offers the prospect of improving current environmental remediation technologies. Engineered nanomaterials (NMs) and magnetic iron oxide (FeO) colloids in particles have been investigated for use in eco-friendly applications, including water treatment and renewable energy sources, due to this worldwide nanorevolution [5].

Nanoparticles (NPs) have the features of both separate and dissolved particle forms. Their surface-to-volume ratio is 35%–45% times greater than those of larger particles or atoms. This distinctive extrinsic attribute of NPs' exceptional SA is a contributing factor to their superior value, and it also induces their various intrinsic attributes such as their outstanding surface reactivity (which is size-related). Generally, these NPs' remarkable characteristics are accountable for their multifunctional properties and increasing use in different sectors [6].

MNPs are of particular interest owing to their minute size, distinct physicochemical characteristics, and inexpensive production. MNPs' physical characteristics and stability, which are highly reliant on their size and structure, are crucial in defining their noncrystalline nature and, in turn, their magnetic moment. Superparamagnetic signal-domain particles occur when the particle size drops below a supposed critical length and its coercivity rises to zero. Research has indicated that MNPs between 10 and 20 nm have the maximum phases of superparamagnetism, which causes them to react swiftly to applied magnetic fields by aligning their magnetic moments in a net alignment. When the external magnetic field (EMF) is removed, the MNPs' magnetic moments become randomly oriented, restoring the net magnetic moment to zero [7].

Since MNPs are so adaptable for *in situ* applications, they are also heavily utilized for the remediation of both carbon-based and inanimate contaminants from green

Table 4.1. The magnetic and physical characteristics of FeO NPs.

Molecular formula	Fe_3O_4	$\gamma - Fe_2O_3$	$\alpha - Fe_2O_3$
Melting point (°C)	1583–1597.	—	1350
Density (g cm^{-3})	5.2.	4.9.	5.3.
Néel temperature (K)	—	—	948 $< T_N >$ 963.
Hardness	5.5.	5.0	6.5
Point of zero charge (pH_{PZC})	—	—	—
Morin temperature (K).	—	7.5	263
Magnetism type	Ferromagnetic	Ferromagnetic	Weakly ferromagnetic and antiferromagnetic
Lattice parameter (Å)	0.8394 (cubic)	0.8346 (cubic)	(Rhombohedral)

matrices. The MNPs most often employed as sorbents or catalysts for contaminant removal are FeO NPs such as maghemite (γ-Fe_2O_3), hematite (α-Fe_2O_3), magnetite (Fe_3O_4), and goethite (α-FeOOH), metals and metal alloys such as iron (Fe), cobalt (Co), nickel (Ni), etc. and spinel ferrites with the universal formula AFe_2O_4 (A = Mn, Co, Ni, Mg, or Zn), such as manganese ferrite ($MnFe_2O_4$), cobalt ferrite ($CoFe_2O_4$), magnesium ferrite ($MgFe_2O_4$), etc. The magnetic and physical properties of FeO NPs [8, 9] are shown in table 4.1. Specifically, γ-Fe_2O_3 and Fe_3O_4 (Fe_3O_4 oxidized form) have remarkable magnetic crystalline forms that can be exemplified as cubic inverse spinel structures with oxygen (O_2) shaping on the face-centered cubic (FCC) structure and Fe cations in the octahedral and tetrahedral sites [5].

MNPs have been applied to resolve ecological issues due to the unconventional economic and ecological efficiency of the processes based on MNPs' superior chemical and physical attributes (extraordinary SA, chemical stability, ease of functionalization, etc.) in addition to MNPs' superparamagnetic properties, which support the separation and recovery stages of the designs employed in multifaceted multiphase systems [2].

Today, MNPs are capable of efficiently eliminating a comprehensive range of contaminants, including dyes, HMs, microalgae, bacteria, and radioactive waste. Due to their use in enhancing the sensitivity and stability of sensors developed and manufactured for the identification of ecological analytes, MNPs have drawn more attention in recent times as labeling materials. In these applications, they are incorporated into transducer materials and suspended in samples, and then an EMF attracts the substance to be quantified to the operational detection surface of the biosensor. There are numerous uses for these MNP-based biosensors in the food, environmental, and therapeutic domains [7]. Figure 4.1 shows the numbers of articles published over the last 14 years (2009–2023) with the title 'Application of Magnetic Sorbents for Heavy Metal Removal from Aqueous Solutions' obtained from SCOPUS. It shows that research into this subject first increased, followed by a decrease in research output in later years. Hence, the objective of this chapter is to review the applicability of MNPs to environmental issues such as HM removal as well as the detection of HMs using MNP-derived sensors.

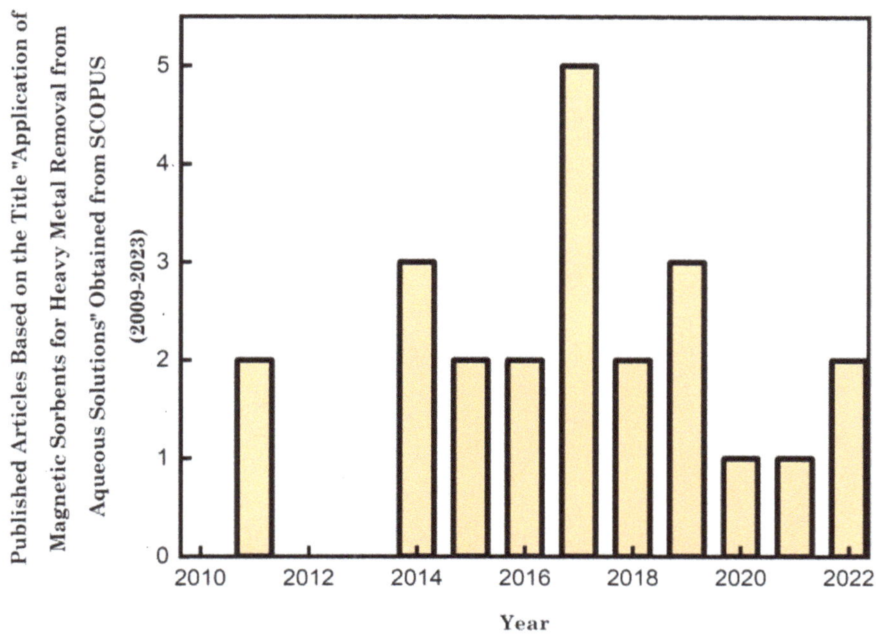

Figure 4.1. Published articles based on the title 'Application of Magnetic Sorbents for Heavy Metal Removal from Aqueous Solutions' obtained from SCOPUS.

4.2 Magnetic nanoparticle synthesis approaches

In the synthesis of MNPs, the three key approaches used are the chemical, biologic, and physical methods. The chemical and biologic methods are jointly referred to as the bottom-up technique, while the physical method is referred to as the top-down method [10]. Various chemical and physical approaches are generally used for MNP preparation and stabilization. The selection of an MNP synthesis approach is critical because NP production procedures such as metal ions' (MIs) kinetic interaction with reducing agents, the stabilizing agent sorption process with MNOs, and the different experimental approaches have strong impacts on MNPs' morphology, stability, and physicochemical features [6]. The physical approach for MNP synthesis uses well-described parameters but suffers from an inability to regulate the particle size at the nanoscale. The chemical approach uses operational parameters such as the pH, reaction time, solvent and precursor types, and temperature optimization, thereby providing control over the key properties (size, structure, and biocompatibility). MNP synthesis using a biological approach substantially regulates the constituent and particle geometry under mild conditions and offers extraordinary yield, stability, low cost, and reproducibility [7].

The various methods employed to produce MNPs are the sol–gel, coprecipitation, hydrothermal, high-temperature pyrolysis, microemulsion, thermal, bionic mineralization, electron beam lithography, sonochemical, and laser-induced pyrolysis approaches, among others [8, 11, 12], as depicted in figure 4.2.

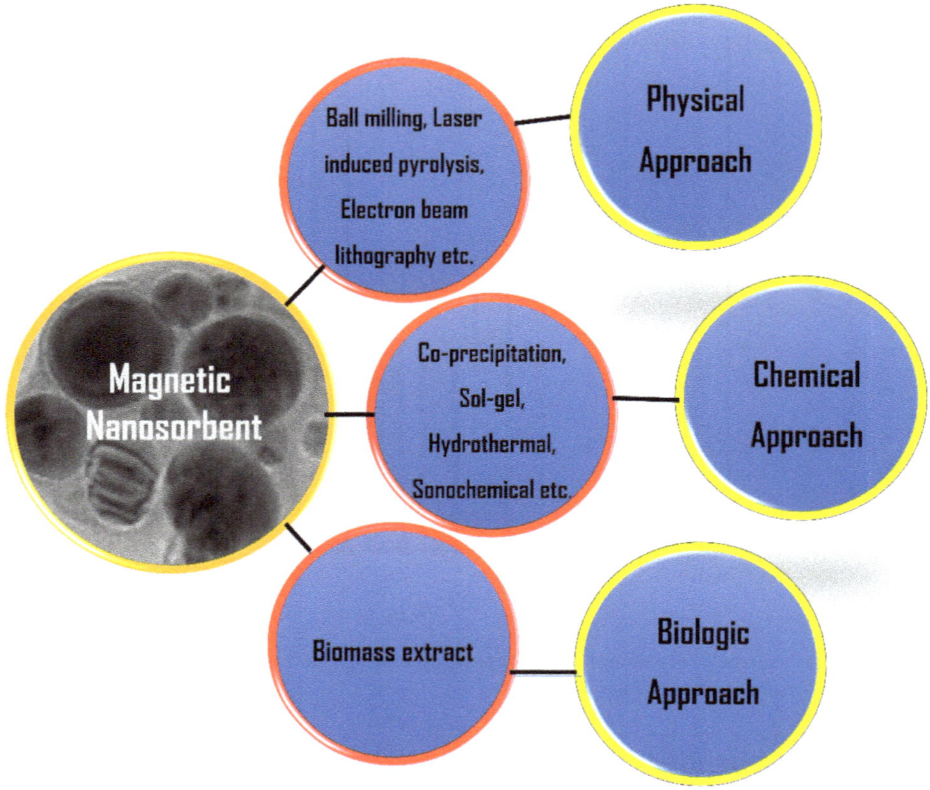

Figure 4.2. Different synthesis approaches employed for MNP production.

The various chemical methods employed for MNP synthesis consist of the coprecipitation, hydrothermal, microemulsion, and thermal decomposition approaches. The hydrothermal and thermal decomposition methods afford improved solutions in terms of control over morphology and size, compared to the other techniques. The chemical approach involves precipitating NPs from a solution to produce monodispersed particles. The precipitation should conform to LaMer and Dinegar's precipitation template (or a similar template) [13]. This template states that when precipitation from a solution occurs at a certain super-saturation phase, a nucleation burst happens, and this progressively increases in size via the diffusion of solute from the solution to the nuclei until the ultimate monodispersed size is attained [14].

Owing to the astonishing benefits of MNPs, various synthesis approaches have been designed to produce them. The approach generally used for MNP synthesis is the coprecipitation approach. The performance of an MNP depends on its characteristic shape, size, and composition, the pH, reaction temperature, the ratio of Fe^{2+}/Fe^{3+} iron salts, and the medium's ionic properties. The coprecipitation approach is a modest and efficient approach for the production of MNPs such as ferric oxides (Fe_2O_3, Fe_3O_4, etc.) and ferritic alloys (Zn–Mn, Ni–Zn, and Co–Zn). For the metal salts of Fe^{2+} and Fe^{3+} to precipitate in this approach, an alkaline solution of ammonia, sodium, and sodium hydroxide is added. It is generally employed because of the straightforward availability of its materials and its simple synthesis approach [8]. MNP synthesis generally involves producing the magnetic material, coating the magnetic core, and detailed characterization of the magnetic core–shell structure. To prepare Fe_3O_4 as a magnetic core, a chemical coprecipitation approach is employed based on the use of Fe^{2+} and Fe^{3+} ions in a water-soluble medium. The production process involves the energetic stirring of a dissolved mixture of $FeCl_3.6H_2O$ and $FeCl_2.4H_2O$ in ultrapure water at 70 °C–85 °C in an atmospheric nitrogen environment with the swift addition of liquid ammonia, leading to the formation of a dark precipitate [15].

The sol–gel approach was first reviewed by Livage *et al* in 1988 based on the chemistry of transition-metal oxides (MeO). This approach involves the mixture of colloid MeO with a solution (sol) comprising matrix-producing species, resulting in gel creation. This approach for MNP production can also be carried out by the direct blending of metal and MeO or NPs within pre-hydrolyzed silica or by metal complexation with silicone and metal reduction before hydrolysis. Metal alkoxide ions and alkoxysilanes are employed as precursors for colloid synthesis. Metal alkoxides are the organometallic precursors for different metals such as aluminum, silica, titanium, and many others and they are nonmiscible in water. At the start of this process, a uniform solution of one or more specified alkoxides is made with the addition of a catalyst to trigger the reaction at a monitored pH. The four steps involved in sol–gel formation consist of hydrolysis, condensation, particle growth, and particle aggregation. In the metal or MeO direct precipitation approach, MeO particles are generally precipitated from a silica sol by a low-temperature heat treatment, and thin films are usually formed via this approach [6].

The microemulsion method is an isotropic, visible, and thermodynamically stable approach. The microemulsion, which has minimal viscidity, consists of an oil (a hydrocarbon), a surfactant (alcohol), and water (a water-soluble electrolyte solution). The reaction types within this approach are categorized as the 'water into oil' and 'oil into water' forms. The reaction space is limited to the micro-reactor droplet. The microemulsion droplet size is on the nanometer (nm) scale, and the individual droplets are separate. The reaction of the reactants with two microemulsions leads to a substance transfer or exchange in the water core caused by micellar particle collisions. This procedure causes chemical reactions in the nucleus that create MNPs. This approach can inhibit the aggregation of MNPs in the synthesis procedure and regulate the particle size [8].

Solvothermal (hydrothermal) synthesis is an approach for the preparation of MNPs and ultrafine powders. These reactions are implemented in a water-soluble medium in reactors or autoclaves at an extreme pressure of >2000 psi and a temperature of >200 °C. This approach includes the utilization of liquid–solid solution reactions (LSSRs), which give outstanding management of the shape and size of the resulting MNPs. It also involves the production of MNPs at the extreme boiling point of a water-soluble solution at elevated vapor pressure [14]. The solvothermal approach is employed for nanophase preparation in the presence of water or other natural chemicals such as ethanol, polyol, or methanol as a solvent. The reaction created in the pressure vessel permits the solvent (alcohol and water) from being heated beyond the boiling point. The crystallization kinetics (formation of crystal) can be enhanced by one to two orders of magnitude using microwave-supported reactions. This is an outstanding method for the production of MeO, metal, semiconducting, rare-earth, and transition-metal magnetic nanocrystals and dielectric, rare-earth fluorescent, and polymeric nanoparticles [6].

The thermal decomposition approach comprises the chemical breakdown of a substance at extreme temperatures. During this approach, chemical bonds are broken. Magnetic nanostructure synthesis uses organometallic compounds such as acetylacetonates in organic solvents with surfactants. The ultimate size and morphology of the resulting magnetic nanostructures are determined by the different precursors involved in the reaction [12, 14].

An emerging NT employed for NP synthesis is the green synthesis approach. These approaches emerged to address the issues encountered with traditional approaches, such as elevated cost and safety problems. This approach features unique strategies for the process of MNP synthesis and various uses of chemical substances to decrease the risk to the human physical condition and the ecosystem. These methods generally comprise clean chemistry, atom reduction, naturally benign chemistry, and benign-by-design chemistry. The biologic techniques used for NP synthesis comprise the application of various isolates and extracts of plant products and microorganisms and their enzymes [16]. These approaches have many advantages over other traditional approaches, as they are cost-efficient, green, and easily scaled up for large-scale manufacture. These approaches do not involve high pressure, temperature, energy, or the use of noxious chemicals. The biosynthesis of MNPs is categorized based on the use of bioreduction (metal ions' (MIs) chemically reduced into a naturally stable form using microorganisms and their enzymes or plant extracts). The NPs created are stable and unreactive in the environment and can be carefully separated from polluted samples. Biosorption is an exceptional approach for NP production, in which metal cations in a water-soluble medium are permitted to bond with an organism's cell wall, which also leads to the creation of stable NPs due to the peptide or cell wall interaction [6, 17].

Another approach used for NP synthesis is microwave-supported NP synthesis, in which microwave frequencies of 300 MHz to 300 GHz are employed, leading to the orientation of polar molecules, such as those of water, with an electric field (EF). The reorientation of the dipolar molecules within an alternating EF is due to molecular friction and energy loss in heat form. It is a useful approach in various fields of chemistry and materials science and is generally employed in the preparation of MNPs from various plant-based extracts [6, 18]. Table 4.2 shows the various approaches utilized for MNP synthesis with their advantages and disadvantages [6, 8, 19].

Table 4.2. A comparison of the various approaches employed for MNP production with their advantages and disadvantages.

Synthesis approach	Advantages	Shortcomings
Coprecipitation	Mild reaction conditions, extreme product purity, rapid process, effective and simple method for creating MNPs at the gram scale. The mean diameter of particles produced via this procedure varies up to 50 nm, and the reaction procedure is quick, with considerable yield and reasonable precursor cost.	Products in the washing, filtering, and drying stages are susceptible to aggregation. A size-sorting procedure is required owing to the reduced crystalline particles created. The polymeric matrix consists of particle masses owing to the application of polymers for Fe_3O_4 colloid stability.
Sol–gel	Modest technique. Particle size and morphology are controlled by efficient monitoring of reaction factors	Moderately high expense of metal alkoxides, high content of alcohol discharged at the time of calcination, ineffective bonding, extraordinary permeability, and elevated safety measures needed for the process.
Hydrothermal/solvothermal	This procedure can create crystalline FeO NPs. Crystalline phases are formed by this technique that are unstable at their melting point. FeO NP size and shape can be controlled in this approach.	A pressure of >2000 psi needs to be maintained in this procedure.
Sonochemical decomposition	Combines homogeneity and crystal growth reduction, which results in acceleration of chemical dynamics. The rate of reaction and particle growth can be manipulated by manipulating polymers, organic agents, etc.	The approach is unable to accomplish the creation of FeO NPs with controllable dispersity and shape. Cavitation is produced in the water-soluble phase owing to the application of ultrasonic treatment.
Microemulsion	The produced MNPs have a size distribution, consistent shape, and absolute dispersion property.	Minimal yield; lots of solvents required for this approach.
Electron beam lithography	Well-preserved interparticle spacing.	High-cost approach due to exceptionally complex machines required.
Laser-induced pyrolysis	A time-saving method, with a regular shaped product. Continuous, direct, and produced a large quantity of MNPs.	Excessive cost, a prerequisite of reagent/laser resonance, and specialized installation required.
Microbial or plant extracts	Cost-effective and sustainable approach, with acceptable reproducibility and scalability, and superior yield at low temperature and energy.	More time is required to achieve the required product

4.3 The functionalization or modification of MNPs

MNPs are usually stable in the solid state, and because of this, they are utilized in various applications such as magnetic recording devices, catalytic materials, magnetic hyperthermia (MH), pigments, coatings, gas sensors, magnetic data storage devices, magnetic resonance imaging (MRI), drug delivery, bioseparation, etc [2, 19]. Reduced NPs have an elevated surface free energy that leads to NP aggregation due to the reduced energy barriers. The formation of aggregation reduces the MNPs' SA, thereby decreasing their confiscation capacity and reactivity and hence reducing the treatment performance. Aggregation also undercuts MNPs' efficiency during remediation owing to mobility loss. The size distribution, particle concentration, solution composition, surface chemistry, and NP magnetism are the various conditions that influence aggregation [20]. Extended chains of interparticle magnetic force can lead to aggregation, and because the blocking temperature is relative to the volume, huge particles are more inclined to aggregation than smaller particles. The main magnetic interactions that cause aggregation in nanoparticulate systems are the result of anisotropic dipole–dipole forces; explicit exchange interactions between particles that are in contact play only a negligible role [8].

The shape and potential to create anisotropic magnetic lattices are entirely typical. To reduce interparticle interactions, MNPs generally have to be segregated from each other by a covering added to the particles through functionalization or modification [19]. MNPs are characterized by their specific SA, excellent dispersion, small particle size, etc. MNPs have the benefits of extraordinary SA, minimal cost, ecological protection, superior sorption efficacy, and straightforward separation and recovery under EMFs [8, 11].

Surface modification and copolymerization are two further ways to modify MNPs. The modification or functionalization of NMs to detect contaminants in the environment has gained significant consideration in current times as a core research area. The most widely utilized sorbents among NMs are MNPs, which can be used to sorb HMs, inorganic salt contaminants, and trace organic contaminants as a method of effluent contamination detection [8]. MNPs have a modest size effect because of their minute particle sizes. MNPs' activity is improved by their ability to sorb more of a chemical of interest thanks to their considerable specific SA. The effective separation of MNPs under an EMF is made possible by their remarkable magnetic responsiveness. The core idea behind MNP detection is that the material to be assessed can be removed from the sample by MNPs through magnetization and concentration under specific conditions, and then the material can be separated and detected by an EMF The application of MNPs in the identification of soil and water contamination has garnered much of interest. Current research hotspots include the modification of MNPs to remediate contamination and the more efficient use of MNP properties in a microbial composition biosensor to directly detect samples' biotic noxiousness [8].

Encapsulating MNPs with inorganic and organic materials is a way to overcome some of the concerns associated with the reactivity and stability of MNPs (as it decreases particle–particle aggregation and degradation). Encapsulated MNPs provide enhanced effectiveness in targeted sorption and dual functionality such as chemo- and fluorogenic sensing [21]. The modification of MNPs with organic materials is done via *in situ*, sorption, and post-synthesis coating approaches. Inorganic and organic materials act as stabilizers by covalently attaching to NPs to assist these particles in attaining advanced-stage magnetic susceptibility. They are transformed with remarkable groups to achieve improved functionalization via the addition of different biologic functional molecules, thus resulting in inorganic and organic complexes that have a core–shell nanostructure for different purposes [12].

Generally, MNPs are coated with biological films such as polymers or surfactants (dextran and polyethene glycol) or mineral films such as metallic elements (platinum or gold), metal oxides (MeO) (aluminum oxide and cobalt oxide), activated carbon (AC), silica, etc. Functionalized surfaces are implemented to improve MNPs' physicochemical stability and give them resistance to spontaneous aggregation, oxidation, and corrosion [2]. When MNPs are modified via ligand exchange, maleimide coupling, or covalent linkage, they become very efficient for contaminant sequestration from wastewater. Ligands have excellent adsorptive chelating features or FGs that can make bonds through their terminal groups on encrusted MNPs to confiscate inorganic or organic contaminants, improve sorption capacity (Q_e), assist with appropriate separation, and avoid releasing noxious NPs into the ecosystem. The four approaches used to perform multiple MNP alterations for effluent decontamination applications are: (i) ligand exchange (ligand molecules on the surface of the NPs are exchanged for another to improve NMP stability and Q_e), (ii) click-reaction (the cycloaddition of Cu (I) catalyzed alkyne-azide under a variety of conditions with substantial stereospecificity and efficiency), (iii) covalent bond development (approaches involve amine, carboxylic (–COOH) and hydroxyl (OH^-) groups added to the NP surface), and (iv) maleimide conjugation (the coupling of thiols with primary amines) [19, 22].

The functionalization approaches include the addition of carboxyl (–COOH) (the materials thus functionalized permit more diverse synthetic reagent selection due to the large number of acids with –COOH groups). No post-modification is needed to functionalize NMPs with –COOH. A good carboxylated reagent to employ for this purpose is ethylenediaminetetraacetic acid (EDTA) due to its extraordinary -COOH content. The key benefit of these functionalized materials is their sorption performance, and their main drawback is their relatively poor material selectivity. A further functionalization approach is the addition of amino groups (achieved by combining functionalized reagents comprising amino groups with magnetic materials in a reaction, specifically in the one-pot and post-functionalization approaches). The commonly utilized amino-functional reagents comprise (3-aminopropyl)triethoxysilane (APTES), triethylenetetramine (TETA), and hexane-1,6-diamine. Mercapto functionalization is usually employed for trace elemental analysis, since the sulfide group has an excellent affinity for a series of elements such as Hg, Pb, Cd, and noble metals. Due to the strong electronegativity of S, the capacity of S-containing sorbents to link to trace elements is greater than that of the corresponding amino-group-containing sorbents [23].

Coating MNPs can have a huge impact on their properties, both for individual NMPs and groups. In virtually all settings, pure Fe NPs open to the atmosphere or water-soluble environments swiftly form a native oxide layer, generally in the inverse spinel structure of Fe_3O_4 and γ-Fe_2O_3, which is initially several nanometers thick. Such a layer limits reactivity, thereby reducing the particle magnetization of the layer owing to the lower saturation magnetization (M_s) of Fe oxides (the M_s of bulk $Fe_3O_4 = 92$ emu g^{-1} and the M_s of bulk Fe = 218 emu g^{-1} at 300 K). This impact is more distinct for reduced particles, since the volumetric fraction of the oxide shell grows. The structured growth of high-quality Fe/FeO core–shell NPs has been implemented to moderate the impacts of oxidation whilst maintaining the superior moment of the core. However, the lack of stability linked with the transformation between oxidation states remains a problem for metal/MeO NPs applied for environmental purposes. Surface impacts manifest as spin canting at the surface that can decrease the magnetization. Functionalization may also alter the surface features at the ferromagnetic core surface, as the sturdy bond created between the binding molecules and metals alters the electronic structure. These interactions may also efficiently pin the spins of the surface molecules over the spherical particle's surface, causing their net magnetization to be zero [21].

4.4 Heavy metals

Intensification in the magnitude of contaminants in surface and subsurface water is an obvious wide-reaching issue due to population increase, the fast development of industry, and the repeated occurrence of drought. The persistent contaminants in industrial seepage consist of mineral compounds, HMs, dyes, natural pollutants, and other problematic compounds [24]. Water contaminants are categorized into point and non-point sources, depending on whether the pollution originates from definite and detectable sources or a multiplicity of sources such as drainage pipes, industrial seepage, smokestacks, factories, power plants, and municipal wastewater treatment plants, according to the United States Environmental Protection Agency (US EPA). Non-point source contaminations do not emanate from an identifiable source and are a blend of contaminants that are distributed over a significant area (grease, oil, animal wastes, runoff, thunderstorms, fertilizer, and pesticides). According to the EPA, the six classification categories based on point and non-point sources are mining and industry, agriculture and forestry, diverse urbanization, waste maladministration, and intrinsic sources [25].

HMs and natural pollutants in effluents are generating a terrifying ecological issue due to their severe toxicity and non-compostable nature. They are also a threat to human health when they accumulate in living bodies. Notwithstanding their toxicity, they are valued production chemicals employed for industrial manufacture. HMs are categorized based on their toxicity into biologically essential HMs such as Zn, copper (Cu), Fe, Ni, and Co and biologically nonessential HMs such as arsenic (As), cadmium (Cd), chromium (Cr), lead (Pb), and mercury (Hg). The biologically essential HMs are not hazardous at low concentrations, while the biologically nonessential HMs are exceptionally lethal even at low concentrations [24].

Table 4.3. Anthropogenic HMs and their maximum concentration levels (MCLs).

HM ion	Anthropogenic origins	Toxic effects	MCL (mg l^{-1})
As	Animal supplements, metallurgy, ceramics, fireworks, geothermal production of electricity, and pesticides	Vascular problems and skin disease	0.05
Cr	Ferroalloy production, pigments, data storage, coloring of textiles and leather	Vomiting, migraines, and carcinogenesis	0.05
Cu	Chemical/pharmaceutical equipment, roofing, alloys, and water pipelines	Sleeping disorders and liver problems	0.25
Cd	Metal coatings, Ni–Cd batteries, coal combustion, and pigments	Renal disease, kidney damage, and carcinogenesis	0.01
Ni	Computer components, glass/ceramic molds, Ni–Cd batteries, and catalysts	Carcinogenesis, breathing disorders, and skin disease.	0.20
Hg	Catalysts, mercury vapor lamps, and metal extraction processes	Kidney failure, circulatory and nervous system conditions	0.000 03
Zn	Zn alloys, rubber/paint industry and polyvinyl chloride.	Fatigue and anxiety	0.80

HMs can travel through the food chain via bioaccumulation, and they can spread throughout the body resulting in major illnesses such as pancreatic disease, heart disease, brain disease, capillary damage, gastrointestinal swelling, and necrotic changes in certain tissues. Because HMs can have harmful effects on ecosystems and living things even at low concentrations, the World Health Organization (WHO) has set limits on their permissible levels [24]. Table 4.3 gives details of anthropogenic sources, the maximum allowed levels (MALs) of HM contaminants, and their effects on human health [19]. Currently, HM sequestration from industrial wastewater is gaining attention, and sorption is seen as a leading technique for improving water and effluent improvement polluted by HMs and biotic and inert compounds. This is because an extensive variety of sorbents have been produced with various compositions. Nanosorbents have recently been considered for contaminant removal, and next-generation nanosorbents are taking the form of NPs [19].

4.4.1 Heavy metal sorption using magnetic nanoparticles

Over five billion Earth inhabitants depend on surface and groundwater systems for a diversity of uses, including manufacture, housing, agriculture, and the supply of clean water. As a well-explored phenomenon, the depletion of water resources can be ascribed to both natural processes (such as climate change, water–rock interactions, and terrestrial factors), the actions of humans (such as urban waste and agricultural practices), and the significant presence of chemical compounds that have existed since the industrial revolution. Notwithstanding this, there are still several situations in which managing surface water and groundwater as reserves is difficult, and pertinent data is still lacking. Geological heterogeneities of rock and soil interact with water in ways other than those caused by human action, affecting intrinsic water cycles and changing water quality in all fields. Such changes may have detrimental outcomes. Furthermore, the impact of both natural and human activities causes fluctuations in water resources' quality, quantity, and accessibility as well as their physicochemical and natural features [25]. Consequently, there is a great demand for innovative solutions that can reuse the produced effluent. For progressing nations, the treatment or remediation of this wastewater is a critical issue, especially when significant considerations such as cost, sustainability, reusability, and recycling are taken into account. HMs are challenging to break down or remove, particularly those originating from the Earth's crust. With the development of novel products, cleanup and treatment methods, and environmental monitoring sensors, NT is believed to have the power to substantially alter our current surroundings [4, 26–28].

In the field of water purification, technology must be able to reduce unsafe pollutants in the environment to a safe level swiftly, effectively, and affordably. Therefore, the creation of novel NMs with enhanced attraction and potential selectivity for HMs and other contaminants is a dynamically expanding area of study in the field of NT. These NMs can be used, for example, to enable extended water reuse, recycling, and desalination through improved filtering, which improves the quality, availability, and sustainability of water. NT offers innovative water treatment solutions based on the use of sorption, electrostatics, reactivity, catalysis, and changeable pore depth. It can also be used to produce highly sensitive sensors and optoelectronics and hydrophilic and hydrophobic interactions. Ease of implementation and reasonably priced water and effluent treatments are made possible by the efficient, adaptive, and flexible nature of NT-based processes. Moreover, abnormal water bodies can be inexpensively cleaned and restored using NT [4]. In general, sorbent performance is regulated by the nature of the surface functional groups and the metal–ligand bond strength [29].

At the laboratory scale, magnetic nanosorbents are becoming remarkably efficient functional materials with quick sorption kinetics and a remarkable capacity to sequester microcontaminants [22]. However, the application of most of these approaches at a significant scale faces numerous downsides such as cost, complexity, lack of effectiveness, and the generation of sludge. Sorption is deemed to be a promising technology due to its ease of design and operation, minimal cost, possibility for renewal, absence of sludge, and extraordinary retention efficiency when performed by a suitable sorbent. MNPs have been applied as nanosorbent resources for HM confiscation due to their outstanding features such as the superior number of operational surface sites, high specific SAs, huge pore volumes, sturdy structural design, extensively interconnected porous networks, and low intraparticle effects in comparison to traditional sorbents, make their application very suitable for contaminated water treatment, decreasing costs and creating less pollution.

Furthermore, recent advances in the production approaches permit the straightforward attachment of various FGs to the nanosorbents' surface, which take part in the sorption process as precise binding sites, enhancing the Q_e and the selectivity of the procedure for particular contamination issues. One of the exceptionally important characteristics of magnetic nanosorbents is their ability to be detached *in situ* from sorption-treated waters in magnetic nanosorbent–sorbate slurry form using a suitably strong magnetic field. Other benefits include their redox activity, surface charge features, affordable cost of production, nontoxicity, outstanding selectivity, binding specificity, and excellent reusability. In addition, their recovery can easily be accomplished using magnetic separators, thereby overcoming the conventional fixed-bed sorption column pressure drop [2, 22].

Contaminants such as HMs can have a variety of characteristics depending on the application type they are intended for, particularly in terms of hydrophilicity or hydrophobicity and whether they are neutral, cationic, or anionic. These characteristics have a considerable effect on how they sorb when exposed to different sorbents [30]. Fe_3O_4, γ-Fe_2O_3, and nano zero-valent iron (nZVI) are MNPs widely used in environmental applications. The reactivities of these Fe-based MNPs in the sequestration of pollutants vary based on their Fe oxidation state, which reveals their chemical characteristics. Fe_3O_4 is a ferromagnetic FeO that contains both Fe (III) and Fe(II) and has been the subject of much research due to the presence of the Fe state, which can act as an electron. Figure 4.3 is a graphical depiction of pollutant removal mechanisms using MNPs [31]. With the appropriate surface modifications, NMs with a magnetic core can be employed as sorbents for HMs, organics,

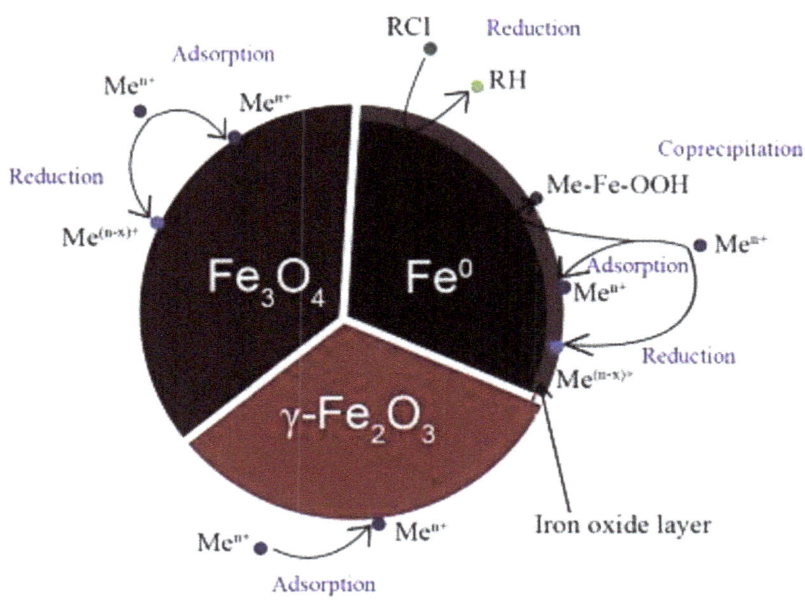

Figure 4.3. A graphical depiction of pollutant removal mechanisms based on MNPs [31].

pesticides, radionuclides, and hydrophobic compound pollutants. The surface sorption and coprecipitation of a broad range of organic and inorganic contaminants can arise in most aqueous ecosystem settings due to the creation of FeO/hydroxide shells. Furthermore, Fe can act as a strong reducing agent, which facilitates the degradation of different inorganic and organic pollutants into less unsafe substances [31, 32].

Fe_3O_4 was synthesized via the thermal decomposition approach and modified with thiol (-SH and -COOH groups) employing meso-2,3-dimercaptosuccinic-acid (DMSA) to create Fe_3O_4/DMSA NCs. These were subsequently utilized for the elimination of Pb^{2+}, Ni^{2+}, and Cd^{2+} ions. A transmission electron microscopy (TEM) analysis showed that the produced nanocomposites (NCs) were exceptionally monodispersed MNPs with a regular particle size of 8.24 ± 1 nm. It was noticed that the sorption of various HMs ions was pH dependent, with an increase in Q_e as the pH increased. The optimal Q_e was reported to occur at pH 6 for Pb^{2+} and Ni^{2+} and at pH 7 for Cd^{2+} (figure 4.4). The optimum sorption capacities (Q_m) for Pb^{2+}, Ni^{2+}, and Cd^{2+} ion sequestration to the NCs were assessed and found to be 116.54, 102.74, and 75.48 mg g^{-1} and 64.5, 53.9, and 27.17 mg g^{-1} for single and ternary MI systems, respectively. The sorption process followed the Langmuir (LNIR) and pseudo-second-order (PSOR) models [33].

The functionalization of Fe_3O_4 NPs with APTES resulted in the attachment of the key amino groups to the NPs and the subsequent grafting of heterocyclic groups (HCGs) on the amino groups through a substitution reaction to form Fe_3O_4/SiO_2/HCGs. These particles were explored for the elimination of Cu^{2+}, Hg^{2+}, Pb^{2+}, and Cd^{2+}. For the HMs studied, an increase in the solution pH led to an enhancement in the percentage of MIs removed, with an optimum removal reported at pH 5. In

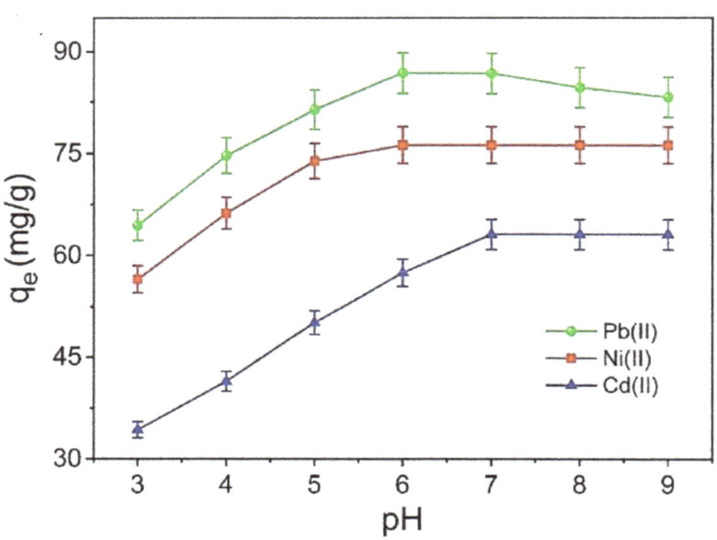

Figure 4.4. The influence of pH on HM (Pb^{2+}, Ni^{2+}, and Cd^{2+}) sorption [33].

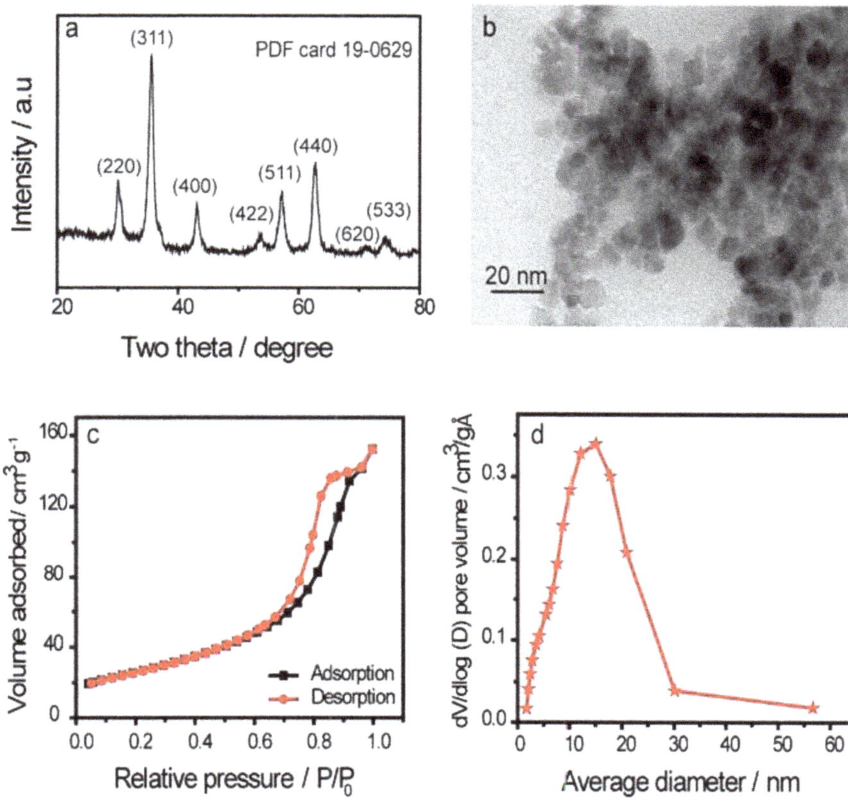

Figure 4.5. (a) XRD pattern, (b) TEM image, (c) N_2 sorption desorption and (d) average pore sizes obtained during Fe_3O_4 NP characterization [35].

terms of the sorption isotherm, the determined Q_m values for Cd^{2+}, Hg^{2+}, Pb^{2+} and Cd^{2+} were 82, 77, 72, and 56 mg g^{-1} due to the two nitrogen atoms which are situated on the same side of the aromatic ring and were useful for HM ions coordination [34]. Ultrafine Fe_3O_4 NPs prepared using the coprecipitation approach were employed for the sorption of Pb^{2+}, Cd^{2+}, Cu^{2+}, and Ni^{2+}. As shown in the x-ray diffraction (XRD) analysis in figure 4.5(a), the diffraction peaks obtained at 2θ were ascribed to the crystal plane of the Fe_3O_4 NPs. The calculated crystalline size based on XRD was determined to be 9 nm using the Debye–Scherrer equation. From a TEM analysis of the image in figure 4.5(b), it was noticed that the particles created were spherical and tended to amass due to the magnetic force between the particles. This was attributed to the lack of a stabilizing surfactant on the surface of the NPs. The typical particle size was revealed to be between 4 and 17 nm. Figure 4.5(c) shows a Brunauer–Emmett–Teller (BET) analysis of the produced NPs, which confirmed an extraordinary SA of 94.43 m^2 g^{-1}. Based on the sorption desorption isotherm, a type VI isotherm was reported, which confirmed the mesoporous structure of the NPs (figure 4.5(d)). The sorption of these HM ions was found to be pH linked, with individual confiscation efficiencies of 98%, 87%, 90%, and 78% reported for Pb^{2+}, Cd^{2+}, Cu^{2+}, and Ni^{2+}. For hybrid Pb^{2+}, Cd^{2+}, Cu^{2+}, and Ni^{2+} ions, the confiscation efficiencies were 86%, 80%, 84%, and 54% for the same river water. The HM sorption by Fe_3O_4 was excellently described by the PSOR and LNIR models, and Q_m values of 85, 79, 83, and 66 mg g^{-1} were obtained for Pb^{2+}, Cd^{2+}, Cu^{2+}, and Ni^{2+}, respectively [35].

Employing the simple coprecipitation approach, pristine silica layered and amino-functionalized $CoFe_2O_4$ NCs were produced. The produced NCs were assessed for Pb^{2+}, Zn^{2+}, and Cu^{2+} ion elimination at a fixed pH of 6.5. The PSOR and LNIR ideally depicted the sorption of these HMs ions, and Q_m values of 277, 255, and 258 mg g^{-1} were obtained for Cu^{2+}, Pb^{2+} and Zn^{2+} ions, respectively. When the reusability of the NCs was measured, it was found that five sorption cycles were possible, with removal efficiencies of 77%, 81%, and 76% for Cu^{2+}, Pb^{2+}, and Zn^{2+}, respectively [36]. Employing a coprecipitation and modified Hummer's approach, GO/Fe_3O_4 NCs were prepared, which resulted in the introduction of amino groups into the NCs. The prepared NCs were assessed for the efficient elimination of various HMs such as Pb^{2+}, Cu^{2+}, Cd^{2+}, Hg^{2+}, Zn^{2+}, Fe^{2+}, Ca^{2+}, Mg^{2+}, Mn^{2+}, and Al^{3+}. Extraordinary removal percentages for many HM ions were observed under slightly acidic conditions. The LNIR and PSOR models optimally defined the sorption process of these NCs. It was also observed in this study that Fe_3O_4/NH_2 prepared using the solvothermal approach, which was sequentially modified with TEOS and APTMS to fix amine groups, was effective in detecting Pb^{2+} in lake water; the cutoff value was 20 ng mL^{-1} and the detection limit was 10 ng ml^{-1} of Pb^{2+} [37]. MNPs functionalized with 3-aminopropyl-triethoxysilane and glutaraldehyde (APTES-GA) were assessed for Cu^{2+} confiscation. The sorption process was found to follow the pseudo-first-order (PFOR) and LNIR models and had a determined Q_m of 19.25 mg g^{-1} [38]. γ-Fe_2O_3 NPs were layered with a silica shell (γ-Fe_2O_3/SiO_2) and further improved by the covalent bonding of derived lysin. The resulting NCs were subsequently used for the remediation of industrial sewage sludge containing Cu, Cr, and Zn. The nanosorbent dosage (2.5–35 mg) impacted the remediation of the industrial sewage sludge: a significant percentage of the individual HM ions was removed as the dosage of the nanosorbent was increased, and the nanosorbent was saturated after 10 mg of the sorbent was used to treat the sewage sludge (figure 4.6). Ten mg of sorbent dosage was chosen as the optimal value to maximize the removal efficacy [39].

Figure 4.7(a) depicts the influence of reaction times varying from 5–120 min on the sorption of various HMs in the sewage sludge. It can be observed that the percentage of HMs eliminated from the sewage sludge swiftly increased with increasing contact time for the first 30 min for Zn and Cu, while the percentage of Cr ions removed swiftly improved with increasing contact time and equilibrium of the sorption process was reached after 120 min. Based on the kinetic studies, various kinetic models (PFOR, PSOR, and IND models) were assessed and the results are represented in figures 4.7(b)–(d). The sorption process of the sewage sludge was ideally defined by the PSOR (Zn and Cu) and the PFOR (Cr).

Table 4.4 summarizes the research carried out into the use of MNPs for the removal of various HMs.

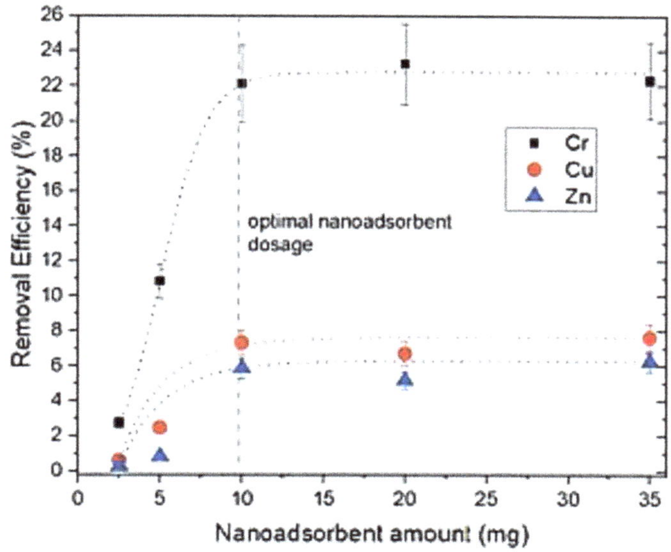

Figure 4.6. The influence of nanosorbent dosage on the sorption of various HMs in industrial sewage sludge. Reproduced from [39]. CC BY 4.0.

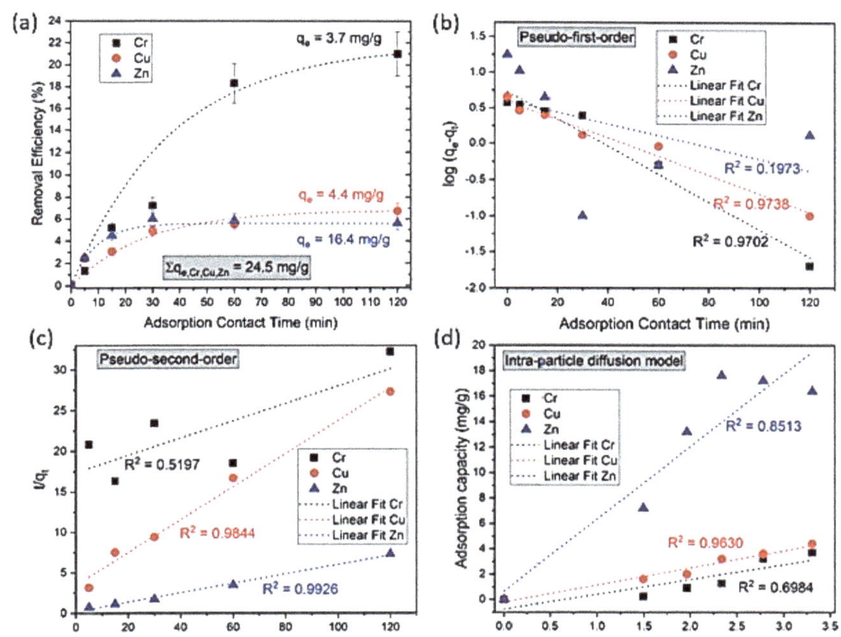

Figure 4.7. (a) The influence of contact time on the sorption of Zn, Cu, and Cr from sewage sludge. (b) The PFOR model fit, (c) PSOR model fit, and (d) IND model fit for the sorption process of sewage sludge by NCs. Reproduced from [39]. CC BY 4.0

Table 4.4. A summary of various studies of the use of MNPs in the sequestration of HMs.

MNPs	Contaminants	pH	Isotherm Model/Q_m	Kinetic model	Thermodynamics	References
Fe_3O_4@SiO_2-S-nNG-SPTZ	Hg^{2+}	7–8	Freundlich (FRCH)/43–67	PSOR.	—	[11]
Fe_3O_4	Zn^{2+}	5.5	—	PSOR.	Endothermic	[40]
Fe_3O_4/SiO_2 nanospheres	Hg^{2+}, Pb^{2+}, and palladium (Pd^{2+})	4	LNIR/303, 270, and 256	PSOR.		[41]
AC/FeO	Cr^{6+} and mordant violet 40 (MV40)	1.6/2.0	FRCH and LNIR/312/833	PSOR.	—	[42]
Fe_3O_4/PAA/CR	Pb^{2+}, Fe^{2+}, Cu^{2+}, Fe^{3+}, and Cd^{2+}	6.5/Pb^{2+}	FRCH and LNIR/223 (Pb^{2+})	PSOR (Pb^{2+}).		[43]

PAA = polyacrylic acid, CR = Congo red.

4.4.2 Factors influencing the sorption of heavy metals by magnetic nanoparticles

The Q_e value or percentage removal is impacted by different factors such as the pH, contaminant concentration, sorbent dosage, contact time, and temperature [44], as depicted in figure 4.8. To realistically design an economical sorption/photoreactor system, it is imperative to comprehend the impact of these factors on the sorption process [45].

A crucial factor in practically all sorption approaches is the contaminant pH. This element influences both the process's efficiency and the adsorbent's capacity. The sorbent's surface charge, competitiveness with coexisting ions in the solution, functional group activity, and the chemistry of the contaminated solution are all impacted by the pH. The characteristics of the sorbent, the sorption procedure, and the dissociation of HM ions can all be impacted by the pH of the aqueous medium. The solution pH can change not just the sorbent but also the contaminants' chemical structure. The sorbed ions' surface charge and level of ionization are modified by the pH [46].

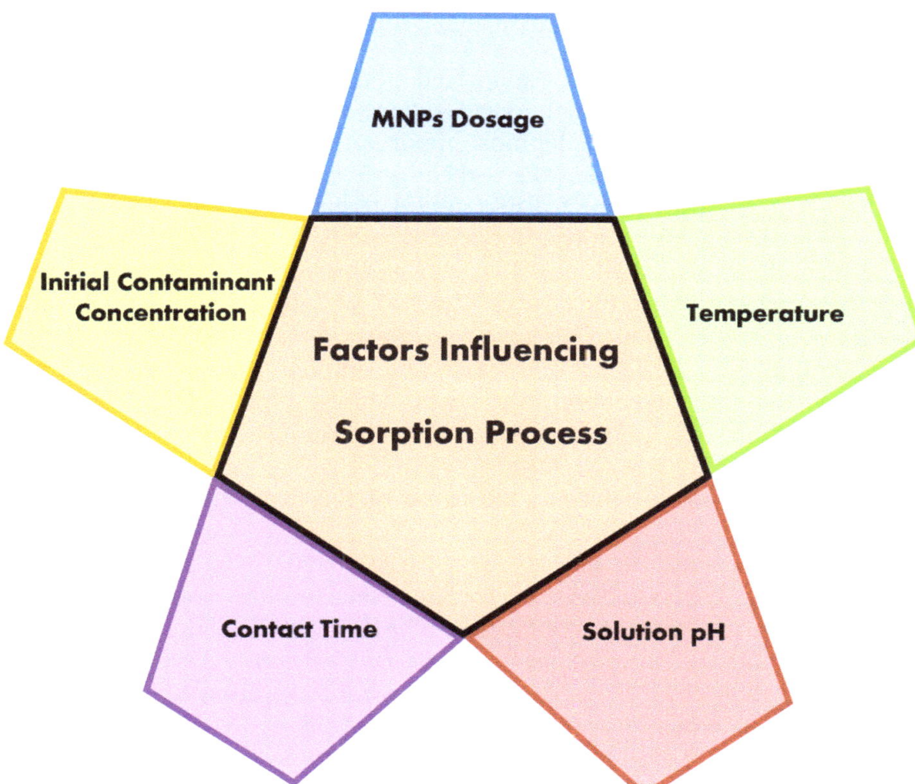

Figure 4.8. Factors influencing the sorption process.

Many H^+ or H_3O^+ ions enter the active sites and form positively charged sorbents when the pH drops below pH_{pzc}. This lowers the elimination effectiveness by decreasing the number of FGs and creating electric repulsion for divalent MIs with positive charges. HM species with positive charges in water (cations) are drawn to negatively charged surfaces of the MNPs by electrostatic forces when the pH is higher than pH_{pzc}. Furthermore, the degree of protonation in the materials decreases as the pH rises, allowing many O_2 and nitrogen-containing sorption sites to readily fill the vacant orbits of HMs ions and substantially boosting the sorption ability of the MNPs [45].

The runoff physical chemistry, including its pH and ionic strength, defines the sorption behavior. It regulates the sorption behavior via electrostatic interaction, modifies the charges of the sorbent and the sorbate, and either promotes or deters charged species contact. By adding or hiding charges, the ionic strength, and pH can alter the surface charge and impact the way in which particles aggregate, either increasing or decreasing the sorption efficiency. In addition, the pH of the solution influences HM speciation, introduces competing reactions among the ionic species in the solution, and interferes with other reactions, including the hydrolysis of ions, either at the interface or in the bulk solution. The size, hydration, and other intrinsic properties of ions alter the surface speciation (inner or outer sphere complex) and the thermodynamics [47].

Kroeker's rule states that as the sorbent mass increases, the specific sorbed volume reduces for a given initial concentration [48]. The dose of the sorbent or photocatalyst is assumed to be the most important factor affecting the process of sorption. It is critical to optimize the sorbent dose during the exclusion process to prevent any excess dose. To be feasible in cost terms, the reaction needs to be as effective as possible while applying the least quantity of sorbent or photocatalyst. Because there are more active accessible sites on the sorbent surface, the exclusion effectiveness increases when the dose is increased. Pollutants are broken down by a sizable amount of OH^- and other reactive oxygen species (ROS) that are created during sorption [45]. On the other hand, the amount sorbed per species (Q_e) decreases as the sorbent concentration rises. This is explained by the fact that when the concentration of the sorbent increases, the sorption isotherm's shape varies. It is likely that selected surface or surface groups are not filled in more concentrated suspensions, which is the cause of the drop in the specific sorbed amount [26].

The initial concentration of MIs is a crucial factor in explaining successful sorption. Due to the saturation of sorption sites on the sorbent surface, an intensification in the initial concentration of MIs leads to a decline in sorption efficiency, although the extraordinary driving force for mass transfer at an elevated initial concentration of HMs resulted in an increase in sorption capacity (mg g^{-1}). This behavior depends upon the potential interactions between the contaminants and the sorbent surface sites. As a result, it is critical to maximize the concentration to ensure that the initial MI concentration and the available active sites of the sorbent bond are practical. The solution's concentration can readily be lowered to approach equilibrium at a lower level when the initial contaminant concentration is low and the sorbent dosage is high. In this scenario, the sorption sites stay unsaturated. As the MI concentrations are increased, a decrease in sorption effectiveness is noticed because of repulsive forces between the sorbed solute molecules in the bulk phase and those on the surface [49].

The effect of contact time must be considered when assessing the sorption/photodegradation of HMs. Published studies indicate that at the beginning of the treatment, the removal capacity improves fast before progressively declining until sorption equilibrium is achieved, at which point the sorption-accessible sites are fully filled. This can be described by the steady decrease in the differential concentration between the sorbent's surface and the bulk phase of the solution as well as the steady saturation of active available sites by sorption on MNPs [45].

Increasing the reaction temperature aids in the sorption of pollutants. An increase in reaction temperature causes more reactive radicals to be generated because of bubble formation during the photoreaction process, which in turn raises the sorption activity [45]. Furthermore, it can change the amount of sorption, which means it has a meaningful impact on it. Depending on the sorbent and the contaminant, temperature can have varying impacts on the sorption efficiency. Raising the sorbate's surface activity and kinetic energy generally advances the sorption of the contaminants; however, it may also weaken the sorbent's structural integrity. Greater temperatures result in above-average sorption efficiency if the sorption procedure is chemisorption (ΔH chemisorption $= 200$ kJ mol^{-1}), since the rate of chemical reaction increases with temperature (although it eventually reaches equilibrium). Conversely, if the sorption process is physical (ΔH physisorption $= 20$ kJ mol^{-1}), higher temperatures have an adverse impact on the sorption. The sorbent, as well as its sorption sites and activity, can be chemically changed by temperature. Sorption processes can be categorized as either endothermic or exothermic. In exothermic processes, the effectiveness of the sorption process reduces as the temperature rises. The decline in HM removal effectiveness can be explained by the fact that when the temperature rises, the sorptive power between the sorbate and the sorbent's active sites is weakened. Since adding heat to a system improves the mobility of the contaminants, exothermic sorption is typically utilized to modulate the diffusion process. In endothermic processes, the sorption process (effectiveness) improves as the temperature rises because of the greater availability of accessible sites due to the activation of the sorbent surface at elevated temperatures [48].

To improve our understanding of the sorption behavior of nanosorbents, models of sorption kinetics, isotherm, and thermodynamics are applied [44]. When analyzing the mechanism of MI sorption onto various sorbents, isotherms are essential. These models provide insight into the intermolecular interactions between sorbed molecules and the sorbent matrix as well as the surface properties of sorbents. Our understanding of the sorption process is aided by isothermic and kinetic models, which depend on several variables such as the structure of the sorbent and the solute's physical and chemical characteristics [50]. The kinetic and isothermic models and the mechanism of the sorption process depend on different factors such as the morphology of the sorbent surface and the magnetic behavior of the sorbent. They are also impacted by the time of irradiation, the concentration of the sorbent, pH, effluent temperature, and the contaminant's initial concentration [51].

Equations in the form of a sorption isotherm, which associate the amount of sorbate held by the sorbent (Q_e) with the concentration in solution at equilibrium (C_e), can be used to quantify sorption at a specific temperature. The Freundlich (FRCH) and LNIR isotherms are the two practical models that are most regularly employed to describe the HMs sorption process at a certain temperature and on diverse MNP materials. They are shown in equations (4.1) and (4.2) (linear forms). In addition, the interactions between toxic contaminants and sorbent materials have been described by the Temkin (TKN), Dubinin–Radushkevich (DR), Redlich–Peterson (RP), Koble–Corrigan (KC), and Toth isotherms. When analyzing the mechanism of MI sorption to diverse sorbents, sorption isotherms are fundamental. These models provide insight into the properties of a sorbent's surface as well as the interactions between the sorbed molecules and the sorbent matrix [50, 52].

$$\frac{C_e}{Q_e} = \frac{1}{Q_m k_L} + \frac{C_e}{Q_m} \tag{4.1}$$

$$\log Q_e = \log k_f + \frac{1}{n} \log C_e \tag{4.2}$$

Q_e = the total amount of HMs sorbed by the sorbent surface in the equilibrium state (mg g^{-1}), C_e = the concentration of the HM solution in the equilibrial condition (mg l^{-1}), k_L = the LNIR isotherm constant, k_f = the sorption capacity of the sorbent, and $1/n$ = sorbent intensity [46].

The LNIR model is used in solid–liquid systems to show that HMs have an equal chance of occupying any sorbent's surface site. Conversely, the FRCH model defines a nonideal process that often involves the creation of several layers on diverse surfaces [50]. Based on the reviewed articles, most sorption processes have generally been described by the LNIR model, followed by the FRCH model.

To optimize the design parameters, such as the reactor dimensions and sorbate residence time, kinetic sorption models are utilized to explain the mechanism of HM ion sorption by MNPs and regulate the rate of sorption during the industrial eradication of HMs from industrial effluent. Several mathematical models, the majority of which are empirical equations, have often been utilized in experimental investigations to assess the elimination effectiveness of HMs by different sorbents and provide a kinetic description of the sorption processes. The popular kinetics models that have been employed to fit the experimental data are the PFOR, the PSOR, the IND kinetic, and Elovich (ELH) models (equations (4.3)–(4.6)). The goodness of fit, as determined by the coefficient of determination (R^2), is often the pivotal factor when choosing between these equations [50].

The Lagergren PFOR, which is typically applicable over the first stage of a sorption process, assumes that the rate of alteration of solute uptake over time unswervingly corresponds to the alteration in saturation concentration and the quantity of solid uptake over time. It has regularly been noted that when sorption happens by diffusion over the area of contact, the kinetics obeys this Lagergren PFO rate equation [47].

The PSOR is based on the supposition that chemical sorption (chemisorption) is the rate limiting step. It forecasts behavior across the complete sorption range. Under these circumstances, the sorption rate depends more on the sorption capacity than the sorbate concentration. The ability to compute the equilibrium sorption capacity directly from the model removes the theoretical need to assess the sorption equilibrium capacity by experimentation, which is one of the main advantages of this model over the Lagergren PFOR. The ELH kinetic model, which is based on the concept that that the surface is energetically diverse, is frequently applied to explore the sorption kinetics and realistically characterize PSOR kinetics [47, 53]. In 1962, Weber and Morris presented the IN mass transfer diffusion model as a means of defining the sorption's diffusion mechanism. The diffusion of the sorbate into pores of varying sizes is known as the physisorption phenomenon, and the IND rate model is normally applied to porous materials. Diffusion via IND is the process that limits the pace at which sorbate molecules or ions are transported from the bulk solution to the sorbent solid surface [54].

$$\log Q_e - Q_t = \log Q_e - \frac{k_1}{2.303}t \tag{4.3}$$

$$\frac{t}{Q_t} = \frac{1}{k_2 Q_e^2} + \frac{1}{Q_e}t \tag{4.4}$$

$$Q_t = k_{ip}\sqrt{t} + C \tag{4.5}$$

$$Q_t = \frac{1}{\beta}\ln \alpha\beta + \frac{1}{\beta}\ln t \tag{4.6}$$

q_t = the sorption capacity of the sorbent at time t (mg g^{-1}), k_1 = the PFOR rate constant (min^{-1}), t = the reaction time (min), k_2 = the PSOR rate constant (g mg^{-1} min^{-1}), k_{ip} = a measure of the diffusion coefficient, C = the IND constant, α = the initial sorption rate and β = the desorption constant [46, 47].

Process modeling is an essential component of any systematic research project. The nature of the sorbate–sorbent interaction and the sorption structure's time dependence are understood using the sorption isotherm and kinetics investigations. Nonlinear regression is used to evaluate distinct factor sets because of the attribute preference that arises from the linearization. This provides a rigorous statistical method for assessing factors using a special version of the equations for the models. Since different versions of the sorption model equations affect the R^2 values during the linear examination, nonlinear analysis is the primary method for preventing these kinds of mistakes. The insufficiency of the linear regression technique in estimating just two variables in an experimental equation is another significant problem. For this reason, it is considered an inappropriate method for determining sorption kinetics and isotherm parameters. To evaluate the best fit of the sorption factors to the experimental results, a more complex and rigorous process for assessing the kinetics and isotherm factors with an error estimation is provided by the nonlinear optimization approach. It is generally accepted that error function consideration is required to estimate whether a model computation is applicable for a given set of experimental data. Numerical equations called error functions are employed to estimate the difference between estimates of actual experimental data and hypothetically projected data. Error functions are primarily useful for measuring the distribution of the sorbent and providing a mathematical assessment of the outcome; nevertheless, their most important use is in verifying the validity of the experimental results, which have produced sorption isotherms and kinetics [55]. Based on the articles reviewed, the LNIR and PSOR models generally describe HM ion sorption by MNPs.

4.5 The environmental applications of sensors

The presence of natural and mineral chemical substances in water may result in a swift decline in water quality and probably leads to adverse impacts on human health. Water quality is generally impacted by contamination sources of chemical and natural (decaying vegetation and microbiota metabolite) origins. The most widespread specification for water quality depends on its acceptability to users. The regulations related to water for human consumption consider whether its odor and flavor are acceptable to its users and do not display unusual changes. The European Union (EU) standard for water quality (EN 1622) includes numeric thresholds for odor and flavor (the threshold odor number (TON) and the threshold flavor number (TFN)). But if water is acceptable to its users, these assessments are not needed, unless indicated otherwise. The TFN and TON standards are qualitative assessments of water quality, and upon request, a quantitative assessment of the compounds of interest can also be performed. Generally, compounds that trigger a sensory response exist in very minimal concentrations in water bodies and their quantitative assessment can be performed using proper and sensitive analytical approaches such as separation approaches, chromatography techniques, and colorimetric analysis [56].

Water quality and environmental cleanup efforts are now being improved with the assistance of NT and NPs. It is also becoming feasible to utilize these as environmental sensors to monitor contamination in the environment. NPs can be employed as sensors to find out whether certain substances, including contaminants such as HMs, are present in the environment. The compact size and broad detection range of nanosensors offer a great deal of versatility in real-world applications. Biological substances including poisons and microbial infections can be detected in marine settings using nanoscale sensors. Contaminants can be made to selectively bind to NPs, which makes it possible to detect them even at minimal concentrations [10].

In another study, a sensitive fluorescence sensor based on fluorophore quenching was created to detect Cu^{2+} ions. The sensor contained Fe_3O_4 core nanostructures tailored with amino silane shells functionalized with pyranine (Pyre)-$Fe_3O_4/SiO_2/NH_2$/Pyr. The magnetic core was found to be strongly superparamagnetic, with an average particle diameter of up to 10 nm, according to the characterization results. The fluorescent response of the $Fe_3O_4/SiO_2/NH_2$/Pyr nanosensors to Cu^{2+} ions exhibited increased and targeted fluorescence quenching. It was discovered that the $Fe_3O_4/SiO_2/NH_2$/Pyre nanosensors were extremely selective for Cu^{2+} ions and did not react to other interfering biomolecules and cations. The limit of detection (LOD) of this nanosensor for Cu^{2+} was assessed to be 6 nM and it can be applied in biologic and ecological fields owing to the novel fluorescence and biocompatibility of the material [57]. To detect Cr^{6+} ions, two distinct Fe_3O_4/SiO_2/TPED/BODIPY and Fe_3O_4/SiO_2/TMPTA/BODIPY nanosensors were created based on the use of Fe_3O_4 fluorescence (TPED = N-[3-(Trimethoxysilyl)propyl]ethylenediamine, and TMPTA = methanol, ethanol, N1-(3-Trimethoxysilylpropyl)diethylenetriamine. The mean diameters of the Fe_3O_4 particles in the fluorescent Fe_3O_4/SiO_2/TPED/BODIPY and Fe_3O_4/SiO_2/TMPTA/BODIPY nanosensors, as determined by SEM characterization studies, were 18.5 and 19 nm, respectively. When exposed to Cr^{6+} ions in a pH 1 medium, the Fe_3O_4 fluorescent Fe_3O_4/SiO_2/TPED/BODIPY and Fe_3O_4/SiO_2/TMPTA/BODIPY nanosensors demonstrated fluorescence quenching responses. Due to these characteristics, Fe_3O_4 fluorescent Fe_3O_4/SiO_2/TPED/BODIPY and Fe_3O_4/SiO_2/TMPTA/BODIPY nanosensors may find use in environmental applications as a novel class of nontoxic sensors [58]. For the specific purpose of detecting Cr_{6+} ions, two distinct Fe_3O_4/SiO_2/TPED/BODIPY and Fe_3O_4/SiO_2/TMPTA/BODIPY nano-sensors were created using Fe_3O_4 fluorescence in this work. The mean particle diameter of the Fe_3O_4 in fluorescent Fe_3O_4/SiO_2/TPED/BODIPY and Fe_3O_4/SiO_2/TMPTA/BODIPY nano-sensors, as determined by SEM characterization studies, was 18.5 and 19 nm. When exposed to Cr_{6+} ions in the pH 1 medium, the Fe_3O_4 fluorescent Fe_3O_4/SiO_2/TPED/BODIPY and Fe_3O_4/SiO_2/TMPTA/BODIPY nano-sensors demonstrated fluorescence quenching responses. Due to these characteristics, Fe_3O_4 fluorescent Fe_3O_4/SiO_2/TPED/BODIPY and Fe_3O_4/SiO_2/TMPTA/BODIPY nano-sensors may find usage in environmental applications as a novel class of non-toxic sensors [59]. In another study, Fe_3O_4/GMP/Tb was designed, synthesized, and explored as a sorbent for Zn^{2+} detection using magnetic solid phase extraction. The prepared Fe_3O_4/GMP/Tb (GMP = Guanosine 5′-monophosphate disodium salt hydrate) was studied in terms of its morphology and structure and was shown to have a network structure that included magnetic Fe_3O_4 NPs. The findings demonstrated that the nanoscale Fe_3O_4/GMP/Tb particles' size and their network structure that integrated Fe_3O_4 gave them a great enhancement ability (the enhancement factor reached 29) and high sensitivity for Zn^{2+}. With an LOD of 0.21 μM, the prepared Fe_3O_4/GMP/Tb nanosensor for Zn^{2+} demonstrated a broad linear range from 0 to 160 μM. The prepared Fe_3O_4/GMP/TbAs was also successfully used as a fluorescent probe for Zn^{2+} in living cells [60].

Lateral flow immunoassays (LFIAs) on paper are an intriguing on-site process for HM detection. One of the hazardous HMs found in contaminated ambient water is Pb^{2+}. The production and classification of three Fe_3O_4-based nanosorbents were explored in a study. When graphene oxide (GO) or an amino group was added, the performance of these nanosorbents varied with respect to Pb^{2+}, Cu^{2+}, Cd^{2+}, Hg^{2+}, As (V), Zn^{2+}, Fe^{3+}, Ca^{2+}, Mg^{2+}, Mn^{2+}, and Al^{3+}. Under somewhat acidic conditions, Fe_3O_4/GO exhibited a greater sorption percentage for heavy MIs in comparison to Fe_3O_4. Four primary divalent MI sorption interactions were favorably facilitated by the negative surface on Fe_3O_4/NH_2, ordered as follows: $Pb^{2+} > Cu^{2+} > Hg^{2+} > Cd^{2+}$. A fivefold increase in sensitivity was achieved in the detection of Pb^{2+} in lake water using a Fe_3O_4/NH_2-based LFIA combined with antibody-functionalized AuNPs that demonstrated strong affinity and specificity toward Pb^{2+}. This cutoff value of this Fe_3O_4@NH_2 and LFIA method was 20 ng ml^{-1} for Pb^{2+} detection in lake water, and the detection results for 10 ng ml^{-1} Pb^{2+} were visible to the unaided eye or a strip reader device. As a very sensitive screening tool, this detection technique might be easily modified for additional small compounds or biomolecules. Fe_3O_4/GO exhibited a greater proportion of sorption of numerous heavy MIs under slightly acidic conditions as compared to Fe_3O_4. [37]. In a further study, GO was combined with mesoporous $MnFe_2O_4$ ($MnFe_2O_4$/GO) to create an electrochemical platform to sense Pb^{2+}. The $MnFe_2O_4$ particles in the prepared mesoporous $MnFe_2O_4$/GO NCs were consistently sized and had a diameter of around 400 nm. The $MnFe_2O_4$/GO particles had a maximum pore width of 4.5 nm, and the addition of GO raised their specific SA from 167 to 286 m^2 g^{-1}. The addition of GO improved the electrochemical activity of $MnFe_2O_4$ because of its enhancement of the specific SA, electrocatalytic activity, and conductivity supported by GO sheets, according to the cyclic voltammograms (CV), electrochemical impedance spectra (EIS), and square wave anodic stripping voltammetry (SWASV) responses to Pb^{2+}. For the electrochemical detection of Pb^{2+}, the $MnFe_2O_4$/GO NCs offered a sensitivity of 33.9 μA μM^{-1} and an LOD of 0.0883 μM. They were best suitable to the selective detection of Pb^{2+} in the presence of Zn^{2+}. Furthermore, an electrode modified by the addition of $MnFe_2O_4$/GO demonstrated satisfactory stability, repeatability, reproducibility, and usability for Pb^{2+} detection. Hence, MNP-based sensors are effective for the recognition of contaminants in the ecosystem [61].

4.6 Conclusions

This chapter reviewed the recent research advancement in MNPs, from their different synthesis and modification/functionalization approaches to the reduction of particle aggregation and the attachment of certain FGs to their surface for their effective use in environmental waste management and detection. Based on the articles reviewed in this chapter, it was noted that the coprecipitation approach was generally employed for the synthesis of most MNPs and that the confiscation of most HMs was found to be generally pH-reliant and was perfectly defined by the LNIR and PSOR models. Functionalized synthesized MNPs were also found to be highly sensitive and selective in HM detection.

References and further reading

[1] Peralta M, Ocampo S, Funes I, Onaga Medina F, Parolo M and Carlos L 2020 Nanomaterials with tailored magnetic properties as adsorbents of organic pollutants from wastewaters *Inorganics* **8** 24

[2] Gómez-Pastora J, Bringas E and Ortiz I 2014 Recent progress and future challenges on the use of high performance magnetic nano-adsorbents in environmental applications *Chem. Eng. J.* **256** 187–204
[3] Mahmood T, Momin S, Ali R, Naeem A and Khan A 2022 Technologies for removal of emerging contaminants from wastewater *Wastewater Treatment* (London: IntechOpen)
[4] Aigbe U, Ukhurebor K, Onyancha R, Osibote O, Darmokoesoemo H and Kusuma H 2021 Fly ash-based adsorbent for adsorption of heavy metals and dyes from aqueous solution: a review *J. Mater. Res. Technol.* **14** 2751–74
[5] Gallo-Cordova A, Streitwieser D, Puerto Morales M and Ovejero G 2021 Magnetic iron oxide colloids for environmental applications *Colloids: Types, Preparation and Applications* (London: IntechOpen)
[6] Jamkhande P, Ghule N, Bamer A and Kalaskar M 2019 Metal nanoparticles synthesis: an overview on methods of preparation, advantages and disadvantages, and applications *J. Drug Deliv. Sci. Technol.* **53** 101174
[7] Jiang B, Lian L, Xing Y, Zhang N, Chen Y, Lu P and Zhang D 2018 Advances of magnetic nanoparticles in environmental application: environmental remediation and (bio) sensors as case studies *Environ. Sci. Pollut. Res.* **25** 30863–79
[8] Zhang K, Song X, Liu M, Chen M, Li J and Han J 2023 Review on the use of magnetic nanoparticles in the detection of environmental pollutants *Water* **15** 3077
[9] Liandi A, Cahyana A, Kusumah A, Lupitasari A, Alfariza D, Nuraini R, Sari R and Kusumasari F 2023 Recent trends of spinel ferrites (MFe_2O_4: Mn, Co, Ni, Cu, Zn) applications as an environmentally friendly catalyst in multicomponent reactions: a review *Case Stud. Chem. Environ. Eng.* **7** 100303
[10] Altammar K 2023 A review on nanoparticles: characteristics, synthesis, applications, and challenges *Front. Microbiol.* **14** 1155622
[11] Chen D, Sawut A and Wang T 2022 Synthesis of new functionalized magnetic nano adsorbents and adsorption performance for Hg (II) ions *Heliyon* **8** E10528
[12] Aigbe U *et al* 2023 Utility of magnetic nanomaterials for theranostic nanomedicine *Magnetic Nanomaterials: Synthesis, Characterization And Applications* (Cham: Springer Nature) 47–86
[13] Aigbe U O, Onyancha R B, Ukhurebor K E *et al* 2023 *Utility of Magnetic Nanomaterials for Theranostic Nanomedicine* Magnetic Nanomaterials: Synthesis, Characterization and Applications (Cham: Springer International) 47–86
[14] Abu-Dief A and Hamdan S 2016 Functionalization of magnetic nano particles: synthesis, characterization and their application in water purification *Am. J. Nanosci.* **2** 26–40
[15] Kanjilal T and Bhattacharjee C 2018 Green applications of magnetic sorbents for environmental remediation *Organic Pollutants in Wastewater I. Methods of Analysis, Removal and Treatment* (Millersville, PA: Materials Research Forum LLC) pp 1–41
[16] Aigbe U and Osibote A 2024 Green synthesis of metal oxide nanoparticles and their various applications *J. Hazard. Mater. Adv.* **13** 100401
[17] Miu B and Dinischiotu A 2022 New green approaches in nanoparticles synthesis: an overview *Molecules* **27** 6472
[18] Kustov L and Vikanova K 2023 Synthesis of metal nanoparticles under microwave irradiation: get much with less energy *Metals* **13** 1714
[19] Khan F, Mubarak N, Khalid M, Walvekar R, Abdullah E, Mazari S, Nizamuddin S and Karri R 2020 Magnetic nanoadsorbents' potential route for heavy metals removal—a review *Environ. Sci. Pollut. Res.* **27** 24342–56

[20] Tang S and Lo I 2013 Magnetic nanoparticles: essential factors for sustainable environmental applications *Water Res.* **47** 2613–32

[21] Pratt A 2014 Environmental applications of magnetic nanoparticles *Frontiers Nanosci.* **6** 259–307

[22] Mudhoo A and Sillanpää M 2021 Magnetic nanoadsorbents for micropollutant removal in real water treatment: a review *Environ. Chem. Lett.* **19** 4393–413

[23] He M, Chen Z, Xu C, Chen B and Hu B 2021 Magnetic nanomaterials as sorbents for trace elements analysis in environmental and biological samples *Talanta* **230** 122306

[24] Aigbe U, Ukhurebor K, Onyancha R, Okundaye B, Aigbe E, Kusuma H *et al* 2023 Applications of magnetic nanomaterials for wastewater treatment *Magnetic Nanomaterials: Synthesis, Characterization And Applications* (Cham: Springer International Publishing) 129–69

[25] Akhtar N, Syakir Ishak M, Bhawani S and Umar K 2021 Various natural and anthropogenic factors responsible for water quality degradation: a review *Water* **13** 2660

[26] Silva J 2023 Wastewater treatment and reuse for sustainable water resources management: a systematic literature review *Sustainability* **15** 10940

[27] Ukhurebor K, Aigbe U, Onyancha R, Nwankwo W, Osibote O, Paumo H, Ama O, Adetunji C and Siloko I 2021 Effect of hexavalent chromium on the environment and removal techniques: a review *J. Environ. Manage.* **280** 111809

[28] Aigbe U, Das R, Ho W, Srinivasu V and Maity A 2018 A novel method for removal of Cr (VI) using polypyrrole magnetic nanocomposite in the presence of unsteady magnetic fields *Sep. Purif. Technol.* **194** 377–87

[29] Molina-Calderón L, Basualto-Flores C, Paredes-García V, Gutierrez-Cutiño M and Venegas-Yazigi D 2023 Magnetic nanoadsorbent functionalized with aminophosphonic acid for NdIII ion extraction from aqueous media *J. Mol. Liq.* **384** 122258

[30] Bunge A, Leoştean C and Turcu R 2023 Synthesis of a magnetic nanostructured composite sorbent only from waste materials *Materials* **16** 7696

[31] Kim I, Yang H, Park C, Yoon I and Sihn Y 2021 Environmental applications of magnetic nanoparticles *Magnetic Nanoparticle-Based Hybrid Materials* (Cambridge: Woodhead Publishing) 529–45

[32] Prabhu S *et al* 2023 Magnetic nanostructured adsorbents for water treatment: structure-property relationships, chemistry of interactions, and lab-to-industry integration *Chem. Eng. J.* 143474

[33] Kothavale V *et al* 2023 Carboxyl and thiol-functionalized magnetic nanoadsorbents for efficient and simultaneous removal of Pb (II), Cd (II), and Ni (II) heavy metal ions from aqueous solutions: studies of adsorption, kinetics, and isotherms *J. Phys. Chem. Solids* **172** 111089

[34] Chen D, Awut T, Liu B, Ma Y, Wang T and Nurulla I 2016 Functionalized magnetic Fe_3O_4 nanoparticles for removal of heavy metal ions from aqueous solutions *e-Polymers* **16** 313–22

[35] Fato T, Li D, Zhao L, Qiu K and Long Y 2019 Simultaneous removal of multiple heavy metal ions from river water using ultrafine mesoporous magnetite nanoparticles *ACS Omega* **4** 7543–9

[36] Saleem M, Hussain H, Shukrullah S, Yasin Naz M, Irfan M, Rahman S and Ghanim A 2024 Study of kinetics and the working mechanism of silica-coated amino-functionalized $CoFe_2O_4$ ferrite nanoparticles to treat wastewater for heavy metals *ACS Omega* **9** 3507–24

[37] Sun M, Li P, Jin X, Ju X, Yan W, Yuan J and Xing C 2018 Heavy metal adsorption onto graphene oxide, amino group on magnetic nanoadsorbents and application for detection of Pb (II) by strip sensor *Food Agric. Immunol.* **29** 1053–73

[38] Kothavale V, Chavan V, Sahoo S, Kollu P, Dongale T, Patil P and Patil P 2019 Removal of Cu (II) from aqueous solution using APTES-GA modified magnetic iron oxide nanoparticles: kinetic and isotherm study *Mater. Res. Express* **6** 106103

[39] Plohl O, Simonič M, Kolar K, Gyergyek S and Fras Zemljič L 2021 Magnetic nanostructures functionalized with a derived lysine coating applied to simultaneously remove heavy metal pollutants from environmental systems *Sci. Technol. Adv. Mater.* **22** 55–71

[40] Shirsath D and Shirivastava V 2015 Adsorptive removal of heavy metals by magnetic nanoadsorbent: an equilibrium and thermodynamic study *Appl. Nanosci.* **5** 927–35

[41] Vojoudi H, Badiei A, Bahar S, Ziarani G, Faridbod F and Ganjali M 2017 A new nanosorbent for fast and efficient removal of heavy metals from aqueous solutions based on modification of magnetic mesoporous silica nanospheres *J. Magn. Magn. Mater.* **441** 193–203

[42] Mohamed S, Yılmaz M, Güner E and El Nemr A 2024 Synthesis and characterization of iron oxide-commercial activated carbon nanocomposite for removal of hexavalent chromium (Cr^{6+}) ions and Mordant Violet 40 (MV40) dye *Sci. Rep.* **14** 1241

[43] Sadak O, Hackney R, Sundramoorthy A, Yilmaz G and Gunasekaran S 2020 Azo dye-functionalized magnetic Fe_3O_4/polyacrylic acid nanoadsorbent for removal of lead (II) ions *Environm Nanotechnol. Monit. Manag.* **14** 100380

[44] Aigbe U, Ukhurebor K, Onyancha R, Osibote O, Kusuma H and Darmokoesoemo H 2020 Measuring the velocity profile of spinning particles and its impact on Cr (VI) sequestration *Chem. Eng. Proces.-Process Intensif.* **178** 109013

[45] Amdeha E 2023 Biochar-based nanocomposites for industrial wastewater treatment via adsorption and photocatalytic degradation and the parameters affecting these processes *Biomass Convers.Biorefinery* 1–26

[46] Panda S, Aggarwal I, Kumar H, Prasad L, Kumar A, Sharma A, Vo D, Van Thuan D and Mishra V 2021 Magnetite nanoparticles as sorbents for dye removal: a review *Environ. Chem. Lett.* **19** 2487–525

[47] Sahoo T and Prelot B 2020 Adsorption processes for the removal of contaminants from wastewater: the perspective role of nanomaterials and nanotechnology *Nanomaterials for the Detection and Removal of Wastewater Pollutants* (Amsterdam: Elsevier) 161–222

[48] Rápó E and Tonk S 2021 Factors affecting synthetic dye adsorption; desorption studies: a review of results from the last five years (2017–2021) *Molecules* **26** 5419

[49] Jadoun S, Fuentes J, Urbano B and Yáñez J 2023 A review on adsorption of heavy metals from wastewater using conducting polymer-based materials *J. Environ. Chem. Eng.* **11** 109226

[50] Raji Z, Karim A, Karam A and Khalloufi S 2023 Adsorption of heavy metals: mechanisms, kinetics, and applications of various adsorbents in wastewater remediation—a review *Waste* **1** 775–80

[51] Qasem N, Mohammed R and Lawal D 2021 Removal of heavy metal ions from wastewater: a comprehensive and critical review *Npj Clean Water* **4** 36

[52] Abebe B, Murthy H and Amare E 2018 Summary on adsorption and photocatalysis for pollutant remediation: mini review *J. Encapsulation Adsorpt. Sci.* **8** 225–55

[53] Lodh B 2021 Biosorbents for heavy metal removal *Microbial Ecology of Wastewater Treatment Plants* (Amsterdam: Elsevier) pp 377–94
[54] Ray S, Gusain R and Kumar N 2020 Adsorption equilibrium isotherms, kinetics and thermodynamics *Carbon Nanomaterial-Based Adsorbents for Water Purification* (Amsterdam: Elsevier) 101–18
[55] Aigbe U, Maluleke R, Lebepe T, Oluwafemi O and Osibote O 2023 Rhodamine 6G dye adsorption using magnetic nanoparticles synthesized with the support of vernonia amygdalina leaf extract (Bitter Leaf) *J. Inorg. Organomet. Polym. Mater.* **33** 1–20
[56] Kekes T, Tzia C and Kolliopoulos G 2023 Drinking and natural mineral water: Treatment and quality–safety assurance *Water* **15** 2325
[57] Shah M, Alveroglu E and Balouch A 2018 Pyranine functionalized Fe_3O_4 nanoparticles for the sensitive fluorescence detection of Cu^{2+} ions *J. Alloys Compd.* **767** 151–62
[58] Bilgic A and Cimen A 2020 Two novel BODIPY-functional magnetite fluorescent nanosensors for detecting of Cr (VI) Ions in aqueous solutions *J. Fluores.* **30** 867–81
[59] Xu S, Zhang L, Zhao Y, Luo Y, Yu B and Zhang W 2021 A magnetic functionalized lanthanide fluorescent sensor for detection of trace zinc ion *Res. Chem. Intermed.* **47** 3487–500
[60] Zhou S, Han X, Fan H, Huang J and Liu Y 2018 Enhanced electrochemical performance for sensing Pb (II) based on graphene oxide incorporated mesoporous $MnFe_2O_4$ nanocomposites *J. Alloys. Compd.* **747** 447–54
[61] Liosis C, Papadopoulou A, Karvelas E, Karakasidis T and Sarris I 2021 Heavy metal adsorption using magnetic nanoparticles for water purification: a critical review *Materials* **14** 7500

Chapter 5

The applications of magnetic sorbents for the sequestration of dyes from aqueous solutions

Sefiu Olaitan Amusat, Oluseyi Salami, Temesgen Girma Kebede, Simiso Dube and Mathew Muzi Nindi

The application of magnetic sorbents (MSs) for the sequestration of dyes from aqueous solutions has gained significant attention due to the environmental and health concerns associated with dye pollution. MSs offer a viable solution for the removal of dyes from aqueous solutions. Their high adsorption capacity, selective adsorption, easy separation, reusability and regeneration, application flexibility, potential for combination with other treatment methods, and scalability make them effective materials in addressing the challenge of dye pollution. This chapter provides a comprehensive overview of the use of MSs as efficient and sustainable materials for the removal of dyes.

5.1 Introduction

Water remains the most vital resource on the Earth's surface. Concerns about water quality, availability, and remediation approaches are directly linked to the physicochemical characteristics of water. The growth of urban regions and practices that can add contaminants to potable water supplies have made water safety increasingly indispensable. Consequently, water treatment specialists work with a variety of water qualities and have access to a broadening choice of treatment methods [1].

Dye is a noxious contaminant that colors water when released into aquatic systems; as a result, water that contains dyes even in small amounts as low as ($1.0\ mg\ l^{-1}$) is indeed unsuitable for human consumption [2, 3]. Dyeing agents are compounds that contain chromophores, the building blocks of color, and are frequently used in polymers, cosmetics, and other products as well as in fabrics [1]. The adherence of dyes to surfaces can occur through covalent interactions, the creation of metal complexes with salts or metal ions, physisorption, or mechanical retention [4]. The molecular makeup or intended use of dyes determines how they are categorized. Over

doi:10.1088/978-0-7503-5909-2ch5

ten thousand distinct synthetic dyes and pigments are made globally each year, amounting to 7×10^5 tons. Two hundred thousand tons of colorants used in the textile industry are dumped into effluents each year. Consequently, these dyes persist in the ecosystem and conventional effluent because of their high stability in different matrices. Dyes are nonbiodegradable because of their intricate structural compositions [5]. Dyes can be categorized based on their origin, application method, the nature of their chromophores, and their solubility in water, as shown in figure 5.1.

Studies have revealed that these dyes are carcinogenic. They reduce photosynthesis rates and dissolved oxygen concentrations in polluted water and can increase allergic responses and anaphylactic shock in the organisms [7–9].

In recent years, eliminating dyes from wastewater in an attempt to mitigate water contamination has received a lot of attention. There are several strategies for removing dye from wastewater, some of which include flocculation, oxidation, and electrolysis. Adsorption has emerged as a simple, efficient, and cost-effective method and is used in several wastewater treatment strategies for removing dye from wastewater [10, 11]. Adsorbents are typically used at smaller particle sizes, as their greater surface area results in the maximum adsorption capacity. However, once the maximum adsorption capacity has been reached, it is extremely difficult to separate them from water. Considering these constraints, the magnetic separation of adsorbents has been one of the most promising methods for wastewater treatment because it produces no contaminants, such as flocculants, and can treat high volumes of effluent in a short period [12, 13]. Due to their tremendous importance in advancing the extraction pace and enhancing the effectiveness of water treatment, the magnetic separation methods based on the use of magnetic composites (MCs)

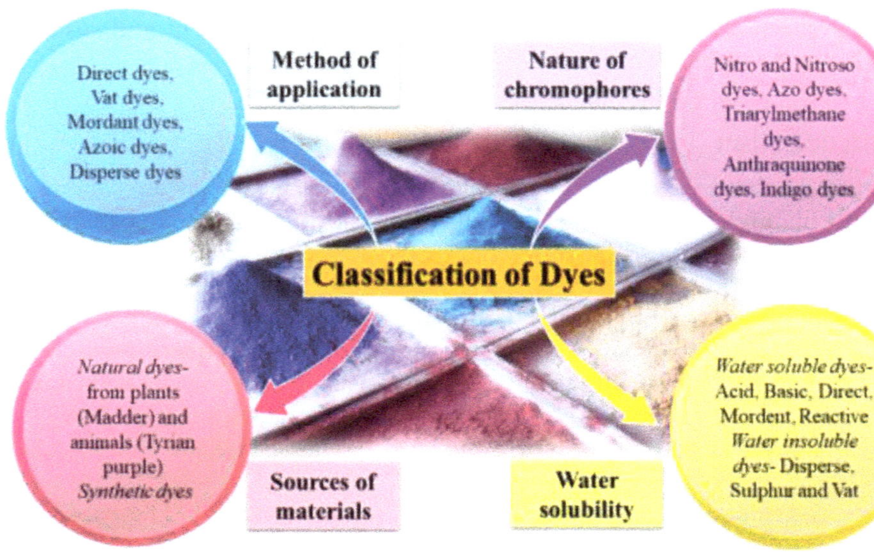

Figure 5.1. The classification of dyes [6].

and magnetic nanocomposites (MNCs) have gained widespread use in the sequestration of dyes [14]. An overview of the utilization of MCs as dye adsorbents is provided in this chapter. We briefly review their categorization and toxicity and several MCs used for the removal of dyes from aqueous solutions.

MSs have emerged as a promising alternative for the efficient sequestration of dyes from aqueous solutions. MSs are composite materials that combine the advantages of magnetic nanoparticles (MNPs) with the adsorption properties of sorbent materials. The incorporation of MNPs imparts magnetic responsiveness to the sorbents, enabling their facile separation from water using an external magnetic field. This feature not only enhances their ease of handling and separation but also enables their repeated use, thus reducing the overall cost and environmental impact [15].

The application of MSs for the sequestration of dyes offers several advantages over traditional sorbents. First, the extensive surface and porosity of the sorbents provide abundant active sites for dye adsorption, leading to enhanced removal efficiency. In addition, their magnetic properties allow for rapid separation and recovery of the sorbents, minimizing the contact time between the sorbents and the dye-containing solution. Furthermore, the reusability of MSs provides an environmentally friendly means of dye elimination, reducing the generation of waste and the consumption of resources [16].

Various types of MSs have been researched for dye sequestration, including MNPs coated with organic polymers, carbonaceous materials, and metal–organic frameworks (MOFs). These sorbents can be tailored to exhibit specific surface properties, such as hydrophobicity or hydrophilicity, to selectively adsorb different types of dyes. Moreover, functionalization of the sorbents with specific functional groups can enhance their affinity and selectivity towards particular dye molecules [17].

5.2 The fundamentals of magnetic sorbents

Since the beginning of society's evolution, the phenomenon of magnetism has played a significant role. The Earth's magnetic field has demonstrated its importance in various environmental applications [18].

Adsorbents that exhibit magnetic characteristics and are affected by a magnetic field are referred to as magnetic adsorbents. By having magnetic species added to them, adsorbents gain magnetic characteristics. Iron, cobalt, nickel, copper, and their oxides are the most common magnetic species. If the adsorbent contains one or more of these metal species, it can attract and separate magnetic adsorbents from a fluid (liquid or gas) under the influence of a magnetic field quickly and effectively [19].

Magnetic adsorbents improve the potential and performance of adsorbents. The possibility of reclaiming the solid adsorbent from the fluid phase utilizing magnets, which is considered to be an efficient separation method, makes the use of magnetic adsorbents feasible. The benefits of employing magnetic adsorbents include ease of separation, cheap production costs, the need for less adsorbent, and high adsorptive capabilities because of their enormous surface areas [20].

The characteristics of MSs collectively contribute to the effectiveness and practicality of MSs in dye sequestration processes, offering a promising approach for the removal of dyes from wastewater or other dye-contaminated solutions. Metal oxides have been widely used to manufacture magnetic adsorbents as a supporting material for both natural and synthetic adsorbents. Adsorbent/catalyst-like magnetic metal-oxide composites and magnetic metal-oxide-supported adsorbents have high adsorption capacities and catalytic activity towards organic and inorganic wastewater produced by chemical industries. The various types of magnetically modified adsorbents may be broadly categorized as magnetic-particle-modified carbon adsorbents, magnetic-particle-modified clay mineral adsorbents, magnetic-particle-modified biopolymer adsorbents, etc [21].

Magnetic adsorbents mostly exhibit superparamagnetic characteristics. Because these particles' magnetization is instantly changed by the application of external magnetic fields, the following operations are made possible after adsorption: the accurate and quick steering and placement of nanoparticles as well as their manipulation by field gradients. Superparamagnetism in nanoparticles is more directly related to their alignment and thermal fluctuation than is the case for individual dipoles in molecules or atoms [22].

5.2.1 Types of magnetic sorbents commonly used in dye sequestration

MSs are widely used in dye sequestration due to their unique properties that allow for simple extraction from a solution using an external magnetic field. The following list shows some types of MSs commonly used for dye sequestration:

- ***Magnetic iron oxide nanoparticles:*** iron oxide nanoparticles, such as magnetite (Fe_3O_4) and maghemite (γ-Fe_2O_3), are frequently employed as MSs. These nanoparticles possess high surface area, good magnetic response, and chemical stability, making them effective in the adsorption of dyes [23].
- ***Magnetic activated carbon:*** activated carbon particles impregnated with magnetic materials, such as iron oxide, are used as MSs. Activated carbon has a porous structure that enhances dye adsorption capacity, while the magnetic component enables easy recovery [24–26].
- ***Magnetic polymer composites:*** magnetic polymer composites combine the advantages of both polymers and magnetic materials. Polymers with MNPs embedded within or attached to their structure can be synthesized to create efficient MSs for dye removal [27–29].
- ***Magnetic MOFs:*** MOFs are a class of porous materials composed of metal ions or clusters connected by organic linkers. Incorporating MNPs into MOFs can result in MSs with high surface area, tuneable pore size, and excellent dye adsorption properties [30–32].
- ***Magnetic hydrogels:*** hydrogels are three-dimensional cross-linked polymer networks that can absorb large amounts of water. The introduction of MNPs into hydrogel matrices allows magnetic hydrogels to be created for dye sequestration. These sorbents offer both high water content and magnetic responsiveness [33–35].

These are just a few examples of MSs used in dye sequestration. Various modifications and combinations of these materials can be explored to tailor sorbents for specific dye removal applications.

5.2.2 The principles of magnetism and its application in sorption processes

5.2.2.1 Magnetic principles

The ability of a magnetic material to attract or repel another material in the presence of a magnetic field is regarded as *magnetism*. Magnetism finds its way into materials science, nanotechnology, ecotoxicology, and wastewater management (WWM), among others, as a useful property for purification, separation, remediation, detoxification, and environmental management. Different materials have different magnetic properties, causing them to be classified as magnetic and nonmagnetic materials or nanomaterials (NMs).

The basic principle of the magnetic properties of magnetic nanomaterials (MNMs) or MNPs, such as MSs, is attributed to their interaction with an external (applied) magnetic field and the way in which their magnetic moments (μ) are distributed in the external magnetic field ($\vec{B}_{\text{external}}$) [36]. Based on this principle, magnetism is broadly classified into five categories, viz. paramagnetism, diamagnetism, ferromagnetism, ferrimagnetism, and antiferromagnetism, as seen in figure 5.2, [38].

Paramagnetism, which is associated with many NMs, is a characteristic property of magnetic materials with one or more unpaired electrons. The magnetization (M) of these materials lies parallel to $\vec{B}_{\text{external}}$ with $\mu > 0$, but has low M, where M is the ratio of μ to the material volume. In contrast, diamagnetism is the fundamental

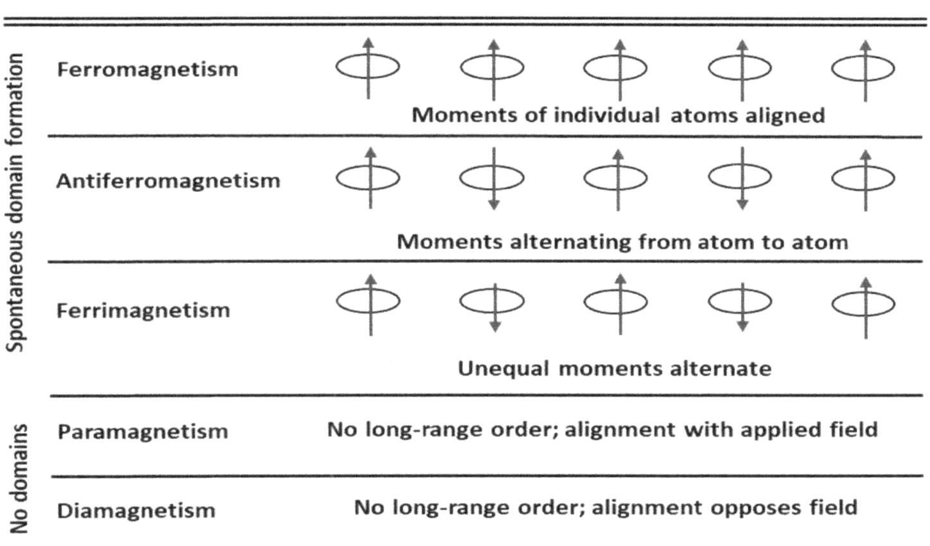

Figure 5.2. Categories of magnetism based on μ and $\vec{B}_{\text{external}}$ [38].

characteristic of most atoms in which the magnetization is very low and in the opposite direction to $\vec{B}_{external}$. Ferromagnetism is characteristic of naturally magnetically organized materials that spontaneously develop magnetization, even in the absence of a field. Ferrimagnetism has the property that different atoms in a material have varied moment strengths but are always ordered below a given critical temperature ($T_{critical}$). As the name implies, antiferromagnetism has properties that are opposite to those of ferromagnetism.

In a similar way, magnetic susceptibility (χ), which is the ratio of the M value of a material to the applied magnetic field intensity H (that is, $\chi = M/H$, where $H = B/\mu - M$), is also used to classify magnetic materials. In this classification system, $\chi < 0$ for diamagnetic materials such as copper (Cu), carbon (C), and lead (Pb), $\chi > 0$ for paramagnets such as iron oxides (Fe_3O_4 and γ-Fe_2O_3), and $\chi \gg 0$ for ferromagnets e.g. nickel (Ni), iron (Fe), and cobalt (Co) [18, 36, 38].

If the magnetic moments of magnetic iron oxide (ferrimagnetic) NMs that are less than 20 nm in size undergo a transition (or equilibrium condition) across (at) the Néel relaxation point due to a substantial size reduction, the ferrimagnetic NMs develop so-called superparamagnetic properties. *Superparamagnetism* is a magnetic property in which the overall μ of a ferrimagnet aligns in the field direction when subjected to an external magnetic field but then quickly returns to zero when the field decreases to zero [18].

As a rule, the magnetic behavior of a material is characterized by the connection between the direction of the magnetic moment (μ) and the associated electrons. This can be measured using the formula for the spin on the magnetic moment (μ_s): $\mu_s = \sqrt{n(n+2)}$, measured in Bohr magneton (BMs), where n is the number of unpaired electrons in a material. According to this measurement, the magnetic moment of a magnetic nanomaterial arbitrarily considered to have only one electron is 1.73 BM [36].

The adsorption capacity and magnetism of MNMs (especially MNPs) are attributed to their high surface area-to-volume ratio (A/V) and their composition (mainly the presence of unpaired electrons), respectively [36, 37]. For instance, the shape effect has been found to influence the volume or corresponding size-related features of MNCs. This gives environmentally friendly and low-cost powder $CuFe_2O_4$ and MnO-Fe_2O_3 nanocomposites a porous structure, a tiny particle size, and a comparatively large surface area and leads to their use in novel magnetic metal-oxide composite adsorbents for dye sequestration [36].

5.2.3 The application of magnetic principles in sorption processes

The application of magnetic principles in sorption follows a simple stepwise procedure stated below and illustrated in figure 5.3 [39]:

- ***MS synthesis:*** this is the first and one of the most crucial steps for magnetic adsorption, and it requires specific synthesis steps to obtain given MS molecules depending on the adsorbates (the target molecules to be sorbed), such as dyes. This step is discussed fully in section 5.3.

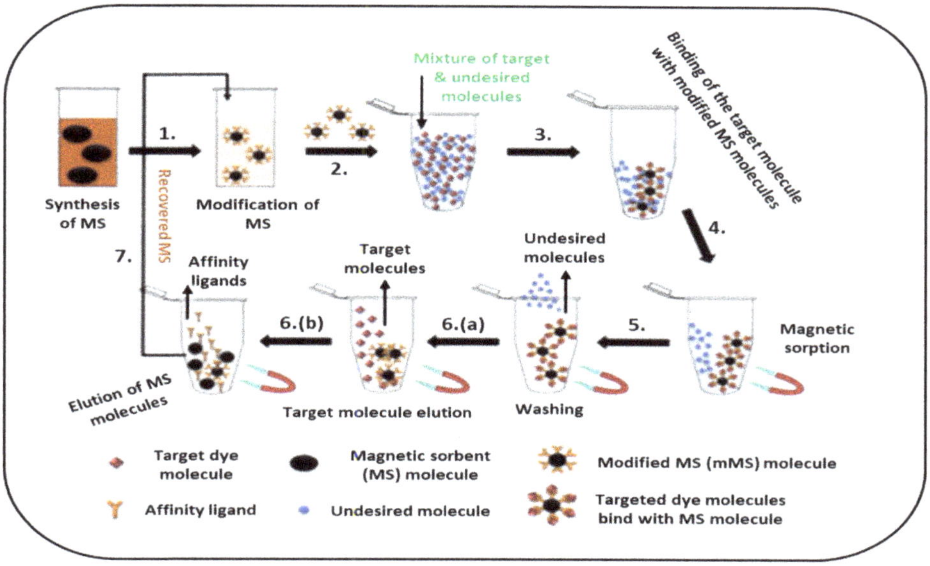

Figure 5.3. The application of magnetic principles in sorption processes [39].

- ***MS modification:*** in this step, the surface of the synthesized MS molecules is functionally modified to inhibit any MS reactions and agglomeration in the aqueous phase and preconditioned for biomedical importance. Such surface modification of MS molecules improves their multifunctional qualities, for example, their biotargeting, colloidal stability, and biocompatibility.
- ***Adsorption step:*** this step requires a thorough mixing of the modified MS (mMS) molecules with the adsorbates present in the sample aqueous solution followed by incubation for a short period. This then makes the mMS molecules adsorb the adsorbates.
- ***Separation step:*** in this step, activation of the magnetic field keeps the mMS molecules and adsorbates together while separating unwanted molecules.
- ***Multiple washing steps:*** after separation, a washing buffer is introduced into the column. Several on–off magnetic field cycles are applied to the mixture. The target-containing mMS molecules are resuspended during the 'off' period, and they are collected from the washing buffer during the 'on' period.
- ***Elution steps:*** following washing, (a) the adsorbates are recovered through the addition of an elution buffer, and then (b) the MS molecules are eluted using affinity ligands.
- ***Recycling step:*** after each adsorption procedure, the MS molecules are incubated with an appropriate solution(s) to create new binding sites for the subsequent sorption processes.

5.3 Fabrication techniques for magnetic sorbents and their modifications

The demand and utilization of various types of MSs in different fields of science and technology have resulted in their modifications and the development of a wide range of fabrication techniques for their preparation. These modifications and fabrication techniques offer MSs some advantages, such as control over their shape, size, composition, size distribution, crystallinity and crystal structure, configurations and conformations, thermal stability, size uniformity, monodispersity, and aggregation tendency, among others [2, 13, 15–17, 36, 37, 39–43].

5.3.1 Fabrication techniques for magnetic sorbents

The different fabrication techniques for MS preparation are presented in figure 5.4 and some are listed (with their pros and cons) in table 5.1. These techniques include physical methods (such as gas-phase deposition and electron-beam lithography) and various chemical methods (such as the solvothermal process, chemical reduction, sol–gel synthesis, coprecipitation, hydrothermal decomposition, and hydrolysis) [16–18, 36, 37, 39, 40, 42].

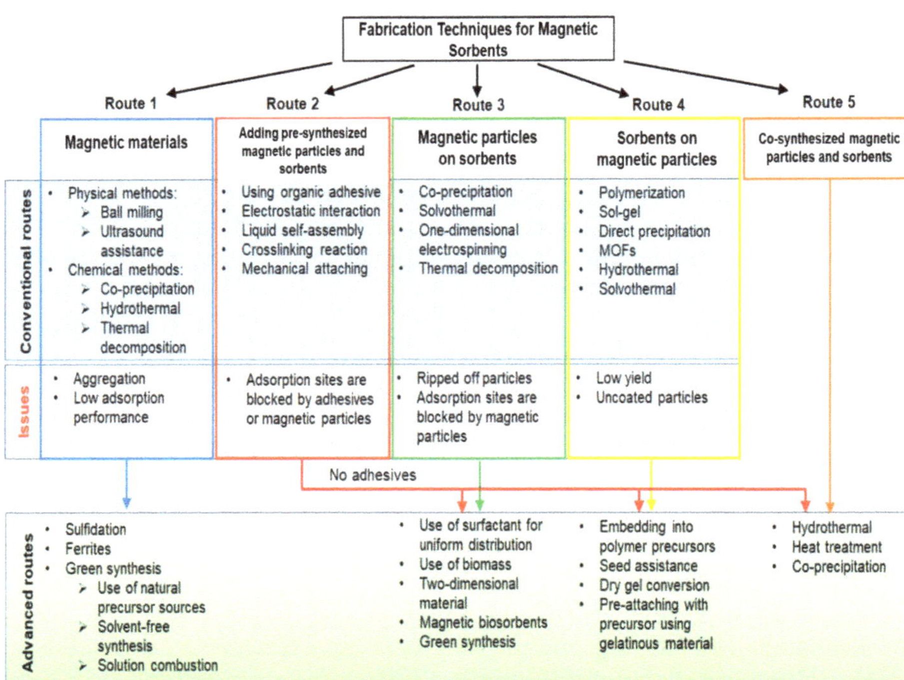

Figure 5.4. The fabrication techniques used to produce MSs. Reprinted from [36], Copyright (2014), with permission from Elsevier.

Physical methods offer some benefits such as good interparticle spacing control and simplicity [39], whereas chemical methods aid in the control of the size, shape, and compositions of MSs under controllable experimental conditions such as solution pH, stoichiometry, reaction temperature, salt nature, and ionic strength, as seen in table 5.1 as adapted from [16, 39, 44, 45].

Table 5.1. The fabrication techniques used to produce MSs, with their pros and cons.

tFabrication technique	Controlling parameters	Pros	Cons
Coprecipitation	Type of meta salts M^{2+}/M^{3+} ratio Temperature pH Precipitating agent	Simple method Functionalization is simple Ecologically cleaner High output Rapid response time	Lack of particle uniformity Low reproducibility
Thermal decomposition	The proportion of precursor to surfactant; the solvent Temperature during annealing Reaction time	Limited size range Length and structure determined A strong magnetic field Improved crystallinity	Higher pressure and temperature required Safety concerns related to reactants Inorganic solvent solubility
Microwave-assisted pyrolysis	Microwave power Radiation time Temperature	Excellent adsorption ability High surface area A high total pore volume Quick crystallization	Special equipment requirement
Hydrothermal	Temperature of a reaction Response pressure Response time Concentration of precursors	Synthesis in a liquid Outstanding magnetic properties Broad size distribution with a high surface area High levels of purity	Slower kinetic speed High pressure and temperature

(*Continued*)

Table 5.1. (*Continued*)

tFabrication technique	Controlling parameters	Pros	Cons
Microemulsion	Oil/water ratio	Strong magnetic abilities Small particle size	Low yield Expensive Large quantity of emulsifier required
Arc discharge	Temperature Pressure Power supply Electrode geometry	Efficient Cheap Eco-friendly Nontoxic	Low efficiency Difficulty of size control Limited industrial use
Polyol	Precursor concentration Reaction temperature	Outstanding magnetic properties Can be used at high temperatures Unique morphology High crystallinity	Manufacture of precious metals Particle size is polydisperse
Chemical reduction	Reducing agent type Precursor concentration	Simple and safe Proceeds at room temperature Easy to perform in the lab	Magnetic metal sorbents' ability to oxidize

M^{2+}/M^{3+} ratio is the ratio of the divalent ionic precursor to the trivalent ionic precursor.

5.3.2 Surface modification methods for magnetic sorbents to enhance dye sorption capacity

The efficiency of the adsorptive property of any adsorbent is primarily influenced by its surface area. However, the readily accessible porous MNMs employed for adsorption have low specific surface areas and so exhibit poor adsorption capacities [46–49]. Therefore, there is a need for MS surface modification or functionalization.

The fabrication technique used to produce MSs is vital in determining their surface-specific properties, since every preparation process has its pros and cons (figure 5.4 and table 5.1). Hence, to control the distinctive MNMs' properties such as porosity, particle size, and crystalline nature, a specific fabrication procedure should be utilized, followed by an efficient postfabrication modification approach [40, 47–49].

Theoretically, MNMs are subject to four different interparticle forces: van der Waals, magnetic, electrostatic repulsion, and steric hindrance forces are all present.

The latter two aid in stabilizing MNMs, whereas the first two tend to aggregate MNMs. Based on the effects of these forces, surface modifications have been developed to add defensive molecules to the nanomaterial's surface to increase the repulsive force between particles. These modifications also help to prevent the oxidation of Fe_3O_4 to γ-Fe_2O_3, reduce surface energy, and enhance hydrophilic/hydrophobic characteristics [50].

Generally, surface modifications of MSs are centered on four major purposes, viz.:

(i) to alter or improve MNM dispersion [18],
(ii) to increase MNM surface activity [51, 52],
(iii) to improve MNMs' physicochemical and mechanical qualities [18], and
(iv) to boost MNM biocompatibility [23].

The two major approaches to surface modifications of MSs are physical and chemical, as described briefly below [50–53].

5.3.2.1 Physical approach

Physical means such as adsorption, coating, wrapping, deposition, and surface interactions with plasma and UV light are employed in the physical approach to surface modification [53].

5.3.2.2 Chemical approach

The chemical approach to the surface modification of MSs changes the MNM surface through chemical processes, and it is the most adopted surface modification technique [13, 15, 23, 37, 47, 49–54]. Chemical modifications involve coating the MNM surface with organic functional groups (such as hydroxyl, carboxyl, sulfonate, amino, and mercapto groups), polymers, ligand-like inorganic substances (such as silicon dioxide (silica), metal oxides (such as TiO_2), Au, Ag, and quantum dots) through chemical reactions to aid the sorbent selectivity, non-agglomeration, and other desired properties of MSs [15, 50, 53, 54].

The most popular chemical modification technique for MSs is surface silanization, which involves coating the surface of naked MNMs with functionalized groups obtained from silane molecules to obtain numerous features, including greater stability, decreased toxicity, inertness to redox reactions, and high responsiveness [23, 39, 53]. This is because the surface silanization procedure may be carried out in both organic and aqueous media at medium temperature without requiring any specific reaction conditions [39].

Some of the various silane molecules commonly used in surface silanization are (3-aminopropyl)triethoxysilane (APTES), (3-mercaptopropyl)trimethoxysilane (MPTMS), and 3-(trihydroxysilyl)propyl methylphosphonate, which react with silica-coated MNMs to install functional groups such as mercapto, amino, *n*-octyl, iminodiacetic acid (IDA), thiosemicarbazide, ethylenediaminetetraacetic acid (EDTA), and phosphoric acid groups on the silica surface of the MNMs [39, 50, 53]. The coating mechanism of SiO_2 (silica) on the surface of Fe_3O_4 NMs (a type of MS) is illustrated in figure 5.5.

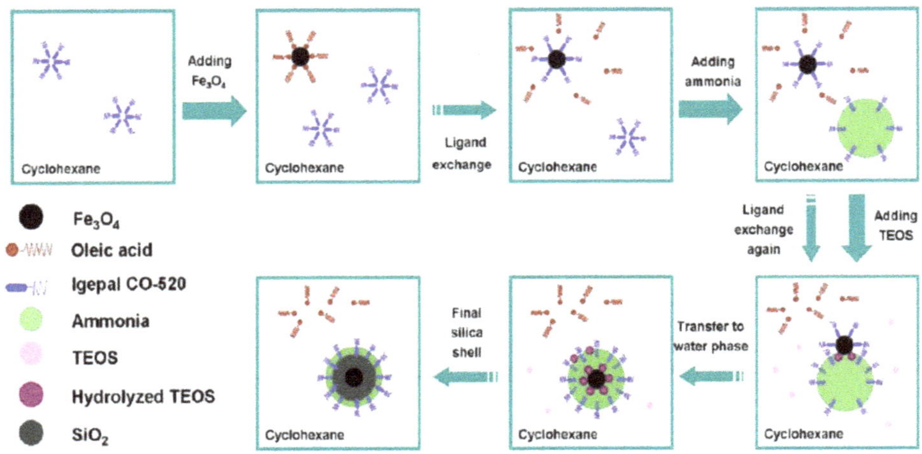

Figure 5.5. A diagram showing how SiO_2 (silica) coats the surface of Fe_3O_4 NMs in an organic solvent (cyclohexane) (where TEOS = tetraethoxysilane) [23].

5.3.3 Tailoring magnetic sorbents for specific dye classes and applications

MNPs or NMs coated with organic polymers, MOFs, and carbon-based materials are only a few of the MSs that have been studied for the removal of dyes. For selective dye adsorption, these sorbents can be tailored to have specific surface characteristics such as hydrophobicity or hydrophilicity. Furthermore, functionalizing the sorbents with particular functional groups can increase their affinity and selectivity for a given dye molecule, as seen in section 5.3.2 [11, 17, 50].

5.3.3.1 Specific dye classes

Two crucial functionalizing components that make up dye molecular structures are *auxochromes*, which increase a dye's affinity for fibers, and *chromophores*, which create the color [55]. These components, in the presence of some other dye components, allow the dyes found in many textile industries to be classed according to their physicochemical properties and whether they are natural or synthetic.

Some of the dye classes are acidic (anionic), basic (cationic), non-ionic (neutral), azo, direct, reactive, sulfur, disperse, and mordant (chrome) dyes, as shown in figures 5.1 and 5.6, and their basic characteristics [6, 56–61] are presented in table 5.2.

5.3.3.2 Various adsorptive methods for sequestering dyes from aqueous solutions

The contamination of aquatic systems with dyes due to waste disposal from the textile, printing, paper, dyeing, and plastics industries has demanded quick intervention and regulation to reduce their detrimental effects, such as disease, the death of aquatic organisms, and the inhibited growth of aquatic plants, as explained in section 5.1.

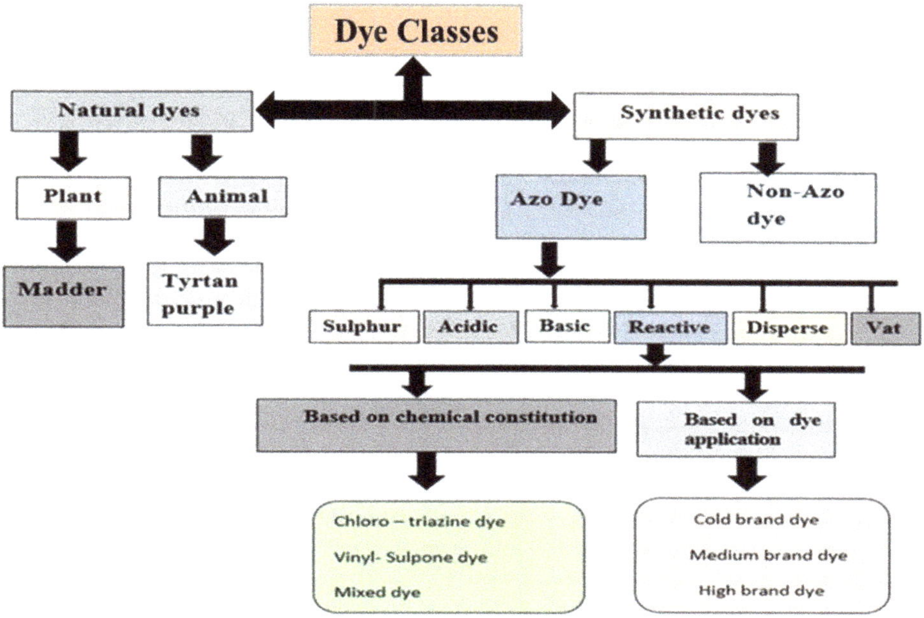

Figure 5.6. A flow diagram showing dye classes based on their physicochemical properties and applications [56] [2023], reprinted by permission of the publisher (Taylor & Francis Ltd, https://www.tandfonline.com).

Due to developments in science and technology and the intervention of health authorities in the issue of the toxicity of dyes in the environment, a wide range of technological dye remediation methods have been developed and applied by the relevant industries to minimize the negative impacts of dye contamination. These methods of dye sequestration are broadly categorized into three groups, viz. physical, chemical, and biological methods [60–62].

(i) The *physical method* includes techniques such as adsorption, ion exchange, nanoremediation, membrane separation, and the photo-Fenton technique, among others.
(ii) The *chemical treatment method* includes chemisorption, flocculation, photochemical reduction, electrochemical reduction, and catalytic reduction.
(iii) The *biological method* involves plant, animal, and microbial-based remediation techniques that use plant and animal matter, bacteria, algae, fungi, and appropriate enzymes (as living catalysts) [60, 62].

A summary of the three broad conventional methods with their advantages and disadvantages is illustrated in figure 5.7.

Table 5.2. The classes, basic characteristics, fiber suitability, and polluting agents of some dyes.

Dye class	Characteristics	Fiber suitability	Typical fixation (%)	Associated pollutants	Example(s)
Acidic	Water-soluble, anionic compounds	Wool, nylon, silk	80–93	Color, organic acids, unfixed dyes	Acid yellow 36, acid violet
Azo or azoic (naphthol)	Water-insoluble, bright, with all-round fastness, possess azo group	Cotton, other cellulosics	*NA*	Alkali, unfixed dyes, naphthol, diazotized base, soaping agents	Blue-red azo dye
Basic (cationic)	Water-soluble, applied in weakly acidic dyebaths; very bright dyes	Acrylic, some polyesters	97–98	Color, organic acids, dyeing auxiliaries	methyl violet, methylene blue, rhodamine B (RB)
Direct (substantive)	Water-soluble, anionic compounds, directly applied to cellulosics *without* mordants (i.e. metals such as Cr and Cu)	Cotton, rayon, other cellulosics	70–95	Color, salt, unfixed dyes, cationic fixing, levelling and retarding agents, surfactant, finish, defoamer, diluents	Indigo, direct orange 26, lichens
Disperse	Water-insoluble, low molecular weight	Polyester, acetate, and other synthetics	90–92	Color, organic acids, carriers, levelling, and dispersing agents	Disperse red 4, disperse blue 27
Reactive	Water-soluble, anionic compounds, major dye class	Cotton, wool, silk, nylon, and other cellulosics	60–90	Color, salt, alkali, unfixed dyes, surfactants, soaping agents	Reactive blue 5, procion H
Sulfur	Sulfur-containing organic compounds, water-insoluble, solubilized on weak alkaline reduction	Cotton, other cellulosics	60–70	Color, sodium sulfide, alkali, unfixed dyes, oxidizing and reducing agents	Sulphur black 1, indophenol
Vat	Water-insoluble, oldest dyes, chemically complex, solubilized on strong alkaline reduction	Cotton, other cellulosics	80–95	Color, alkali, oxidizing, and reducing agents, sodium hydrosulfite, soaping agents	Vat blue 4 (indanthrene)

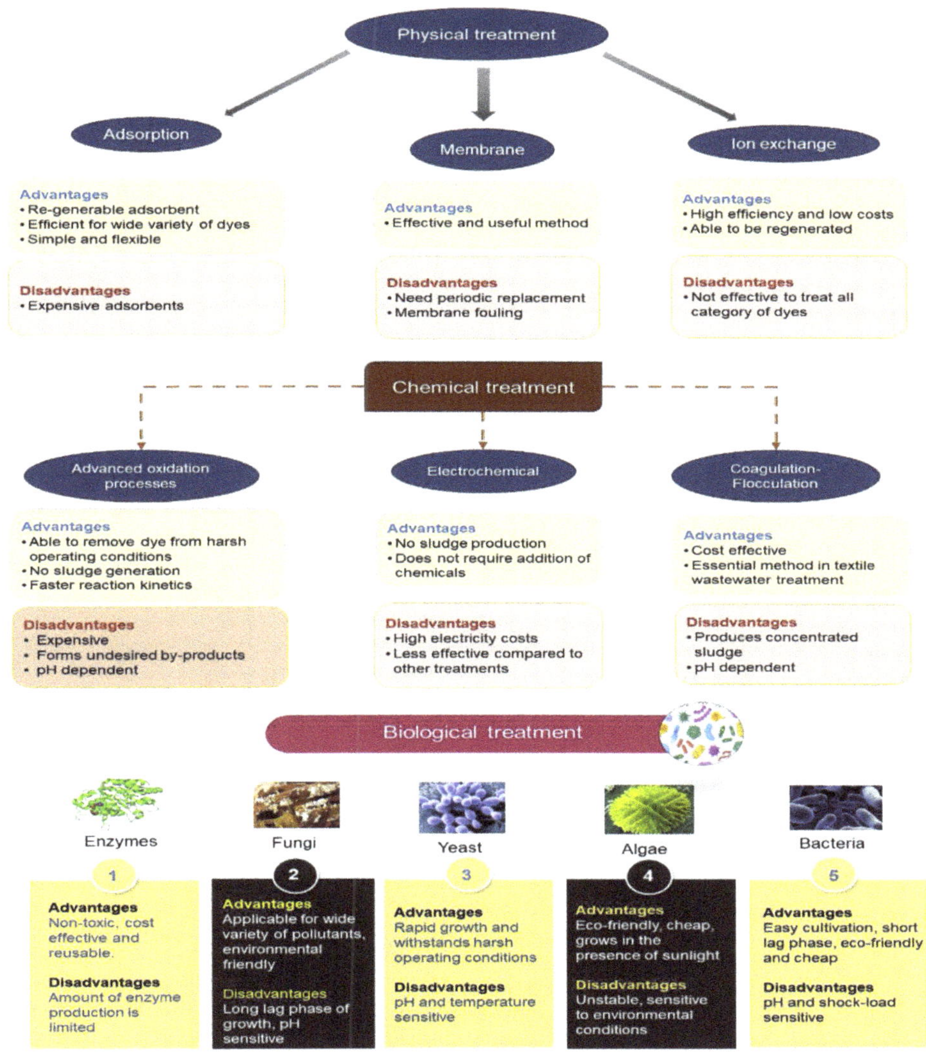

Figure 5.7. Various methods for the sequestration of dyes from aqueous solutions. Reprinted from [62], Copyright (2020), with permission from Elsevier.

5.3.3.3 The application of magnetic sorbents for sequestering dyes from aqueous solutions

The application of MSs for removing dyes from aqueous solutions depends on the characteristics of the dye (an adsorbate) and the adsorbent surface chemistry [6, 55].

MSs available for dye sequestration from aqueous solutions can be categorized into various groups based on their sources, compositions, synthesis methods, modifications, and effectiveness. Some of these groups are explained below, and examples are presented in table 5.3.

Table 5.3. The sequestration of dyes (adsorbates/pollutants) from aqueous solutions using various MSs: preparation, operating conditions, maximum adsorption capacity, applicable isotherm, and kinetics models.

Magnetic sorbent (MS)	Preparation method of MS	Sequestered dye(s) (adsorbate(s)/ pollutant(s))	Optimum operating conditions (time/pH/ sorbent dosage/initial conc.)	Maximum adsorption capacity, q_{max} (mg g^{-1})	Applicable isotherm and kinetic models	References
Fe_3O_4/banana peels	Thermochemical precipitation	Bromophenol blue	140 min/2.0/0.02 g/50 ppm	8.1	Freundlich and PFO	[65]
Nanocomposite of carbon nanotubes and iron oxide	*NA*	Methylene blue (MB), neutral red (NR), brilliant cresyl blue (BCB)	24 h/7.0/NA/1.4–37.4 ppm	MB: 15.74; NR: 20.33; BCB: 23.55	Freundlich and PSO	[66]
Magnetic SiO_2/*Bacillus subtilis*	Functionalization of MNPs and substitution	MB	3 h/6.8/25 mg/100 ppm	59.0	Freundlich and PFO	[67]
Sodium dodecyl sulfate-modified maghemite	*NA*	BCB, Janus green (JG) B, thionine (Th)	5 min/6.0/NA/1–400 ppm	BCB: 166.7; Th: 200; JG: 172.4	Langmuir	[68]
Iron oxides/coconut shells	Simultaneous activation and hydrolysis of $FeCl_3.6H_2O$	Sunset yellow	75 min/4.51/NA/NA	22.3	Langmuir and PSO	[24]
Mesoporous magnetic biochar (corn straw) composite	Chemical reduction and sonication	Malachite green	20 min/6.0/0.25 g l^{-1}/50–150 ppm	515.7	Langmuir and PSO	[69]
Fe_3O_4/pomelo peels	Carbonization, NaOH activation, and coprecipitation	Reactive red 21	60 min/3.0/2 g l^{-1}/300 ppm	26.8	Langmuir and Sips, and PFO	[70]
$CoFe_2O_4$/activated *Nerium oleander* leaf waste	Chemical treatment, pyrolysis, and autocombustion	Acid violet	55 min/6.5/1.0 g l^{-1}/100–200 ppm	83.33	Langmuir and PSO	[71]
Fe_3O_4/*Forsythia suspensa* leaf powders	One-step ultrasonication	Congo red (CR) and rhodamine B (RB)	200 min/low pH/20 mg/100 ppm	39.7	Langmuir and PSO	[72]

Magnetic chicken bone biochar	Pyrolysis and coprecipitation	RB	500 min/10.0/NA/100 ppm	113.31	Freundlich and MSM	[73]
Iron oxides/walnut shells	Carbothermal and microwave-assisted	MB	30 min/NA/NA/NA	130.0	Langmuir	[26]
Fe_3O_4/*Sorghum* husks	*In situ* coprecipitation	Crystal violet (CV) and MB	4 h/4.1/1.0 g/50 ppm	CV: 18.9; MB: 30.0	Langmuir and PSO	[74]
Magnetic chitosan nanoparticles	Carboligation and precipitation reactions	Metanil yellow (MY) and reactive black 5 (RB5)	1020 min/3.0/NA/0.67 mmol l^{-1}	MY: 620 RB5: 2549	Langmuir and PSO	[75]
Iron oxides/egg white wastes	Hydrothermal carbonization and coprecipitation	MB	24 h/4.0/15 mg/200 ppm	236.9	Langmuir and PSO	[25]
Fe_3O_4/peach gum	Concurrent creation of MNPs and peach gum cross-linkage	MB	1 h/7.0/50 mg/400 ppm	231.5	Langmuir and PSO	[76]
Fe_3O_4/moss biomass	Microwave synthesis	Thioflavin T	240 min/6.0/2.5 g l^{-1}/313 µmol l^{-1}	483.0 µmol g^{-1}	Langmuir and PSO	[77]
$CoFe_2O_4$/rice husk biosilica	Biosilica extraction and precipitation	MB	120 min/9.0/8 mg/100 ppm	253.6	Langmuir and PSO; Freundlich and Elovich	[78]
Magnetic alginate beads	Mechanical mixing	Methyl orange (MO) and MB	48 h/6.7 ± 0.2/NA/0.4–20 mmol l^{-1}	MO: 0.002 mmol g^{-1}; MB: 0.059 mmol g^{-1}	Langmuir and PSO	[12]
Sodium alginate @$CoFe_2O_4$–polydopamine bead	Solvothermal and sonication	MB, CV, malachite green (MG)	90 min/4.0–9.0/2.0 g l^{-1}/1000 ppm	MB: 466.60; CV: 456.52; MG: 248.78	Freundlich and PSO	[79]
Magnetically modified spent grain	*NA*	Aniline blue, Bismarck brown (BB) Y, CV, etc.	3 h/NA/30 mg/NA	44.7	Langmuir	[80]
Magnetic-fluid-modified peanut husk composite	Mechanical and rotary mixture	BB, acridine orange (AO), CV, and safranin O (SO)	180 min/NA/3.0 g l^{-1}/NA	BB: 95.3; AO: 71.4; CV: 80.9; SO: 86.1	Langmuir	[81]

(Continued)

Table 5.3. (*Continued*)

Magnetic sorbent (MS)	Preparation method of MS	Sequestered dye(s) (adsorbate(s)/ pollutant(s))	Optimum operating conditions (time/pH/ sorbent dosage/initial conc.)	Maximum adsorption capacity, q_{max} (mg g^{-1})	Applicable isotherm and kinetic models	References
Peanut husk-Fe_3O_4-iminodiacetic acid-Al (PN– Fe_3O_4-IDA-Al)	Copre cipitation	CR	6 h/6.26/1.0 g l^{-1}/100 ppm	79.0	Freundlich and Elovich	[82]
$CuFe_2O_4$/sawdust nanomagnetic composite	*In situ* coprecipitation of $CuFe_2O_4$ on the material surface	Cyanine acid blue	15 min/2.0/0.1 g/70 ppm	178.6	Langmuir and PSO; exothermic process	[83]
κ-carrageenan-coated Fe_3O_4 nanoparticles	Copre cipitation and sonication	MB	5 min/9.0 /NA/120 ppm	185.3	PFO and PSO	[84]
Carbon dots with magnetic $ZnFe_2O_4$	Electrothermal and hydrothermal	MO	40 min/5.0/NA/20 ppm	181.2	Langmuir and PSO	[85]
Magnetite rGO/ chitosan nanocomposite	One-step solvothermal synthesis	Acid red 22 (AR22) and Remazol black (RBB)	30–180 min/1.0–8.0/5–25 mg/10–50 ppm	AR22: 99.46 RBB: 95.32	Langmuir, Elovich, and Blanchard PSO	[86]
Mag-γ-Fe_2O_3 and cross-linked chitosan composite	Microemulsion process	MO	100 min/4.0/1.0 g l^{-1}/30 ppm	29.50	PSO	[87]
Fe_3O_4/MIL-100(Fe)	Catalytic liquid-phase acetylation and hydrothermal	RB	30 min/4–8/20 mg/NA	28.36	Freundlich and PSO	[31]
Cu-MOFs/Fe_3O_4	Solvothermal	MG	65 min/NA/50 mg/4.64 and 46.35 ppm	113.67	Freundlich, D–R and PSO	[30]

$MgFe_2O_4$@MOF	Simple solvothermal	RB and rhodamine 6G (Rh6G)	5 min/(RB: 3.0; Rh6G: 4.0)/10 mg/(RB: 50.03 ppm; Rh6G: 50.15 ppm)	RB: 219.78 Rh6G: 306.75	Langmuir and PSO	[88]
$Fe_3O_4/Cu_3(BTC)_2$	Solvothermal and sonication	MB	3 h/2.0–11.0/1000 mg l^{-1}/300ppm	245	Freundlich and PSO	[32]
Fe(III)-MOFs	Microwave-assisted ball milling method	Orange II (OII), CR, MB	300 min/NA/100 mg/20 ppm	OII: 97.7%, CR: 99.7%; MB: 97.5%	PFO	[89]
Fe_3O_4/MIL-101(Cr)	Simple one-step reduction–precipitation	Acid red 1 (AR1) and orange G (OG)	120 min/(AR1: 5.0; OG: 3.0)/30 mg/(AR1: 10 ppm; OG: 40 ppm)	AR1: 142.9 OG: 200.0	Langmuir and PSO	[90]
Fe_3O_4@SiO_2@UiO-66-urea	Solvent-assisted ligand exchange (SALE)	MO and MB	60 min/(MO: 3.0; (MO +MB): 7.0; MB: 11.0)/50 mg l^{-1}/1000 ppm	MO: 183 MB: 121	Langmuir and PSO	[91]
Modified zeolite MNPs (MHY and M13X)	Chemical synthesis before modification	MB	5–65 min/(MHY: 9.0; M13X: 8.93)/(MHY: 1.186 g l^{-1}; M13X: 1.198 g l^{-1})/(MHY: 10.23 ppm; M13X: 10.05 ppm)	MHY: 99.85% M13X: 95.9%	Langmuir and PSO	[92]

NA = not available; PFO kinetic model = pseudo-first-order kinetic model; PSO kinetic model = pseudo-second-order kinetic model; D–R isotherm= Dubinin–Radushkevich isotherm; MNPs = magnetic nanoparticles; MSM = multilayer sorption mechanism; rGO = reduced graphene oxide.

MIL = Matérial Institute Lavoisier; BTC = benzene-1,3,5-tricarboxylate; MHY = magnetic zeolite HY (with Si/Al ratio > 1.5); M13X = magnetic zeolite 13X (with 1 < (Si/Al ratio) < 1.5)

5.3.3.3.1 Magnetic biosorbents

The most common method for creating magnetic biosorbents is to combine previously synthesized magnetite (Fe_3O_4) in powder form or suspension with the biosorbent (biomass) in an organic solvent. As a result, magnetic particles are added to the pores or surface of the biosorbent [41]. Magnetic biosorbents are efficient in the elimination of dyes from aqueous solutions and have greener properties, such as good selectivity, high specific surface area, physicochemical stability, affordability, effective adsorbent modifications/functionalization, easy recovery, and eco-friendliness [37, 47, 54, 55]. Magnetic biosorbents include animal-based magnetic biosorbents (such as chitosan-based magnetic adsorbents), plant-based magnetic adsorbents, and microbe-based magnetic biosorbents, among others [16, 17, 37, 41, 49, 50, 54, 55, 63, 64].

5.3.3.3.2 Carbon-based magnetic adsorbents

Carbon-containing adsorbents (CCAs) that are functionally modified with magnetic substances (mainly iron oxides) have been developed for dye sequestration at greater adsorption capacities. Most CCAs are biological in origin. Carbon-based magnetic adsorbents (CBMAs) include magnetic graphene (such as graphene oxide (GO)/iron oxide nanocomposites), magnetic β-cyclodextrin–chitosan/GO (MCCG) adsorbents, magnetic activated carbon, and magnetic biochar (magnetic BC), among others [24–26, 41].

Other categories of MSs for sequestering dyes from aqueous solutions include magnetic polymer composites [15, 27–29], MOFs [15, 32, 59], silica-based magnetic adsorbents [41], inorganic clay-based magnetic composites (ICMCs) [15], zeolite-based magnetic adsorbents [41], and magnetic hydrogels (MHs) [33–35].

5.4 The factors, mechanisms, and isotherm–kinetic models that influence the sequestration of dyes by magnetic sorbents

The adsorption process of MSs (or any adsorbents) depends on the optimum operating factors (OOFs) that influence the rate and efficiency of the process, its mechanisms, and the thermodynamic parameters governing the process (figure 5.8), as described below.

5.4.1 The factors that influence magnetic sorbent–dye interactions

Several physicochemical factors have been studied that influence the adsorption capacity (efficiency) of MSs for dyes within their aqueous solution systems. These factors include temperature, concentration, pH, contact time, ionic strength, sorbent type, sorbent surface area and pore structure, particle size, pressure, sorbent surface modifications and activation, adsorption operation mode, and the presence of competing ions. Regarding the dyes (adsorbates), their polarity, size, molecular weights, and molecular structure should also be considered [59, 60, 64].

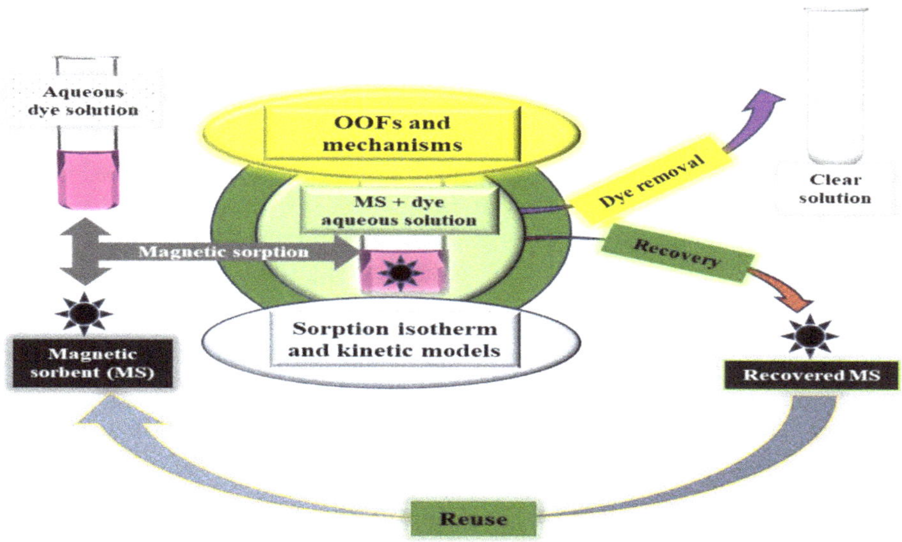

Figure 5.8. Dye sequestration by an MS using OOFs, sorption mechanisms, and sorbent isotherm–kinetic models.

5.4.1.1 The effect of temperature

The temperature of the adsorbate solution (such as an aqueous dye solution) greatly affects the adsorption effectiveness of an adsorbent (such as an MS), and it is significant within both endothermic or exothermic processes [60]. This may be attributed to the dye mobility rate and the amount of the adsorbent's functional or active adsorptive sites [55]. Hence, an increase in temperature may increase the adsorption rate, and vice versa.

Under endothermic conditions, the efficiency of adsorption increases with temperature due to an increase in dye mobility and the number of active sites for adsorption [55, 59, 60]. From a research report, an endothermic process has been observed during the adsorption of methylene blue by $Fe_3O_4/Cu_3(BTC)_2$ [32].

In the exothermic state, an increase in temperature causes the adsorbent's active sites to become less effective at adsorbing dye molecules, and this results in a reduction in adsorption capacity [55, 59, 60]. This is a basis for diffusion control in the adsorption of dye molecules [60]. According to the literature, the adsorption of xylenol orange by MIL-101(Cr) occurs through an exothermic process [93].

Ceteris paribus, as a rule, better adsorption at higher temperatures may signify endothermic activity, whereas exothermic processes are more prevalent at lower temperatures.

5.4.1.2 The effect of the initial dye concentration

The initial dye concentration has an indirect impact on the efficiency of dye sequestration in aqueous systems through changes in the availability of MS surface binding sites. This is because the initial dye concentration aids the mass transfer of the dye molecules or ions to the MS surface until an equilibrium is established and the sorbent is fully saturated [61]. (This is analogous to the effect of the initial concentration of a substrate in the formation of an enzyme–substrate complex according to the principles of enzyme kinetics.)

It has been reported that the adsorption rate becomes higher if the initial dye concentration is low, and vice versa [60, 61].

In the aqueous solution system of a dye in the presence of a sorbent, the dye removal efficiency (E) and the dye adsorption capacity (q, in mg g^{-1}) at equilibrium are related to the initial dye concentration (C_i, in ppm), as stated in equations (5.1) and (5.2):

$$\%E = \frac{C_i - C_f}{C_i} \times 100 \tag{5.1}$$

$$q = \frac{C_i - C_f}{W} \times V, \tag{5.2}$$

where C_f (ppm) is the final dye concentration, V (litres, l) is the volume of dye aqueous solution, and W is the sorbent dosage (g).

5.4.1.3 The effect of the magnetic sorbent dosage

The adsorption efficiency and dye sequestration increase with an increase in MS dosage at a fixed dye concentration in aqueous solution systems. This is because an increase in sorbent dosage makes more active surface area and active adsorption sites available for adsorption [59, 60].

With some exceptions, most research studies have revealed a negative association between sorbent dosage and adsorption capacity but a positive relationship with removal efficiency [64]. A drop in the adsorption capacity of a dye that occurs while increasing the dosage of an MS is attributed to (i) the binding of a smaller amount of dye to the excess sorbent's active sites for dye sequestration, and (ii) active site accumulation because of intense collisions between excess sorbent particles [64].

As a rule, in the equilibrium state, the amount of dye sorbed decreases with an increase in the sorbent dosage, and vice versa. Thus, for research purposes, small amounts of MS are utilized to obtain a higher efficiency of dye sequestration and to minimize research costs and post-adsorption sludge [59].

5.4.1.4 The effect of the dye solution pH

The pH of a dye's aqueous solution plays a vital role in the sequestration of dyes in the presence of MSs. Based on the research literature, the dye solution pH affects: (i) the sorbent capacity, surface charge, and its functional group activity; (ii) the adsorption efficiency and mechanism; (ii) dye sequestration and the solubility, chemical structure, and the dissociation of its molecules; (iv) competing ions or molecules in the solution; and (v) the ionization and surface charge of adsorbed dye ions [59, 60, 64].

In practice, the binding of anionic dyes to a sorbent is favored by an acidic medium (pH $<$ 7) due to the protonation of the sorbent surface, which facilitates an electrostatic attraction between the sorbent and the dye, while the binding of cationic dyes is more effective in a basic medium (pH $>$ 7). On the other hand, a basic medium deprotonates the surface of the sorbent, resulting in an electrostatic repulsion between the sorbent and anionic dyes, and vice versa [60].

5.4.1.5 The effect of contact time

The amount of time required for the MS and dye to be in contact with each other in the aqueous dye solution is a crucial factor in the adsorption process. The adsorption rate, adsorption capacity, and removal efficiency are higher for longer contact times, and vice versa [61, 64]. The chemical kinetics responsible for the adsorption reaction rate require a longer time for the adsorbent to be in contact with the adsorbate and for both to reach equilibrium and attain a maximum efficiency for dye sequestration [61]. Factors such as a larger surface area, more binding sites, good porosity, favorable temperature, and optimum pH reduce the time needed to complete the adsorption process [41, 61].

5.4.1.6 The effect of sorbent particle size

The small particle sizes of MSs result in a higher specific surface area (measured in $m^2\ g^{-1}$) and adsorption capacity due to the increased availability of active binding sites, and vice versa [60]. The dependence of the adsorption capacity of a dye on particle size is attributed to: (i) the ionic charge of the dye, (ii) its capacity to produce hydrolyzed species, and (iii) the adsorbent's intrinsic properties (such as porosity, crystallinity, and polymeric chain rigidity) [59, 60].

In addition, a decrease in the particle size of the adsorbent results in an increase in its Brunauer–Emmett–Teller (BET) surface area. However, depending on the type of adsorbent, if the particle size is too small, the adsorption capacity may be decreased, since the lighter particles float and cannot make effective contact with the solution, and it can be difficult to separate these tiny particles from the water after biosorption [60].

5.4.2 The adsorption mechanisms involved in dye sequestration by sorbents

The mechanisms of the adsorption process provide comprehensive information about the way in which adsorbates are adsorbed by adsorbents in their aqueous solutions. According to the research literature, dyes are adsorbed by sorbents based on their physicochemical properties, functional groups, and the nature of the aqueous solution systems [59]. The common adsorption mechanisms reported in the literature include electrostatic attraction, π–π stacking, hydrogen bonding, complexation, van der Waals forces, chelation, ion exchange, Lewis acid acid–base interactions, and physical adsorption [59, 94–96].

5.4.2.1 The electrostatic attraction mechanism

The most common adsorption mechanism is electrostatic attraction, as shown in figure 5.9. Cationic MSs easily attract anionic (acidic) dyes in aqueous solutions by electrostatic forces to form chemical bonds. Using the same principle, anionic and cationic dyes in the same aqueous solution attract each other and form chemical bonds via electrostatic attraction involving electron transfer (a redox reaction) [59, 64]. Most dyes are adsorbed by MSs using this mechanism [64].

5.4.2.2 The π–π interaction mechanism

The π–π interaction (also known as π–π stacking) is the process of creating chemical bonds between two interacting aromatic rings, and it can be used as an adsorption mechanism, as illustrated in figure 5.9. This kind of adsorption mechanism has been reported for the sequestration of dyes from aqueous solutions by MSs, although the mechanism is not as strong as electrostatic interactions [59, 67]. A few of the reported adsorption processes that involve the π–π interaction mechanism are those between iron oxides/egg white waste composites and methylene blue [25], a Cd-based MOF (TMU-8) and reactive black 5 dye [97], iron oxides/coconut shell composites and sunset yellow dye [24], and a Cu-based MOF and methyl orange, methylene blue, and rhodamine B (RB) dyes [98].

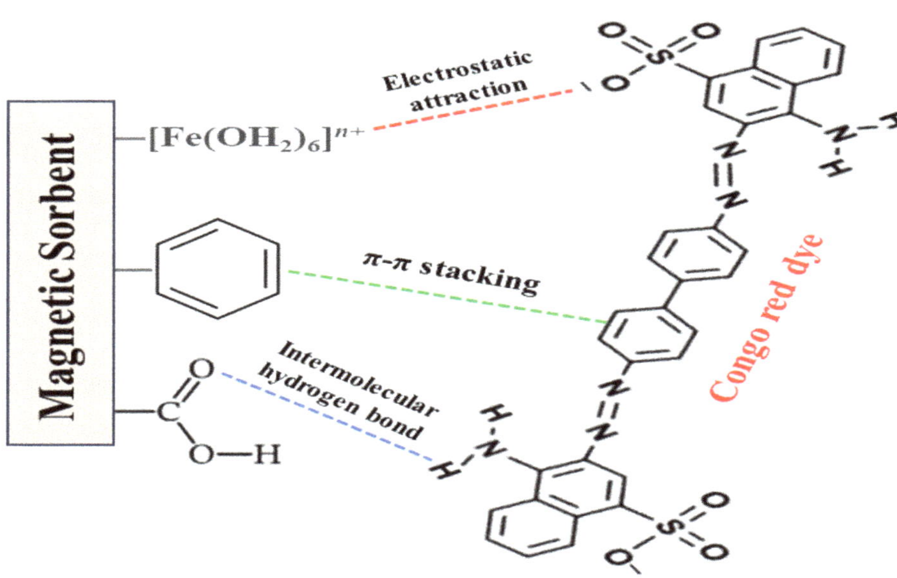

Figure 5.9. The adsorption mechanisms involved in dye sequestration by MSs. Reprinted from [64], Copyright (2022), with permission from Elsevier.

5.4.2.3 The hydrogen bonding mechanism
The formation of hydrogen bonds between a dye and an MS is the basis for the hydrogen bonding mechanism between MSs and dyes, as illustrated in figure 5.9. This may be attributed to the formation of intermolecular hydrogen bonds between their functional groups. According to the research literature, hydrogen bonding mechanisms have been found to occur in the adsorption process between iron oxides/coconut shells and sunset yellow dye [24], $CoFe_2O_4$/rice husk biosilica and methylene blue [78], and magnetic SiO_2/*Bacillus subtilis* and methylene blue [67].

5.4.2.4 The ion-exchange mechanism
The adsorption mechanism based on ion exchange between dyes and MSs involves the transfer of ions or ligands between their structures [59]. According to some studies, an ion-exchange mechanism has been reported to be responsible for the sequestration of crystal violet and acid green 9 by magnetic anion-exchange microbeads (MAMs) and magnetic cation-exchange microbeads (MCMs) from their individual and combined aqueous solutions [99].

5.4.2.5 The Lewis acid–base interaction mechanism
In some magnetic MOFs, adsorption mechanisms based on Lewis acid–base interactions have been reported, although this is an uncommon adsorption process [59]. Since a Lewis acid is a lone-pair electron acceptor, and a Lewis base is a donor of lone-pair electrons, there is the possibility that an MS with a Lewis acid component can use this mechanism to sequester dyes from their aqueous solutions. Typical examples are found to occur between nickel-based MOF/graphene oxide (GO) composites and Congo red dye (in which the Lewis acid Ni^{2+} facilitates the adsorption process) [100] and between a magnetic SiO_2/*B. subtilis* sorbent and methylene blue [67].

5.4.2.6 The physical adsorption mechanism
A physical adsorption (physisorption) mechanism has been reported to take place between dyes and MSs in their aqueous solutions [101], and it usually occurs on the sorbent's crystal surfaces [101]. For instance, the adsorption of malachite green dye by Cu-MOFs/Fe_3O_4 (a magnetic MOF) was reported to occur by physical adsorption in which the dye molecules were a suitable size to fit into the sorbent's mesopores, thus compensating for its smaller surface area [30].

5.4.3 The adsorption isotherm model governing dye sorption by magnetic sorbents

The isotherm model of the adsorption process is the model that governs the interaction between the adsorbate molecules and the adsorbent surface (or between the sorbate molecules and the sorbent surface), their behavior, and the adsorption capacity of the adsorbent [37, 55, 102, 104, 109]. At the isotherm equilibrium state for dye sequestration, the correlation between the amount of dye sorbed at equilibrium, q_e (mg g^{-1}), and the equilibrium dye concentration in an aqueous solution, C_e (mg l^{-1}), with the aid of equation (5.2), serves as a good indicator for depicting the sorption system [102, 109].

Based on the adsorption isotherm model, isotherm plots can be generated from the data obtained for q_e and C_e at the same temperature; these are then used for sorption (or adsorption) analyses performed using the Langmuir (monolayer), Freundlich (multisite), Temkin, Harkins, Dubinin–Radushkevich (D–R), Redlich–Peterson (R–P), Sips, and BET (multilayer) isotherm models, or Henry's law and statistical physics [102, 104], as shown in table 5.4.

5.4.4 Kinetic models used to describe the dye sorption process

The dynamics of the dye sorption process are usually investigated using the sorption kinetics in terms of the order of the rate constant (k_n, where $n = 1$ or 2) [36, 37]. The adsorption rate is taken into consideration when choosing a material to be utilized as an MS. High sorption (or adsorption) capacity and a quick rate of sorption are required of the sorbent (such as an MS).

In this case, the most used adsorption kinetic models as seen in most research are the pseudo-first-order (PFO) and pseudo-second-order (PSO) kinetic models [36, 37, 41, 102, 104], as stated in table 5.5.

5.4.5 Thermodynamic conditions used to describe the dye sorption process

The thermodynamic measurement of the energy changes and the entropic state of any sorption process at a measured temperature T (in kelvin, K) is an essential concept in studying MS–dye interaction in an aqueous solution [107]. To study the thermodynamic conditions governing the sorbent–sorbate interaction in any sorption process, data gathered at known temperatures are used, and the thermodynamic variables, mainly the Gibbs free energy change (ΔG^o), the enthalpy change (ΔH^o), and the entropy change (ΔS^o) are also used to assess the spontaneity (or feasibility) and energy flow governing the sorption process [107, 108, 110, 111].

The thermodynamic equilibrium constant (K_e^o) is obtained from q_e and C_e values measured at the same temperature [104, 107, 108, 110], and the linear plot that can be generated using the van 't Hoff equation serves as an experimental and statistical plot of the sorption process that allows the determination of ΔH^o and ΔS^o, which are then used to obtain ΔG^o at a particular temperature [108, 111], as shown in table 5.5.

5.5 A comparison between magnetic and nonmagnetic adsorbents for the sequestration of synthetic dyes from aqueous solutions or wastewater

The demand for affordable, fast-acting, and more efficient adsorbents for the sequestration of dyes from aqueous solutions has opened the door for magnetic separation technology, which has grown more and more in recent years [41, 112]. Due to the simplicity, enhanced surface area, reduced diffusional resistance, high efficiency, eco-friendliness, economic reusability, diverse applications, and

Table 5.4. Expressions and characteristics of sorption isotherm models governing dye sorption by MSs.

Sorption isotherm model	Expression	Parameters	Characteristics	References
Langmuir	$q_e = \frac{q_{max} K_L C_e}{1 + K_L C_e}$, or $\frac{1}{q_e} = \frac{1}{q_{max}} + (\frac{1}{K_L q_{max}})\frac{1}{C_e}$ $R_L = \frac{1}{1 + K_L C_o}$	q_{max} is the maximum adsorption capacity (mg g^{-1}), K_L is the Langmuir constant (l mg^{-1}), R_L is the separator factor, and C_o is the initial sorbate (dye) concentration (mg l^{-1}).	Most common isotherm models. Based on *an ideal monolayer-surface-phase assumption*: an ideal sorbent containing only a fixed number of energetically homogeneous sorption sites accommodates only one sorbate (dye) molecule on each side to form a mono-layer surface phase. Isotherm linear plot of $\frac{1}{q_e}$ against $\frac{1}{C_e}$, with q_{max} and K_L values estimated from the intercept (on the $\frac{1}{q_e}$-axis) and the slope of the plot, respectively. Adsorption process conformity to the Langmuir model: $R_L = 0$ (irreversible monolayer adsorption process); $R_L = 1$ (linear); $0 < R_L < 1$ (favorable); and $R_L > 1$ (unfavorable).	[55, 102]
Freundlich	$q_e = K_F$, r $\log q_e = \text{lo}K_F + \frac{1}{n}\log C_e$	K_F is the Freundlich constant (mg g^{-1}), and n is the adsorption intensity.	• Based on multisite/multilayer adsorption process on heterogeneous surfaces. • Adsorption content is enhanced infinitely with high concentration. • Isotherm linear plot of $\log q_e$ against the $\log C_e$, with K_F and n values estimated from the intercept (on the $\log q_e$-axis) and the slope of the plot, respectively.	[41, 102]

(*Continued*)

Table 5.4. (*Continued*)

Sorption isotherm model	Expression	Parameters	Characteristics	References
			• The value of n predicts the sorbent site's heterogeneity and distribution energy, and $n > 1$ or $0 < \frac{1}{n} < 1$ (signifies a favorable adsorption process).	
Temkin	$= B \ln (K_T C_e)$ a) $= B \ln K_T + B \ln C_e$ $B = \frac{RT}{b}$	*sub* – *Tis*K_T is the Temkin isotherm equilibrium binding constant (l g^{-1}), B is the adsorption enthalpy (kJ mol^{-1}) linked constant, b is the Temkin isotherm constant, R is the universal gas constant (8.314 J mol^{-1} K^{-1}), and T is the absolute temperature (in kelvin, K).	– Assumes that the adsorption enthalpy of sorbate molecules falls linearly with the growth in covering of the adsorbent surface for a physical adsorption process. – Isotherm linear plot of q_e against $\ln C_e$, with K_T and B values are estimated from the intercepts and the slope of the plot. – $B < 8$ kJ mol^{-1} indicates physical adsorption (physiosorption), otherwise chemisorption (chemical adsorption).	[102, 103]
Dubinin–Radushkevich (D–R)	$q_e = q_{max} e^{-\beta\varepsilon^2}$, or $\ln q_e = \ln q_{max} - \beta\varepsilon^2$ $\varepsilon = RT \ln (1 + \frac{1}{C_e})$ $E = \frac{1}{\sqrt{2\beta}}$	ε is the Polanyi potential beta being the D–R isotherm constant (mol^2 kJ^{-2}), and E is the apparent adsorption energy.	• Assumes that the adsorbent is temperature-independent and comparable to the micropore size. • Isotherm linear plot of $\ln q_e$ against ε^2, with q_{max} and β values estimated from the intercept (on the $\ln q_e$-axis) and the slope, respectively. • Aids in E estimation: $E < 8$ kJ mol^{-1} indicates physical adsorption.	[104]
Henry's law	$q_e = K_H C_e$	K_H is the Henry constant (L g^{-1}).	Based on the adsorption process on a monotonous surface at low enough concentrations. Isotherm linear plot of q_e against C_e, with K_H value estimated from the plot slope.	[105]

Note: Sips and Brunauer-Emmett-Teller (BET) isotherm models with appropriate cited references have been included after Table 5.4 as additional information to make the sorption isotherm models more comprehensive to the reader.

Table 5.5. Expressions and characteristics of adsorption kinetic models and thermodynamic conditions governing dye sorption by MSs

Adsorption kinetic model	Expression	Parameters	Characteristics	References
PFO (Lagergren's kinetic model)	$\frac{dq_t}{dt} = k_1(q_e - q_t)$, $\ln(q_e - q_t) = \ln q_e — k_1 t$, or $\log(q_e - q_t) = \log q — \frac{k_1 t}{2.303}$	q_t (mg g^{-1}) is the adsorption capacity at time t (min), and k_1 (min^{-1}) is the PSO rate constant.	Assumes that one active site binds one adsorbate (dye) molecule. Kinetic linear plot of $\log(q_e - q_t)$ against t, with k_1 and q_e values estimated from the slope and the $\log(q_e - q_t)$-axis intercept of the plot, respectively.	[41, 106]
PSO (Ho's kinetic model)	$\frac{dq_t}{dt} = k_2(q_e - q_t)^2$, or $\frac{t}{q_t} = \frac{1}{k_2 q_e^2} + \frac{t}{q_e}$	$k_2 q_e^2$ is the initial adsorption rate, and k_2 (g (mg min)$^{-1}$) is the PSO rate constant.	Assumes that two active sites bind one adsorbate (dye) molecule. Kinetic linear plot of t/q_t against t, with k_2 and q_e values estimated from the intercept (on the t/q_t-axis) and the slope of the plot, respectively.	[41, 106]
Thermodynamic conditions	$\Delta G^o = \Delta H^o — T\Delta S^o = -RT \ln K_e^o$ $\ln K_e^o = \frac{\Delta S^o}{R} — (\frac{\Delta H^o}{R})\frac{1}{T}$ (The van 't Hoff equation) $K_e^o = \frac{q_e}{C_e}$	ΔG^o is the standard Gibbs free energy change (kJ mol^{-1}), ΔH^o is the standard enthalpy change (kJ mol^{-1}), ΔS^o is the standard entropy change (kJ K^{-1} mol^{-1}), and K_e^o is the equilibrium constant.	Shows the effect of energy change during the adsorption process. For the adsorption process: $\Delta G^o = 0$ (equilibrium); $\Delta G^o < 0$ (feasible or spontaneous); and $\Delta G^o > 0$ (non-feasible or non-spontaneous). The van't Hoff linear plot of ln $\ln K_e^o$ against $\frac{1}{T}$ aids in estimating ΔH^o and ΔS^o values respectively from the slope and intercepts (on the ln $\ln K_e^o$ - axis) of the plot, such that $\Delta H^o < 0$ (signifies exothermic), $\Delta H^o > 0$ (endothermic), $\Delta S^o < 0$ (↓chaos), and $\Delta S^o > 0$ (↑chaos).	[107, 108]

affordability of MSs, this novel separation technology has allowed magnetic adsorption to overtake nonmagnetic adsorption techniques and replace complex industrial dye sequestration methods such as centrifugation and filtration [41, 50, 113].

5.6 Magnetic sorbents in combination with advanced technologies

Today, multiple techniques are employed to eliminate dyes from aqueous solutions (such as wastewater), either individually or in combination, but each has its flaws [61, 62]. Chemical oxidation, coagulation, membranes, and microbial degradation are notable examples of these technologies. These technologies fall into one of three categories: physical, chemical, or biological, as shown in figure 5.7, which also depicts some of their potential benefits and drawbacks [62, 114]. Adsorption is typically seen as being preferable to alternative decontamination methods, and the removal of dyes from aqueous solutions by adsorption, catalysis, and hybrid methods (such as photocatalytic membrane reactors (PMRs) and membrane bioreactors (MBRs)) in combinations for large-scale implementation, is an excellent application for MSs and other adsorbents based on their physicochemical properties and surface chemistry [41, 62–64, 94].

5.7 Challenges and future perspectives

The utilization of MSs for the removal of dyes from aqueous solutions has gained considerable attention due to their simplicity, cost-effectiveness, easy accessibility, broad applicability, eco-friendliness, efficient nature, etc. These materials, typically consisting of MNPs functionalized with adsorptive surfaces, offer a promising solution with which to address dye contamination in water bodies. However, this field faces various challenges and holds intriguing prospects. Among the challenges are:

- *Dye variability:* the vast structural diversity of dyes with variations in their chemical properties, size, and charge poses a challenge in designing MSs with broad-spectrum adsorption capabilities to efficiently capture a wide range of dyes.
- *Competing ions:* coexisting ions in water such as salts and heavy metals can hinder the adsorption of dyes by MSs. In this situation, strategies to address selectivity and improve the sorbent's performance in complex matrices are needed.
- *Regeneration and reusability:* the regeneration and reusability of MSs are essential for cost-effectiveness and sustainability. Developing methods to efficiently desorb adsorbed dyes and restore the sorbents' adsorption capacity is a critical challenge in most cases.

Overcoming these challenges requires an interdisciplinary effort from the fields of materials science, chemistry, and engineering. Based on these challenges, one way forward involves continued research into nanomaterial innovations, such as the development of novel MSs with core–shell structures and improved composite materials that offer the promise of improving adsorption capacities and selectivity to produce novel enhanced dye-removal MSs. Tailoring the surface chemistry of MSs through functionalization with specific ligands or groups can enhance their affinity for target dyes. This approach can aid in the selective removal of dyes belonging to a particular class. Hybrid technologies, in which MSs are integrated with other treatment methods such as photocatalysis, advanced oxidation processes, or the use of molecularly imprinted polymers (MIPs), can lead to synergistic effects and improved dye removal efficiency. These hybrid approaches offer a comprehensive solution that can be used to address dye contamination challenges. In addition, autonomous devices equipped with MSs could offer remote and *in situ* removal strategies that could be deployed to tackle dye contamination in rivers, lakes, and reservoirs. Future research must include a comprehensive assessment of the environmental impact of MSs by considering factors such as sorbent disposal, the possible toxicity caused by the potential leaching of nanoparticles, and long-term effects on ecosystems. The establishment of sustainable practices and responsible disposal methods will also mitigate any unintended consequences, making the process eco-friendly. Hence, addressing issues related to dye variability, competing ions, and regeneration and exploring innovative nanomaterials and hybrid technologies will create a bright future for the use of MSs in water treatment. Ensuring their sustainability and minimizing environmental impact will be crucial steps as we strive for cleaner and safer water resources.

5.8 Conclusions

Magnetic sorption has emerged as a promising and innovative technique for the efficient removal of dyes from aqueous solutions. This technology leverages the unique properties of MSs, which typically consist of MNPs functionalized with adsorptive surfaces, to address the pressing challenges of dye contamination in water bodies. The application of MSs for dye sequestration offers several advantages, including high adsorption capacity, rapid and efficient separation, and the potential for regeneration and reuse. It is a sustainable and eco-friendly approach that has the potential to revolutionize water purification and environmental protection efforts. As we look to the future, the prospects for magnetic sorption in dye removal are bright. Advancements in nanomaterials science will lead to the development of sorbents with enhanced properties and selectivity for specific dye classes. Tailoring sorbents for dyes and exploring green and sustainable practices in sorbent preparation are key directions for future research. Moreover, the application of MSs in industrial settings and their integration into remote-controlled systems for *in situ* dye removal hold great promise. These developments are poised to make a significant impact in industries that generate dye-containing wastewater and address dye contamination in natural water bodies. However, it is crucial to remain vigilant about the environmental impact of MSs and conduct comprehensive assessments to ensure responsible practices. This includes evaluating the disposal of sorbents, potential nanoparticle leaching, and their long-term effects on ecosystems. In summary, the magnetic sorption of dyes from aqueous solutions represents a sustainable and efficient solution to a critical environmental challenge. With continued research, innovation, and a commitment to responsible practices, MSs are poised to play a pivotal role in safeguarding our water resources and contributing to a cleaner and more sustainable future.

Acknowledgments

This chapter was compiled based on the full support received from The World Academy of Sciences–National Research Foundation (TWAS–NRF) African Renaissance Ph. D. Fellowship Program under grant numbers 116111 and 139178. The University of South Africa (UNISA) and its College of Science, Engineering and Technology (CSET) are also acknowledged.

References

[1] Papagiannaki D, Belay M H, Gonçalves N P F, Robotti E, Bianco-Prevot A, Binetti R *et al* 2022 From monitoring to treatment, how to improve water quality: the pharmaceuticals case *Chem. Eng. J. Adv.* 100245

[2] Kumar A, Khan M, Zeng X and Lo I M C 2018 Development of g-C_3N_4/TiO_2/Fe_3O_4@SiO_2 heterojunction via sol-gel route: a magnetically recyclable direct contact Z-scheme nano photocatalyst for enhanced photocatalytic removal of ibuprofen from real sewage effluent under visible light *Chem. Eng. J.* **353** 645–56

[3] Velusamy S, Roy A, Sundaram S and Kumar Mallick T 2021 A review on heavy metal ions and containing dyes removal through graphene oxide-based adsorption strategies for textile wastewater treatment *Chem. Rec.* **21** 1570–610

[4] Lin L, Yang H and Xu X 2022 Effects of water pollution on human health and disease heterogeneity: a review *Front. Environ. Sci.* **10** 975

[5] Chequer F M D, De Oliveira G A R, Ferraz E R A, Cardoso J C, Zanoni M V B and de Oliveira D P 2013 Textile dyes: dyeing process and environmental impact *Eco-Friendly Textile Dyeing and Finishing* ed M Günay (London: IntechOpen) 6 151–76

[6] Yadav S, Yadav A, Bagotia N, Sharma A K and Kumar S 2021 Adsorptive potential of modified plant-based adsorbents for sequestration of dyes and heavy metals from wastewater: a review *J. Water Process. Eng* **42** 102148

[7] Kamal T, Ul-Islam M, Khan S B and Asiri A M 2015 Adsorption and photocatalyst assisted dye removal and bactericidal performance of ZnO/chitosan coating layer *Int. J. Biol. Macromol.* **81** 584–90

[8] Ali F, Khan S B, Kamal T, Anwar Y, Alamry K A and Asiri A M 2017 Anti-bacterial chitosan/zinc phthalocyanine fibres supported metallic and bimetallic nanoparticles for the removal of organic pollutants *Carbohydr. Polym* **173** 676–89

[9] Kubra K T, Salman M S, Znad H and Hasan M N 2021 Efficient encapsulation of toxic dye from wastewater using biodegradable polymeric adsorbent *J. Mol. Liq.* **329** 115541

[10] Shuang C, Li P, Li A, Zhou Q, Zhang M and Zhou Y 2012 Quaternized magnetic microspheres for the efficient removal of reactive dyes *Water Res.* **46** 4417–26

[11] Wawrzkiewicz M 2012 Anion exchange resins as effective sorbents for acidic dye removal from aqueous solutions and wastewaters *Solvent Extr. Ion Exch.* **30** 507–23

[12] Rocher V, Siaugue J M, Cabuil V and Bee A 2008 Removal of organic dyes by magnetic alginate beads *Water Res.* **42** 1290–8

[13] Luo X and Zhang L 2009 High effective adsorption of organic dyes on magnetic cellulose beads entrapping activated carbon *J. Hazard. Mater.* **171** 340–7

[14] Qadri S, Ganoe A and Haik Y 2009 Removal and recovery of acridine orange from solutions by use of magnetic nanoparticles *J. Hazard. Mater.* **169** 318–23

[15] Sharma A, Mangla D and Chaudhry S A 2022 Recent advances in magnetic composites as adsorbents for wastewater remediation *J. Environ. Manage.* **306** 114483

[16] Osman A I, El-Monaem E M A, Elgarahy A M, Aniagor C O, Hosny M, Farghali M *et al* 2023 Methods to prepare biosorbents and magnetic sorbents for water treatment: a review *Environ. Chem. Lett.* **21** 1–62

[17] Wang M and You X yi 2021 Critical review of magnetic polysaccharide-based adsorbents for water treatment: synthesis, application and regeneration *J. Clean. Prod.* **323** 129118

[18] Nisticò R 2017 Magnetic materials and water treatments for a sustainable future *Res. Chem. Intermed.* **43** 6911–49

[19] Cui H, Liu Y and Ren W 2013 Structure switch between α-Fe_2O_3, γ-Fe_2O_3 and Fe_3O_4 during the large scale and low-temperature sol–gel synthesis of nearly monodispersed iron oxide nanoparticles *Adv. Powder Technol.* **24** 93–7

[20] Chen Y H and Li F A 2010 Kinetic study on removal of copper (II) using goethite and hematite nano-photocatalysts *J. Colloid Interface Sci.* **347** 277–81

[21] QU J 2008 Research progress of novel adsorption processes in water purification: a review *J. Environ. Sci.* **20** 1–13

[22] Motornov M, Roiter Y, Tokarev I and Minko S 2008 Colloidal Systems on the Nanometer Length Scale *Handbook of Surface and Colloid Chemistry* (Boca Raton: CRC Press) 3rd edn 131–54

[23] Zhu N, Ji H, Yu P, Niu J, Farooq M U, Akram M W *et al* 2018 Surface modification of magnetic iron oxide nanoparticles *Nanomaterials* **8** 810

[24] Cazetta A L, Pezoti O, Bedin K C, Silva T L, Paesano Junior A, Asefa T *et al* 2016 Magnetic activated carbon derived from biomass waste by concurrent synthesis: efficient adsorbent for toxic dyes *ACS Sustain Chem Eng* **4** 1058–68

[25] Vahdati-Khajeh S, Zirak M, Tejrag R Z, Fathi A, Lamei K and Eftekhari-Sis B 2019 Biocompatible magnetic N-rich activated carbon from egg white biomass and sucrose: preparation, characterization and investigation of dye adsorption capacity from aqueous solution *Surf. Interfaces* **15** 157–65

[26] Salem S, Teimouri Z and Salem A 2020 Fabrication of magnetic activated carbon by carbothermal functionalization of agriculture waste via microwave-assisted technique for cationic dye adsorption *Adv. Powder Technol.* **31** 4301–9

[27] Lu F, Dong A, Ding G, Xu K, Li J and You L 2019 Magnetic porous polymer composite for high-performance adsorption of acid red 18 based on melamine resin and chitosan *J. Mol. Liq.* **294** 111515

[28] Yadav S, Asthana A, Singh A K, Chakraborty R, Vidya S S, Susan M A B H *et al* 2021 Adsorption of cationic dyes, drugs and metal from aqueous solutions using a polymer composite of magnetic/β-cyclodextrin/activated charcoal/Na alginate: Isotherm, kinetics and regeneration studies *J. Hazard. Mater.* **409** 124840

[29] Das P, Nisa S, Debnath A and Saha B 2022 Enhanced adsorptive removal of toxic anionic dye by novel magnetic polymeric nanocomposite: optimization of process parameters *J. Dispersion Sci. Technol.* **43** 880–95

[30] Shi Z, Xu C, Guan H, Li L, Fan L, Wang Y *et al* 2018 Magnetic metal-organic frameworks (MOFs) composite for removal of lead and malachite green in wastewater *Colloids Surf. A: Physicochem. Eng. Asp* **539** 382–90
[31] Liu H, Ren X and Chen L 2016 Synthesis and characterization of a magnetic metal–organic framework for the adsorptive removal of Rhodamine B from aqueous solution *J. Ind. Eng. Chem.* **34** 278–85
[32] Zhao X, Liu S, Tang Z, Niu H, Cai Y, Meng W *et al* 2015 Synthesis of magnetic metal-organic framework (MOF) for efficient removal of organic dyes from water *Sci. Rep.* **5** 11849
[33] Zhang J, Huang Q and Du J 2016 Recent advances in magnetic hydrogels *Polym. Int.* **65** 1365–72
[34] Li K, Yan J, Zhou Y, Li B and Li X 2021 β-cyclodextrin and magnetic graphene oxide modified porous composite hydrogel as a superabsorbent for adsorption cationic dyes: adsorption performance, adsorption mechanism and hydrogel column process investigate *J. Mol. Liq.* **335** 116291
[35] Mahdavinia G R, Massoudi A, Baghban A and Shokri E 2014 Study of adsorption of cationic dye on magnetic kappa-carrageenan/PVA nanocomposite hydrogels *J. Environ. Chem. Eng.* **2** 1578–87
[36] Sivashankar R, Sathya A B, Vasantharaj K and Sivasubramanian V 2014 Magnetic composite an environmental super adsorbent for dye sequestration—a review *Environ Nanotechnol. Monit Manag* **1** 36–49
[37] Nithya R, Thirunavukkarasu A, Sathya A B and Sivashankar R 2021 Magnetic materials and magnetic separation of dyes from aqueous solutions: a review *Environ. Chem. Lett.* **19** 1275–94
[38] Mathew D S and Juang R S 2007 An overview of the structure and magnetism of spinel ferrite nanoparticles and their synthesis in microemulsions *Chem. Eng. J.* **129** 51–65
[39] Fatima H and Kim K S 2017 Magnetic nanoparticles for bioseparation *Korean J. Chem. Eng.* **34** 589–99
[40] Liu J, Yu Y, Zhu S, Yang J, Song J, Fan W *et al* 2018 Synthesis and characterization of a magnetic adsorbent from negatively-valued iron mud for methylene blue adsorption *PLoS One* **13** e0191229
[41] 2021 *Advanced Magnetic Adsorbents for Water Treatment* ed L Meili and G L Dotto (Cham: Springer) Environmental Chemistry for a Sustainable World 61
[42] Pansambal S, Roy A, Mohamed H E A, Oza R, Vu C M, Marzban A *et al* 2022 Recent developments on magnetically separable ferrite-based nanomaterials for removal of environmental pollutants *J. Nanomater.* **2022** 1–15
[43] Ambashta R D and Sillanpää M 2010 Water purification using magnetic assistance: a review *J. Hazard. Mater.* **180** 38–49
[44] Phouthavong V, Yan R, Nijpanich S, Hagio T, Ichino R, Kong L *et al* 2022 Magnetic adsorbents for wastewater treatment: advancements in their synthesis methods *Materials (Basel)* **15** 1053
[45] Stanicki D, Vander Elst L, Muller R N and Laurent S 2015 Synthesis and processing of magnetic nanoparticles *Curr. Opin. Chem. Eng* **8** 7–14
[46] Tang S C N and Lo I M C 2013 Magnetic nanoparticles: essential factors for sustainable environmental applications *Water Res.* **47** 2613–32

[47] Mahmoodi N M 2015 Surface modification of magnetic nanoparticle and dye removal from ternary systems *J. Ind. Eng. Chem.* **27** 251–9
[48] Keyhanian F, Shariati S, Faraji M and Hesabi M 2016 Magnetite nanoparticles with surface modification for removal of methyl violet from aqueous solutions *Arab J. Chem.* **9** S348–54
[49] Goswami L, Kushwaha A, Kafle S R and Kim B S 2022 Surface modification of biochar for dye removal from wastewater *Catalysts* **12** 817
[50] Hu B, He M and Chen B 2019 Magnetic nanoparticle sorbents *Solid-Phase Extraction* ed C F Poole (Amsterdam: Elsevier) 235–84
[51] Ojemaye M O, Okoh O O and Okoh A I 2017 Surface modified magnetic nanoparticles as efficient adsorbents for heavy metal removal from wastewater: progress and prospects *Mater. Express.* **7** 439–56
[52] Manyangadze M, Chikuruwo N H M, Chakra C S, Narsaiah T B, Radhakumari M and Danha G 2020 Enhancing adsorption capacity of nano-adsorbents via surface modification: a review *South African J. Chem. Eng* **31** 25–32
[53] Zuo B, Li W, Wu X, Wang S, Deng Q and Huang M 2020 Recent advances in the synthesis, surface modifications and applications of core-shell magnetic mesoporous silica nanospheres *Chem. Asian J.* **15** 1248–65
[54] Reddy D H K and Lee S M 2013 Application of magnetic chitosan composites for the removal of toxic metal and dyes from aqueous solutions *Adv. Colloid Interface Sci.* **201** 68–93
[55] Salleh M A M, Mahmoud D K, Karim W A W A and Idris A 2011 Cationic and anionic dye adsorption by agricultural solid wastes: a comprehensive review *Desalination* **280** 1–13
[56] Iqbal A, Yusaf A, Usman M, Hussain Bokhari T and Mansha A 2023 Insight into the degradation of different classes of dyes by advanced oxidation processes; a detailed review *Int. J. Environ. Anal. Chem.* 1–35
[57] Chavan R B 2011 Environmentally friendly dyes *Handbook of Textile and Industrial Dyeing* ed M Clark (Cambridge: Woodhead Publishing) 515–61
[58] 1997 *Profile of the Textile Industry: Sector Notebook* EPA 310-R-97-009 EPA National Service Center for Environmental Publications https://nepis.epa.gov/Exe/ZyPDF.cgi/50000HE9.PDF?Dockey=50000HE9.PDF
[59] Uddin M J, Ampiaw R E and Lee W 2021 Adsorptive removal of dyes from wastewater using a metal-organic framework: a review *Chemosphere* **284** 131314
[60] Rápó E and Tonk S 2021 Factors affecting synthetic dye adsorption; desorption studies: a review of results from the last five years (2017–2021) *Molecules* **26** 5419
[61] Katheresan V, Kansedo J and Lau S Y 2018 Efficiency of various recent wastewater dye removal methods: a review *J. Environ. Chem. Eng.* **6** 4676–97
[62] Samsami S, Mohamadizaniani M, Sarrafzadeh M H, Rene E R and Firoozbahr M 2020 Recent advances in the treatment of dye-containing wastewater from textile industries: overview and perspectives *Process Saf. Environ. Prot.* **143** 138–63
[63] Soares S F, Fernandes T, Trindade T and Daniel-da-Silva A L 2020 Recent advances in magnetic biosorbents and their applications for water treatment *Environ. Chem. Lett.* **18** 151–64
[64] Tee G T, Gok X Y and Yong W F 2022 Adsorption of pollutants in wastewater via biosorbents, nanoparticles and magnetic biosorbents: a review *Environ. Res.* **212** 113248

[65] Akpomie K G and Conradie J 2020 Efficient synthesis of magnetic nanoparticle-Musa acuminate peel composite for the adsorption of anionic dye *Arab. J. Chem.* **13** 7115–31

[66] Gong J L, Wang B, Zeng G M, Yang C P, Niu C G, Niu Q Y *et al* 2009 Removal of cationic dyes from aqueous solution using magnetic multi-wall carbon nanotube nanocomposite as adsorbent *J. Hazard. Mater.* **164** 1517–22

[67] Tural B, Ertaş E, Enez B, Fincan S A and Tural S 2017 Preparation and characterization of a novel magnetic biosorbent functionalized with biomass of bacillus subtilis: kinetic and isotherm studies of biosorption processes in the removal of methylene blue *J. Environ. Chem. Eng.* **5** 4795–802

[68] Afkhami A, Saber-Tehrani M and Bagheri H 2010 Modified maghemite nanoparticles as an efficient adsorbent for removing some cationic dyes from aqueous solution *Desalination* **263** 240–8

[69] Eltaweil A S, Mohamed H A, Abd El-Monaem E M and El-Subruiti G M 2020 Mesoporous magnetic biochar composite for enhanced adsorption of malachite green dye: characterization, adsorption kinetics, thermodynamics and isotherms *Adv. Powder Technol.* **31** 1253–63

[70] Nguyen V H, Van H T, Nguyen V Q, Dam X V, Hoang L P and Ha L T 2020 Magnetic Fe_3O_4 nanoparticle biochar derived from pomelo peel for reactive Red 21 adsorption from aqueous solution *J. Chem.* **2020** 1–14

[71] Suba V, Rathika G, Ranjith Kumar E and Saravanabhavan M 2018 Influence of magnetic nanoparticles on surface changes in $CoFe_2O_4$/nerium oleander leaf waste activated carbon nanocomposite for water treatment *J. Inorg. Organomet. Polym. Mater.* **28** 1706–17

[72] Geng J and Chang J 2020 Synthesis of magnetic forsythia suspensa leaf powders for removal of metal ions and dyes from wastewater *J. Environ. Chem. Eng.* **8** 104224

[73] Oladipo A A and Ifebajo A O 2018 Highly efficient magnetic chicken bone biochar for removal of tetracycline and fluorescent dye from wastewater: two-stage adsorber analysis *J. Environ. Manage.* **209** 9–16

[74] Adeogun A I, Akande J A, Idowu M A and Kareem S O 2019 Magnetic tuned sorghum husk biosorbent for effective removal of cationic dyes from aqueous solution: isotherm, kinetics, thermodynamics and optimization studies *Appl. Water Sci.* **9** 1–17

[75] Tarhan T, Tural B, Boga K and Tural S 2019 Adsorptive performance of magnetic nanobiosorbent for binary dyes and investigation of comparative biosorption *SN Appl. Sci.* **1** (1) 11

[76] Li C, Wang X, Meng D and Zhou L 2018 Facile synthesis of low-cost magnetic biosorbent from peach gum polysaccharide for selective and efficient removal of cationic dyes *Int. J. Biol. Macromol.* **107** 1871–8

[77] Pipíška M, Zarodňanská S, Horník M, Ďuriška L, Holub M and Šafařík I 2020 Magnetically functionalized moss biomass as biosorbent for efficient Co^{2+} ions and thioflavin T removal *Materials (Basel)* **13** 3619

[78] Olusegun S J, Freitas E T F, Lara L R S and Mohallem N D S 2021 Synergistic effect of a spinel ferrite on the adsorption capacity of nano bio-silica for the removal of methylene blue *Environ. Technol.* **42** 2163–76

[79] Li X, Lu H, Zhang Y, He F, Jing L and He X 2016 Fabrication of magnetic alginate beads with a uniform dispersion of $CoFe_2O_4$ by the polydopamine surface functionalization for organic pollutants removal *Appl. Surf. Sci.* **389** 567–77

[80] Safarik I, Horska K and Safarikova M 2011 Magnetically modified spent grain for dye removal *J. Cereal Sci.* **53** 78–80

[81] Safarik I and Safarikova M 2010 Magnetic fluid modified peanut husks as an adsorbent for organic dye removal *Phys. Procedia* **9** 274–8

[82] Aryee A A, Dovi E, Han R, Li Z and Qu L 2021 One novel composite based on functionalized magnetic peanut husk as adsorbent for efficient sequestration of phosphate and congo red from solution: characterization, equilibrium, kinetic and mechanism studies *J. Colloid Interface Sci.* **598** 69–82

[83] Hashemian S and Salimi M 2012 Nano composite a potential low cost adsorbent for removal of cyanine acid *Chem. Eng. J.* **188** 57–63

[84] Salgueiro A M, Daniel-da-Silva A L, Girão A V, Pinheiro P C and Trindade T 2013 Unusual dye adsorption behavior of κ-carrageenan coated superparamagnetic nanoparticles *Chem. Eng. J.* **229** 276–84

[85] Shi W, Guo F, Wang H, Liu C, Fu Y, Yuan S *et al* 2018 Carbon dots decorated magnetic $ZnFe_2O_4$ nanoparticles with enhanced adsorption capacity for the removal of dye from aqueous solution *Appl. Surf. Sci.* **433** 790–7

[86] Sheshmani S and Mashhadi S 2018 Potential of magnetite reduced graphene oxide/chitosan nanocomposite as biosorbent for the removal of dyes from aqueous solutions *Polym. Compos.* **39** E457–62

[87] Zhu H Y, Jiang R, Xiao L and Li W 2010 A novel magnetically separable γ-Fe_2O_3/crosslinked chitosan adsorbent: preparation, characterization and adsorption application for removal of hazardous azo dye *J. Hazard. Mater.* **179** 251–7

[88] Tian H, Peng J, Lv T, Sun C and He H 2018 Preparation and performance study of $MgFe_2O_4$/metal–organic framework composite for rapid removal of organic dyes from water *J. Solid State Chem.* **257** 40–8

[89] Chen D, Feng P fei and Hua Wei F 2019 Preparation of Fe (III)-MOFs by microwave-assisted ball for efficiently removing organic dyes in aqueous solutions under natural light *Chem. Eng. Process. Intensif* **135** 63–7

[90] Wang T, Zhao P, Lu N, Chen H, Zhang C and Hou X 2016 Facile fabrication of Fe_3O_4/MIL-101 (Cr) for effective removal of acid red 1 and orange G from aqueous solution *Chem. Eng. J.* **295** 403–13

[91] Huang L, He M, Chen B and Hu B 2018 Magnetic Zr-MOFs nanocomposites for rapid removal of heavy metal ions and dyes from water *Chemosphere* **199** 435–44

[92] Majid Z, AbdulRazak A A and Noori W A H 2019 Modification of zeolite by magnetic nanoparticles for organic dye removal *Arab J. Sci. Eng.* **44** 5457–74

[93] Chen C, Zhang M, Guan Q and Li W 2012 Kinetic and thermodynamic studies on the adsorption of xylenol orange onto MIL-101 (Cr) *Chem. Eng. J.* **183** 60–7

[94] Hassan M, Naidu R, Du J, Liu Y and Qi F 2020 Critical review of magnetic biosorbents: their preparation, application, and regeneration for wastewater treatment *Sci. Total Environ.* **702** 134893

[95] Kumari R, Mohanta J, Sambasivaiah B, Qaiyum M A, Dey B, Samal P P *et al* 2023 Dye sequestration from aqueous phase using natural and synthetic adsorbents in batch mode: present status and future perspectives *Int. J. Environ. Sci. Technol.* 1**20**

[96] Eltaweil A S, Elshishini H M, Ghatass Z F and Elsubruiti G M 2021 Ultra-high adsorption capacity and selective removal of congo red over aminated graphene oxide modified Mn-doped UiO-66 MOF *Powder Technol.* **379** 407–16

[97] Hazrati M and Safari M 2020 Cadmium-based metal–organic framework for removal of dye from aqueous solution *Environ. Prog. Sustain. Energy* **39** e13411
[98] Khan M S, Khalid M, Ahmad M S, Shahid M and Ahmad M 2019 Three-in-one is better: exploring the sensing and adsorption properties in a newly designed metal–organic system incorporating a copper (II) ion *Dalton Trans* **48** 12918–32
[99] Lu C, Yang J, Khan A, Yang J, Li Q and Wang G 2022 A highly efficient technique to simultaneously remove acidic and basic dyes using magnetic ion-exchange microbeads *J. Environ. Manage.* **304** 114173
[100] Zhao S, Chen D, Wei F, Chen N, Liang Z and Luo Y 2017 Removal of Congo red dye from aqueous solution with nickel-based metal-organic framework/graphene oxide composites prepared by ultrasonic wave-assisted ball milling *Ultrason. Sonochem.* **39** 845–52
[101] Zhang J, Li F and Sun Q 2018 Rapid and selective adsorption of cationic dyes by a unique metal-organic framework with decorated pore surface *Appl. Surf. Sci.* **440** 1219–26
[102] Jegede M M, Durowoju O S and Edokpayi J N 2021 Sequestration of hazardous dyes from aqueous solution using raw and modified agricultural waste *Adsorpt. Sci. Technol.* **2021** 1–21
[103] Shikuku V O and Mishra T 2021 Adsorption isotherm modeling for methylene blue removal onto magnetic kaolinite clay: a comparison of two-parameter isotherms *Appl. Water. Sci.* **11** 103
[104] Sarojini G, Babu S V and Rajasimman M 2022 Adsorptive potential of iron oxide based nanocomposite for the sequestration of congo red from aqueous solution *Chemosphere* **287** 132371
[105] 2017 *Adsorption Processes for Water Treatment and Purification* ed A Bonilla-Petriciolet, D I Mendoza-Castillo and H E Reynel-Ávila (Berlin: Springer) 256
[106] Qiu H, Lv L, Pan B C, Zhang Q J, Zhang W M and Zhang Q X 2009 Critical review in adsorption kinetic models *J. Zhejiang Univ.* A **10** 716–24
[107] Hussain S, Kamran M, Khan S A, Shaheen K, Shah Z, Suo H *et al* 2021 Adsorption, kinetics and thermodynamics studies of methyl orange dye sequestration through chitosan composites films *Int. J. Biol. Macromol.* **168** 383–94
[108] Lima E C, Hosseini-Bandegharaei A, Moreno-Piraján J C and Anastopoulos I 2019 A critical review of the estimation of the thermodynamic parameters on adsorption equilibria. Wrong use of equilibrium constant in the Van't Hoof equation for calculation of thermodynamic parameters of adsorption *J. Mol. Liq.* **273** 425–34
[109] Sabzehmeidani M M, Mahnaee S, Ghaedi M, Heidari H and Roy V A L 2021 Carbon based materials: a review of adsorbents for inorganic and organic compounds *Mater. Adv.* **2** 598–627
[110] Anastopoulos I and Kyzas G Z 2016 Are the thermodynamic parameters correctly estimated in liquid-phase adsorption phenomena? *J. Mol. Liq.* **218** 174–85
[111] Atkins P and De Paula J 2006 *Physical Chemistry* (London: Macmillan) 8th edn1
[112] Wang S, Zhai Y Y, Gao Q, Luo W J, Xia H and Zhou C G 2014 Highly efficient removal of acid red 18 from aqueous solution by magnetically retrievable chitosan/carbon nanotube: batch study, isotherms, kinetics, and thermodynamics *J. Chem. Eng. Data* **59** 39–51
[113] Debrassi A, Corrêa A F, Baccarin T, Nedelko N, Ślawska-Waniewska A, Sobczak K *et al* 2012 Removal of cationic dyes from aqueous solutions using N-benzyl-O-carboxymethyl-chitosan magnetic nanoparticles *Chem. Eng. J.* **183** 284–93
[114] Ghoreishi S M and Haghighi R 2003 Chemical catalytic reaction and biological oxidation for treatment of non-biodegradable textile effluent *Chem. Eng. J.* **95** 163–9

IOP Publishing

Environmental Applications of Magnetic Sorbents

Kingsley Eghonghon Ukhurebor and Uyiosa Osagie Aigbe

Chapter 6

The application of magnetic sorbents for the sequestration of pesticides, pharmaceuticals, and perfluoroalkyl and polyfluoroalkyl substances from aqueous solutions

Kingsley Obodo

Pesticides, pharmaceuticals, and per- and polyfluoroalkyl substances (PFASs) are anthropogenic pollutants found in aqueous solutions. They have negative effects on humans because of their persistence. The application of magnetic sorbents for their sequestration has drawn wide attention. Magnetic sorbents offer unique advantages in these pollutants' sequestration. This chapter systematically evaluates the application of magnetic sorbents for the removal of pesticides, pharmaceuticals, and PFASs. The advantages and disadvantages of using magnetic sorbents are highlighted in terms of their improved efficiency in sorption, selectivity, and regeneration as well as the challenges, roles, sorption mechanisms, and factors affecting their efficiency.

6.1 Introduction

In recent times, large amounts of emerging contaminants of natural and anthropic origins have become a matter of serious concern in various aqueous systems. The major driver for aqueous pollution is rapid and large-scale economic and industrial development. Some sources of aqueous system pollution arise from pesticides, pharmaceuticals, and PFASs, and these raise significant concerns for public health and the environment [1–3].

Pesticides are chemical substances used in agricultural pest management to increase the overall agricultural yield by offering protection from pests which might affect the gross agricultural yield. These can pose risks to nontarget organisms because of excessive and improper usage, disposal, other direct causes (spills and leaks), and indirect causes (runoff and drift), which result in contamination and the

doi:10.1088/978-0-7503-5909-2ch6

pervasive presence of pesticides in water bodies. The polluting effect of pesticides poses a risk to the environment not only in terms of ground water quality but also air quality.

Pharmaceuticals are broadly referred to as substances used for the purpose of diagnosing, treating, or preventing diseases. They play an important role in the restoration, correction, or alteration of physiological functions within the body. Various pharmaceuticals applied in the treatment of human and animal diseases are found to constitute a persistent class of pollutants in the environment, particularly in water bodies [4]. Some therapeutic drugs commonly present in water bodies include: (i) antidepressants such as benzodiazepines; (ii) antiepileptic drugs such as carbamazepine; (iii) lipid-lowering medications such as fibrates; (iv) beta-blockers such as atenolol, propranolol, and metoprolol; (v) anti-inflammatories such as diclofenac, paracetamol, acetylsalicylic acid, and ibuprofen; (vi) antihistamines and antiulcer drugs such as famotidine and ranitidine; (vii) antibiotics such as imidazole, macrolides, sulfonamides, tetracyclines, penicillin, quinolones, beta-lactams, fluoroquinolones, and chloramphenicol and its derivatives; (viii) as well as other miscellaneous substances, which may include heroin, opiates, cocaine, methadone, amphetamines, barbiturates, and other narcotics [5]. 'Active pharmaceutical ingredients' (APIs) is a term used by the pharmaceutical industry to describe drugs that exhibit pharmacological activity, resist degradation, display high persistence in water, and have the potential to negatively impact both aquatic life and human health. Pharmaceutical compounds, which often have a molecular mass of less than 500 Da, differ from conventional industrial pollutants due to their diverse molecular characteristics and ionizability, which depend on their pH, polarity, moderate water solubility, lipophilicity, and persistence, which can extend for years [6]. Micropollutant concentrations of ng l^{-1} to $\mu g\ l^{-1}$ of these pharmaceuticals have potential adverse effects on humans, aquatic organisms, and the environment [7, 8]. Clay-based adsorbents have been shown to have improved properties for the removal of pharmaceuticals such as antibiotics from water sources and have associated advantages such as affordability, availability, ease of application, etc [9].

PFASs are synthetic aliphatic compounds which contain the $–C_nF_{2n+1}$ moiety [10]. They have a unique structure with a hydrophilic 'head' and a long hydrophobic 'tail' that create amphiphilic characteristics, resulting in the ability to interact with both water and nonpolar substances. As such, they can be found in most industries and consumer products, where they are applied in the manufacture of fluoropolymer coatings as well as products that resist oil, heat, grease, stains, and water. Examples of products that include fluoropolymer coatings include food packaging, clothing, heat-resistant nonstick cooking surfaces, furniture, adhesives, food packaging, the insulation on electrical wires, etc [11]. PFASs have widespread application in the anthropogenic era and their hydrophobic characteristics are very strong in aqueous solutions and drinking water. Thus, epidemiological research has linked exposure to certain PFASs with various health-related issues, such as immune and thyroid disruptions, liver problems, irregularities in lipid and insulin levels, kidney ailments, adverse effects on reproduction and development, and cancer [12].

Several studies of the techniques, such as adsorption techniques, for environmental remediation and their cost implications have been carried out [13–15]. Adsorption, a widely used technique, has been shown to effectively remove inorganic and organic contaminants from wastewater. The regeneration of spent adsorbents using magnetic separation, filtration, and chemical desorption was explored by Baskar *et al* [16], who highlighted existing drawbacks and suggested future directions for industrial-scale application, considering economic and environmental feasibility [16]. Obodo *et al* [17] discussed the different kinds, properties, and management options of commonly used adsorbent materials in relation to environmental sustainability. They showed that adsorption techniques offer distinct advantages because commonly available sorbents are applied and can be easily recycled under appropriate conditions at a reasonable cost [18].

Recently, the importance of using magnetic sorbents in the treatment of aqueous systems is an area that has gaining much attention due to the possibility of its application in various areas of environmental remediation, such as wastewater, air pollution, etc. Magnetic sorbents are systems that can adsorb pollutants or unwanted systems from aqueous solutions. These magnetic sorbents are readily recovered through the application of an external magnetic field. They also possess a host of other properties that will be discussed in the next section. Magnetic polymeric sorbents have been created using hyper-crosslinked polystyrene (HPS). These sorbents maintain their unique features, including an expanded specific surface area as well as a micro–mesoporous configuration. Due to their magnetic properties, they can be employed as sorbents that are easily separable using magnets, as well as serving as platforms for the active phase in heterogeneous catalytic systems [19]. Zhai *et al* [20] used a facile technique to synthesize an ordered magnetic nanostructured shell with a core of manganese oxide, which had a strong ability to adsorb organic pollutants, a high capacity when used in wastewater treatment, and a regenerative capacity via combustion when used to sequester Congo red dye.

As discussed above, the presence of pesticides, pharmaceuticals, and PFASs in water bodies and aqueous solutions can lead to adverse effects on aquatic ecosystems, wildlife, and human health. Pesticides can harm nontarget organisms and disrupt the ecological balance, while pharmaceuticals and PFASs have been associated with potentially severe health issues that have long-term detrimental effects in humans. Thus, the removal of these pollutants from aqueous solutions is crucial from scientific, environmental, and health perspectives.

6.2 The challenges and role of magnetic sorbents in aqueous solutions treatment

Various challenges persist in relation to the removal of pesticides, pharmaceuticals, and PFASs, which are anthropogenic pollutants, from aqueous solutions. These are associated with several factors such as their low concentrations, diverse chemical structures, and recalcitrant nature. Some of the key challenges include [18]:

(i) **Low concentrations:** the trace amounts of pesticides, pharmaceuticals, and PFASs in water and aqueous solutions pose a challenge to their detection and removal.
(ii) **Chemical diversity:** the pollutants present in aqueous solutions are found to belong to very diverse chemical classes. This implies that different remedial and treatment approaches are required for each specific target pollutant. This poses a challenge for the remediation of aqueous solutions.
(iii) **Persistence:** some pesticides, pharmaceuticals, and PFASs are long-lasting in the environment. This makes their removal problematic using conventional treatment methods.
(iv) **Compatibility of treatment technique:** the application of conventional aqueous solution treatment methods might be ineffective in the removal of micropollutants associated with PFASs, pharmaceuticals, and pesticides from aqueous solutions. The outcome is that they may still be present in some treated aqueous solutions.
(v) **By-product formation:** the inadequate treatment of PFASs, pesticides, and pharmaceuticals can result in the formation of significantly more toxic derivatives. These pose an even greater risk to human health and the environment.
(vi) **Efficiency and cost implication:** the implementation of novel and innovative treatment techniques for the removal of trace and anthropogenic micropollutants results in a competition between cost and efficacy.

These challenges posed by the efficient removal of these pollutants at low concentrations without the generation of harmful by-products require novel techniques that are cost-effective and efficient. Several studies [21–23] have shown the potential application of magnetic sorbents in aqueous solutions and water treatment for the removal of pollutants such as pesticides, pharmaceuticals, and PFASs. Thus, the integration of magnetic particles with sorbents for water treatment is attracting significant interest due to their exceptional physical and chemical properties. Magnetic sorbents are composed of sorbents with magnetic nanoparticles, magnetic coatings, functional groups, etc. that allow them to remove specific pollutants from aqueous solutions, rendering them effective in efficient pollutant sequestration [24]. The application of an external magnetic field enables the easy separation of pollutants from the aqueous solution. As such, magnetic sorbents offer a highly efficient and practical approach for treating wastewater [25].

According to a study by Hu *et al* [26], a magnetic sorbent is ideal if it possesses certain advantageous characteristics, namely: (i) *strong magnetism* for rapid magnetic separation, ensuring efficient and quick removal of the sorbent from the solution; (ii) *excellent dispersion properties,* which promote enhanced adsorption/desorption kinetics for better performance; (iii) *a large specific surface area and appropriate porosity,* allowing for easy modification and providing abundant adsorption sites (this leads to increased adsorption capacity and improves the extraction efficiency and recovery of target compounds); (iv) *good selectivity and anti-interference capabilities,* which enable the method to handle complex matrices effectively; (v) *good stability,* to withstand challenging conditions such as acidic and alkaline environments, stirring, oscillation, and ultrasound treatment; (vi) *reusability,* ensuring reversible adsorption for repeated applications; (vii) *mild operating conditions*—the ability to operate under mild adsorption and desorption conditions, avoiding harsh treatment; (viii) *ease of preparation,* which allows for high reproducibility of the sorbent; (ix) the availability of *low cost, readily available raw materials* and minimal sorbent consumption during the extraction process; (x) *environmentally friendly processing* low reagent consumption during production and extraction, contributing to sustainable practices. The application of magnetic sorbents in the treatment of various aqueous solutions offers various advantages and disadvantages as presented in table 6.1.

Magnetic sorbents can play an active role in the efficient sequestration and removal of pollutants such as pesticides, pharmaceuticals, and PFASs, resulting in ease of separation in aqueous solutions. The costs, synthesis challenges, and dependency of magnetic sorbents on magnetic fields for optimal performance can have potential downsides. The selective adsorption properties and rapid treatment potential of magnetic sorbents offer an upside for widespread applications in large-scale polluted water treatment; as shown in table 6.1, the advantages outweigh the disadvantages. The aim and objectives of this chapter are to deeply examine and highlight the associated environmental implications of these pollutants, to demonstrate the difficulties encountered in the abstraction of these anthropogenic pollutants from aqueous solutions using conventional means, and to show the need for novel techniques such as magnetic sorbents. The advantages and disadvantages of using magnetic sorbents are highlighted in terms of the improved efficiencies of sorption, selectivity, and regeneration.

6.3 Magnetic sorbents: types and synthesis methods

6.3.1 Types of magnetic sorbents

The classification of the different types of magnetic sorbents is based on the nanoparticle size, nanocomposites, and polymeric material applied in their synthesis. The types of magnetic sorbent can be broadly categorized into: (i) magnetic nanoparticles, (ii) functionalized polymeric materials, (iii) metals and metal oxides, (iv) modified biosorbents, (v) magnetic composite sorbents, etc. Thus, different synthesis techniques applied in the production of these materials can impact their properties, as discussed in detail in the following section. Kubrakova *et al* [27] used a microwave technique to prepare and modify nanomagnetite with organic compounds and showed that they are stable. The different types of magnetic sorbents can be functionalized for various applications, including the treatment and removal of contaminants such as pesticides, pharmaceuticals, and PFASs in aqueous solutions.

The functionalization and doping of magnetic sorbents can be carried out to improve their selectivity for specific pollutants and their adsorption capacity [17, 28, 29]. Prasad *et al* [30] showed that a bioinspired approach involving the green synthesis of reduced graphene oxide/iron oxide magnetic nanoparticles from *Murraya koenigii* leaves extract holds promise for the eco-friendly remediation of

Table 6.1. The advantages and disadvantages of magnetic sorbents in the treatment of various aqueous solutions.

Advantages	Disadvantages
Efficient adsorption: magnetic sorbents exhibit high adsorption capacities due to their large surface area, making them effective in removing contaminants from water	Cost: the synthesis and functionalization of magnetic sorbents can be expensive, impacting their practical application
Easy separation: these sorbents can be easily recovered from treated solutions using a magnetic field, simplifying the separation process and enabling their reuse	Complex synthesis: developing efficient magnetic sorbents with the desired properties involves complex synthesis methods
Selective removal: magnetic sorbents can be engineered to target specific pollutants, offering selective removal of contaminants	Dependency on magnetic field: effective separation relies on the availability of a strong magnetic field, which may not be practical in all settings
Rapid treatment: the use of magnetic fields for separation accelerates the treatment process, enabling faster removal of pollutants from water	Limited maturity: Despite their effectiveness, the full-scale application of magnetic sorbents in water treatment is still developing
Environmental benignity: ideal magnetic sorbents are environmentally friendly and demonstrate high selectivity and capacity for adsorption	Regeneration challenges: the regeneration of magnetic sorbents after multiple cycles of use can be challenging and may require additional treatments
Adaptability for large-scale applications: magnetic sorbents can be integrated into existing water treatment systems, making them suitable for large-scale and industrial applications.	
Potential for hybrid treatment approaches: magnetic sorbents can be used in combination with other water treatment technologies to achieve more comprehensive and efficient removal of pollutants.	

aqueous solutions containing Pb(II) ions through extraction. Cai *et al* [31] successfully synthesized a unique Fe_3O_4@TA@P(NVP-co-NIPAM) magnetic microsphere with a core–shell structure through copolymerization directed by tannic acid (TA), facilitating the extraction of polyphenols. Vallez-Gomis *et al* [32] showed in their review article that the utilization of magnetic sorbents has contributed to the advancement of downsized microextraction methods, a development with particular significance in the field of bioanalysis due to occasional limitations in sample volumes. Thus, different types of magnetic sorbents play key roles in aqueous solution remediation.

6.3.2 Synthesis methods for magnetic sorbents

The method of synthesizing magnetic nanoparticles involves three key phases: (i) the creation of a magnetic core, typically composed of magnetite or maghemite, through the synthesis of suitable precursors; (ii) the alteration of the core's surface, involving the formation of one or more shells; and (iii)—if deemed necessary—the further enhancement of surface properties, potentially through multistage functionalization [33]. The surface of the magnetic nanoparticles [34] (MNPs) may feature modifying shells crafted from diverse materials such as gold, silica, surfactants, or organic compounds. A range of functional groups, such as ligands, peptides, radioactive labels, and antibodies, can also be incorporated to enable specific interactions or binding with the target species. It is essential to underscore the significance of the modification and functionalization stages in ensuring that the resulting MNPs possess the precise characteristics required for their intended applications. The synthesis of magnetic sorbents with distinctive characteristics can be achieved using various methods, such as:

(i) Microwave-assisted synthesis allows nanosized magnetic materials to be produced with precise properties [33]. The microwave method can be paired with the hydrothermal method to synthesize small porous systems in zeolites [34].

(ii) The reduction of iron salts using polyhydric alcohols at high temperatures in an inert atmosphere has been utilized to create polymeric sorbents with magnetic attributes. These polymeric sorbents find applications as magnetically separable sorbents and as supports for heterogeneous catalytic systems.

(iii) Ion-imprinted adsorbents enhance the selectivity and adsorption capacity of magnetic sorbents for specific contaminants, such as the removal of lead from aqueous solutions, etc.

(iv) The hydrothermal method involves using water, which exists in both the vapor and liquid phases at high pressure and temperature in the chemical synthesis of nanoparticles and crystals. This method is largely applied in the synthesis of zeolites [35] as well as nanocomposites with exceptional adsorption capabilities [36].

(v) The coprecipitation method entails the simultaneous precipitation of two or more components from a solution to obtain a solid product. This method is largely applied to produce magnetic nanoparticles such as magnetite, substituted ferrites, maghemite, etc [37–39].

(vi) The microemulsion method involves the use of organometallic precursors in the synthesis of monodispersed nanoparticles at elevated temperatures. This method yields nanoparticles with exceptional crystallinity, precisely controlled sizes, and clearly defined shapes [40]. The decomposition of organometallic precursors occurs in the presence of organic surfactants, facilitating the production of nanoparticles with the desired size and shape [41].

(vii) The oxidation method: the oxidation method for magnetic nanoparticle synthesis is a chemical approach used to produce magnetic nanoparticles by inducing the oxidation of precursor materials. This entails controlled oxidation of metal precursors resulting in the formation of oxide nanoparticles with magnetic properties. The process can be applied for the synthesis of various magnetic oxide nanoparticles, such as magnetite (Fe_3O_4), maghemite (γ-Fe_2O_3), or other metal oxides with magnetic properties [42].

(viii) The hydrolysis method: this entails breaking down water molecules via chemical reactions to decompose large molecules, such as metal salts, leading to the formation of magnetic nanoparticles. Iida *et al* [43] synthesized Fe_3O_4 nanoparticles using the hydrolysis of solutions containing different iron salts and explored the influence of the Fe ions' valency on the formation of Fe_3O_4 nanoparticles.

(ix) The thermal decomposition method: this entails controlled metal precursor decomposition at very high temperatures. It is very useful in obtaining a narrow particle size distribution and excellent control of the magnetic nanoparticle size [34].

(x) The ultrasonic method (or sol–gel method): this entails the formation of gel at room temperature from metal alkoxides using hydrolysis and polycondensation reactions [44].

Several other techniques for the preparation of magnetic sorbents exist, depending on the target pollutants. These techniques can be a combination of various methods as discussed in Osman *et al* [45], which showed the different techniques used in the preparation of bio- and magnetic sorbents for water remediation and treatment (see Osman *et al* [45]). Magnetic sorbents and their synthesis hold the potential to play a significant role in addressing water pollution challenges and advancing sustainable water management practices.

6.4 Magnetic sorbents: their adsorption mechanism for pesticides, pharmaceuticals, and PFASs and factors affecting their efficiency

The magnetic sorbent adsorption mechanism for pesticides, pharmaceuticals, and PFASs takes advantage of the materials chemistry of both the absorbent and absorbate. Overall, there are several adsorption mechanisms for these pollutants (pesticides, pharmaceuticals, and PFASs) in aqueous solutions, such as covalent interactions, π–π interactions, electron exchange, ion exchange, surface interactions, pore-filling mechanisms, etc. The different modes of adsorption by magnetic adsorbents are driven either by electrostatic or non-electrostatic interactions. The interaction depends on the pH of the solution as well as the charge/polarity of the adsorbate–adsorbent system. The main modes of adsorption are the following:

(i) **Electrostatic interaction:** this interaction occurs between the adsorbate and adsorbent surfaces and is driven by a difference in charge. Similar charges result in decreased adsorption due to electrostatic repulsion and vice versa.
(ii) **π–π interactions:** these arise from the aromatic groups of the adsorbate–adsorbent system, and their strength mainly depends on the attached functional groups.
(iii) **Van der Waals force:** this is a minute intermolecular attractive force that increases with increasing distance between the adsorbate and adsorbent systems. This force field holds the adsorbate molecules on the surface before they are desorbed.
(iv) **Hydrogen bonds:** this bonding occurs when hydrogen from a molecule and a highly electronegative atom from another molecule give rise to an electrostatic attraction. Various functional groups on magnetic adsorbents possess varying amounts of oxygen-containing functional groups that can lead to the formation of hydrogen bonds.
(v) **Surface complexation:** this arises when the pollutant molecules/adsorbates form complexes with the functional groups on the magnetic adsorbents.
(vi) **Hydrophobic interactions:** these arise when hydrophobic groups of the target adsorbates undergo an interaction with the hydrophobic parts of the magnetic adsorbent surface.

The non-selective nature of various adsorbents for different pollutants due to the materials chemistries involved also determines the different mechanisms applied in the absorption process. Thus, the mechanism employed depends on the type of pollutant to be extracted (either inorganic or organic), the reaction conditions, the characteristics of the adsorbent, and the way in which the adsorbate and adsorbent interact [45]. Parameters such as the ionic strength, contact time, pH, adsorbent dosage, etc. affect the sorption efficiency; thus, the optimization of these conditions can improve the specific surface area and the porous structure of the adsorbent material [46]. An in-depth overview of the benefits and limitations of different adsorbent classes in the water treatment process can be found in a review article by Osman *et al* [45].

6.4.1 The sorption mechanisms of pesticides on magnetic sorbents

The sorption mechanisms of pesticides on magnetic sorbents involve several interactions, which include van der Waals interactions, electrostatic interactions, hydrophobic interactions, and ion exchange. Pesticides are generally amphiphilic compounds, which are made of both hydrophobic and hydrophilic moieties. This implies that that a surface interaction with magnetic sorbents can occur hydrophobically and that the sorbate's functional groups interact with charged sites on the surface of the sorbent. Furthermore, ion exchange can take place between the charged pesticides and magnetic sorbent surface functional groups.

Yin *et al* [47] synthesized and characterized Fe_3O_4@SiO_2–C_{18}, which took advantage of van der Waals and electrostatic interactions to adsorb weakly polar target pesticides. Liu *et al* [46] demonstrated that the Freundlich adsorption model provided a superior fit for the static adsorption data in comparison with the Langmuir model. Utilizing optimized conditions, metal–organic framework ZIF-8/magnetic multiwalled carbon nanotubes (M-M-ZIF-8) exhibited effective removal of eight different organophosphorus pesticides from both environmental water and soil samples.

6.4.2 The sorption behavior of pharmaceuticals on magnetic sorbents

The extraction of pharmaceuticals from aqueous solutions can be challenging due to the generally low levels of these pollutants. Pharmaceutical compounds find their way into natural water bodies following excretion, either in their original form or as potent metabolites originating from wastewater treatment plant effluents [48]. Tarachuk *et al* [49] carried out a review of the application of magnetic sorbents for pharmaceutical removal. They considered two groups of pharmaceuticals: antibiotics (ciprofloxacin, tetracycline, levofloxacin, etc.) and nonsteroidal anti-inflammatory drugs (diclofenac, ibuprofen, etc.) due to their high prevalence in aqueous solutions. These groups of pharmaceuticals are adsorbed mainly via ion exchange, electrostatic interactions, surface complexation, hydrogen bonds, π—π interactions, etc.

Various parameters, such as the contact time, pH, adsorbate concentration, temperature, adsorbent dose, etc. have been shown to affect the adsorption efficiency and behavior of these magnetic sorbents. For dissociated pharmaceutical adsorbates in aqueous solutions with protonated adsorbate, electrostatic interactions are prevalent. However, other interactions can occur, such as hydrogen bonds, π–π interactions, van der Waals forces, etc. as shown in figure 6.1.

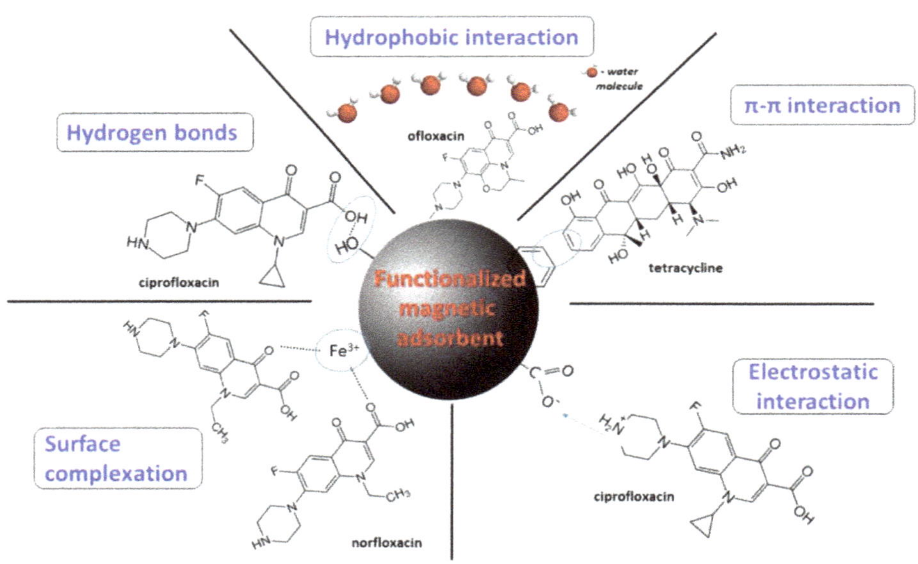

Figure 6.1. Different adsorbent–adsorbate interactions that take place on the surfaces of magnetic adsorbents.

6.4.3 The sorption behavior of PFASs on magnetic sorbents

PFASs are known to be persistent organic pollutants that are widely used and have negative effects on the environment. The adsorption behavior of magnetic sorbents offers a unique avenue for the removal of PFASs. This is because the adsorption mechanism involves various interactions such as hydrophobic interactions, electrostatic interactions, ion-exchange interactions, etc. depending on the magnetic sorbent and the nature of the aqueous solution (polar or nonpolar).

Lei *et al* [50] showed that the mechanism for PFAS sorption in an aqueous solution includes ligand exchange, ion exchange, electrostatic attraction/repulsion, and hydrophobic interaction, which are all dependent on the chemistry of the solution. Sorengard *et al* [51] showed using principal component analysis that sorption via electrostatic forces dominates for shorter-chained PFASs and hydrophobic sorption dominates for longer-chained PFASs. The physiochemical properties of PFASs can affect their sorption rates to a greater extent than the physiochemical properties of other micropollutants affect their respective sorption rates [52]. Different PFASs have a huge variation in their sorption characteristic, which depends on the chain length and functional group of the perfluorocarbon chain. For long-chain PFASs, geometric changes, stearic hindrance, and micelle formation (the self-assembly of amphiphilic molecules) can affect the sorption behavior of the magnetic sorbents [51].

6.5 Factors affecting magnetic sorption efficiency

Different factors and challenges exist that can affect the magnetic sorption efficiency of the extraction of pesticides, pharmaceuticals, and PFASs from aqueous solutions. Three physicochemical phenomena govern the pollutant transfer in aqueous solutions and are crucial for the optimization of the removal process, namely: (i) the two-phase thermodynamic equilibrium which determines the process limit, (ii) the adsorption kinetics, and (iii) the competition between various pollutants [53].

The impact of the solution pH, temperature, sorbent dosage, adsorbent mass, particle size, and coexisting substances are factors that can influence these phenomena and consequently the performance of magnetic sorbents. The above factors can be categorized into adsorbent properties and solution chemistry. The adsorbent properties are factors related to the particular adsorbent used for the sorption process. These properties are the particle size, specific surface area, and pore size, which can be improved for different magnetic sorbents to improve the efficiency of the sorption process. Increasing the pore size as well as the specific surface area can lead to higher adsorption capacity for the same material and improve adsorption kinetics. The solution chemistry is related to the pH of the solution and the effect of coexisting substances such as ions, organic matter, etc. The pH has a strong influence on the surface of the adsorbent as well as the dissociation of the specific pollutants in the solution. Under acidic conditions ($pH < 7$), minerals, activated carbons, and polymeric adsorbents absorb H^+ from the surrounding aqueous solution, resulting in protonation and subsequent positive charge acquisition. This electrostatic alteration renders the sorbents more efficient in attracting negatively charged species, resulting in improved sorption and vice versa. The ions present in the aqueous solution can affect the sorption process because there is an electrostatic interaction between the positively charged surface of the adsorbent and the negatively charged adsorbate, which can decrease the adsorption of anionic pollutants such as PFASs and vice versa. Adsorption can be influenced by electrostatic attraction as well as electrostatic repulsion, depending on the solution chemistry [50].

6.6 Adsorption kinetics and isotherm models

The overall adsorption kinetics and isotherm models are crucial in understanding the adsorption behavior of pesticides, pharmaceuticals, and PFASs on magnetic sorbents. The adsorption kinetics and isotherm model determine the correlation, functional performance, and multifaceted nature of the adsorption method for the adsorbate–adsorbent combination. The kinetic model is a rate equation, which describes the rate at which the solute is transferred/adsorbed by the surface of the adsorbent as well as the time taken for the adsorbate to achieve equilibrium at the solid–liquid interface. The isotherm model, on the other hand, describes the solute concentration in the media and the quantity of the solute adsorbed on the adsorbent surface at a given temperature under equilibrium conditions.

6.6.1 Kinetic models

Several kinetic models exist, such as the Avrami, Elovich, layer diffusion, and Bangham models, the pseudo first-order and pseudo second-order models, the Boyd model, the kinetic model of intraparticle diffusion, etc [53].

The Avrami kinetic model: this model assumes that the reactions occur on the surface active sites of the solid support (equation (6.1)).

$$\theta_t = \theta_{av}(1 - \exp(Kt^n)) \tag{6.1}$$

θ_{av} = the Avrami theoretical adsorption amount (mg g^{-1}), θ_t = the amount of adsorbate in the adsorbent at a given time t (mg g^{-1}), t = time, K = the Avrami constant rate, and n = the Avrami order model.

The Elovich kinetic model: this model describes the adsorption of pollutants from aqueous solution as well as the adsorption kinetics of gaseous species on solids (equation (6.2)).

$$\theta_t = \beta \ln(\alpha\beta) - \ln(t) \tag{6.2}$$

θ_t = the fraction of the surface with adsorbed species at time t (mg g^{-1}), t = the contact time (min), α = the initial adsorption rate (mg g^{-1} min), and β = the sites available for adsorption.

The layer diffusion model: this model involves diffusion through a series of layers with different concentrations. A representative equation is presented below (equation (6.3)):

$$\frac{\mathrm{d}C}{\mathrm{d}t} = \frac{D}{r^2}\frac{\mathrm{d}}{\mathrm{d}r}\left(r^2\frac{\mathrm{d}C}{\mathrm{d}r}\right), \tag{6.3}$$

where C = concentration, t = time, D = diffusion coefficient, and r = the radius or thickness of the diffusion layer.

The Bangham kinetic model: this model accounts for the phase of the adsorption process characterized by sluggish diffusion through the pores. This equation is important for processes that are diffusion controlled (equation (6.4)).

$$\frac{\mathrm{d}\theta}{\mathrm{d}t} = k \cdot 1 - \theta^n \tag{6.4}$$

θ = the fraction of the surface with adsorbed species, t = time, α and β = rate constants that can also be represented by equation (6.5).

$$\log\left(\log\left(\frac{Ct}{Ci}\right) - Q_t m\right) = \log K_0 + Q_t \log\,(t) \tag{6.5}$$

C_t = the fixed bed outlet concentration of the solution at time t, C_i = the initial concentration of the adsorbate (mg l^{-1}), Q_t = the quantity of adsorbate on the adsorbent at a given time t (mg g^{-1}), m = the adsorbent mass in a given litre of adsorbate (g l^{-1}), K_0 = Bangham's model rate constant, and t = contact time (min).

The pseudo first-order kinetic model: the model most widely used to understand adsorption in liquid–solid systems is the Lagergren kinetic model. This can be expressed as (equation (6.6)):

$$\frac{\mathrm{d}Q}{\mathrm{d}t} = K_1 \cdot \left(Q_e - Q_t\right) \tag{6.6}$$

Alternatively, taking an integral of the differential equation and using the boundary conditions $qt = 0$ at $t = 0$ and $qt = qt$ at $t = t$ gives equation (6.7):

$$\ln Q_e - Q_t = \ln Q_e - K_1 t \tag{6.7}$$

Q_e = the quantity of adsorbate on the adsorbent at equilibrium (mg g^{-1}), Q_t = the quantity of adsorbate on the adsorbent at a given time t (mg g^{-1}), and K_1 = Lagergren's first-order rate constant.

The pseudo second-order kinetic model: this model accounts for an adsorption process that occurs on two surface sites. This model, also referred to as Blanchard's model, is expressed as a second-order differential equation (equation (6.8)).

$$\frac{\mathrm{d}Q}{\mathrm{d}t} = K_2 \cdot \left(Q_e - Q_t\right)^2 \tag{6.8}$$

Alternatively, taking an integral of the differential equation and using the boundary conditions $qt = 0$ at $t = 0$ and $qt = qt$ at $t = t$ gives equation (6.9):

$$t/Q_t = \frac{1}{k_2 Q_e^2} - \frac{1}{Q_e}t \tag{6.9}$$

Q_e = the quantity of adsorbate on the adsorbent at equilibrium (mg g^{-1}), Q_t = the quantity of adsorbate on the adsorbent at a given time t (mg g^{-1}), and K_2 = the pseudo second-order rate constant.

6.6.2 Isotherm models

Several isotherm models exist, such as the Langmuir, Freundlich, Bohart–Adams, Brunauer–Emmett–Teller, Dubinin–Radushkevich, Flory–Huggins, Frenkel–Halsey–Hill, Khan, Koble–Corrigan, MacMillan–Teller, Radke–Prausnitz, Redlich–Peterson, Sips, Temkin, Toth, Wolborska, Yoon–Nelson, Harkin–Jura, Halsey, and Elovich–Larionov models, etc [53]. The two isotherms most widely applied for the sequestration of pesticides, pharmaceuticals, and PFASs from aqueous solutions using magnetic sorbents are discussed below:

(a) **Langmuir's isotherm:** Langmuir's isotherm is widely used when considering environmental adsorption due to its capability to evaluate low concentrations of adsorbates and ease of coupling the model with transport phenomena equations for computational simulations. The model considers a solid absorbent of finite adsorption capacity in which each active site is uniform and capable of binding only one molecule of solute. No interaction takes place amongst the adsorbed molecular species (equation (6.10)).

$$Q_e = \frac{Q_L K_L C_e}{1 + K_L C_e} \tag{6.10}$$

Q_L = the maximum coverage capacity of the monolayer (mg g^{-1}), K_L = the Langmuir isotherm constant (l mg^{-1}); C_e = the equilibrium concentration (mg l^{-1}), and Q_e = the quantity of adsorbate in the adsorbent at equilibrium (mg g^{-1}).

(b) **The Freundlich isotherm:** this model characterizes adsorption on heterogeneous surfaces, considering the different energy levels of the adsorption sites. The model assumes a multilayer adsorption process in which the adsorbate molecules bind to the surface, forming layers. The surface energies are nonuniform, which has implications for the density of the sites as well as the interactions responsible for molecular adsorption. The isotherm can be expressed by equation (6.11):

$$Q_e = k_F (C_e)^{\frac{1}{n}} \text{ or } \ln Q_e = \ln K_F + \frac{1}{n}\ln(C_e) \tag{6.11}$$

K_F = the Freundlich isotherm constant (mg g^{-1}) (l g^{-1}); C_e = the equilibrium concentration (mg l^{-1}); and Q_e = the quantity of adsorbate in the adsorbent at equilibrium (mg g^{-1}).

6.7 Conclusions

The application of magnetic sorbents for the sequestration of pesticides, pharmaceuticals, and PFASs from aqueous solutions is promising. Magnetic sorbents have gained significant attention as potential candidate sorbent materials for environmental sustainability and remediation. This chapter provided a comprehensive evaluation of the different magnetic sorbents, synthesis methods, challenges, roles, sorption mechanisms, and factors influencing efficiency, contributing to a deeper understanding of their potential in addressing stringent environmental and health standards for different anthropogenic pollutants. The adsorption kinetics, isotherm models, and various adsorption mechanisms, including electrostatic, π-π interactions, van der Waals forces, hydrogen bonds, surface complexation, and hydrophobic interactions were also discussed. An ideal magnetic sorbent should be strongly magnetic and have excellent dispersion properties, a large specific surface area, appropriate porosity, good selectivity, stability, reusability, ease of preparation, low cost, and environmental friendliness. Notwithstanding the advantages of magnetic sorbents, certain disadvantages were thoroughly discussed in this chapter.

References and further reading

[1] Zhang D Q, Zhang W L and Liang Y N 2019 Adsorption of perfluoroalkyl and polyfluoroalkyl substances (PFASs) from aqueous solution—a review *Sci. Total Environ.* **694** 133606

[2] Gil A *et al* 2021 A review of organic-inorganic hybrid clay based adsorbents for contaminants removal: synthesis, perspectives and applications *J. Environ. Chem. Eng.* **9** 105808

[3] Chaukura N, Gwenzi W, Tavengwa N and Manyuchi M M 2016 Biosorbents for the removal of synthetic organics and emerging pollutants: opportunities and challenges for developing countries *Environ. Dev.* **19** 84–9

[4] Ortúzar M, Esterhuizen M, Olicón-Hernández D R, González-López J and Aranda E 2022 Pharmaceutical pollution in aquatic environments: a concise review of environmental impacts and bioremediation systems *Front. Microbiol.* **13** 1–25

[5] Rivera-Utrilla J, Sánchez-Polo M, Ferro-García M Á, Prados-Joya G and Ocampo-Pérez R 2013 Pharmaceuticals as emerging contaminants and their removal from water. a review *Chemosphere* **93** 1268–87

[6] Lipinski C A, Lombardo F, Dominy B W and Feeney P J 1997 Experimental and computational approaches to estimate solubility and permeability in drug discovery and development settings *Adv. Drug Deliv. Rev.* **23** 3–25

[7] Jaafar M, Alatabe A and Hussein A A 2021 Review paper. utilization of low-cost adsorbents for the adsorption process of chromium ions *IOP Conf. Ser. Mater. Sci. Eng.* **1076** 012095

[8] Quesada H B, Baptista A T A, Cusioli L F, Seibert D, de Oliveira Bezerra C and Bergamasco R 2019 Surface water pollution by pharmaceuticals and an alternative of removal by low-cost adsorbents: a review *Chemosphere* **222** 766–80

[9] Hacıosmanoğlu G G, Mejías C, Martín J, Santos J L, Aparicio I and Alonso E 2022 Antibiotic adsorption by natural and modified clay minerals as designer adsorbents for wastewater treatment: a comprehensive review *J. Environ. Manage.* **317** 115397

[10] Buck R C *et al* 2011 Perfluoroalkyl and polyfluoroalkyl substances in the environment: terminology, classification, and origins *Integr. Env. Assess. Manag* **7** 513–41

[11] Gluge J *et al* 2020 An overview of the uses of per- and polyfluoroalkyl substances (PFAS) *Environ. Sci. Process. Impacts* **22** 2345

[12] Fenton S E *et al* 2021 Per-and polyfluoroalkyl substance toxicity and human health review: current state of knowledge and strategies for informing future research *Environ. Toxicol. Chem.* **40** no 606–30

[13] Shreya , Verma A K, Dash A K, Bhunia P and Dash R R 2021 Removal of surfactants in greywater using low-cost natural adsorbents: a review *Surf. Interfaces* **27** 101532

[14] Sabzehmeidani M M, Mahnaee S, Ghaedi M, Heidari H and Roy V A L 2021 Carbon based materials: a review of adsorbents for inorganic and organic compounds *Mater. Adv.* **2** 598–627

[15] Laskar N and Kumar U 2022 Application of low-cost, eco-friendly adsorbents for the removal of dye contaminants from wastewater: current developments and adsorption technology *Environ. Qual. Manag.* **32** 209–21

[16] Baskar A V *et al* 2022 Recovery, regeneration and sustainable management of spent adsorbents from wastewater treatment streams: a review *Sci. Total Environ.* **822** 153555

[17] Obodo K and Aigbe U O 2023 The types, characteristics, and management options (reusability/recyclability/final disposal) of commonly used adsorbents in environmental sustainability *Adsorption Applications for Environmental Sustainability* (Bristol: IOP Publishing) pp 2-1–2-25

[18] Obodo K and Aigbe U O 2023 Cost and environmental evaluations and comparisons of commonly used sorbents *Adsorption Applications for Environmental Sustainability* (Bristol: IOP Publishing) pp 14-1–14-13

[19] Manaenkov O and Kislitsa O 2022 Synthesis of polymeric sorbents with magnetic properties *Int. J. Chem. Eng. Mater.* **1** 25–9

[20] Zhai Y, Zhai J, Zhou M and Dong S 2009 Ordered magnetic core-manganese oxide shell nanostructures and their application in water treatment *J. Mater. Chem.* **19** 7030–5

[21] Asgharinezhad A A, Karami S, Ebrahimzadeh H, Shekari N and Jalilian N 2015 Polypyrrole/magnetic nanoparticles composite as an efficient sorbent for dispersive micro-solid-phase extraction of antidepressant drugs from biological fluids *Int. J. Pharm.* **494** 102–12

[22] Giakisikli G and Anthemidis A N 2013 Magnetic materials as sorbents for metal/metalloid preconcentration and/or separation. a review *Anal. Chim. Acta* **789** 1–16

[23] Mhd. Haniffa M A C *et al* 2021 Cellulose supported promising magnetic sorbents for magnetic solid-phase extraction: a review *Carbohydr. Polym.* **253** 117245

[24] Anwar S *et al* 2019 Magnetic iron oxide nanoparticles: synthesis, characterization and functionalization for biomedical applications in the central nervous system ' *Magn. Iron Oxide Nanoparticles Synth. Charact. Funct. Biomed. Appl. Cent. Nerv. Syst.* **12** 465

[25] Moosavi S, Lai C W, Gan S, Zamiri G, Akbarzadeh Pivehzhani O and Johan M R 2020 Application of efficient magnetic particles and activated carbon for dye removal from wastewater *ACS Omega* **5** 20684–97

[26] Hu B, He M and Chen B 2019 Magnetic nanoparticle sorbents *Solid-Phase Extraction* (Amsterdam: Elsevier) pp 235–84

[27] Kubrakova I V, Koshcheeva I Y, Pryazhnikov D V, Martynov L Y, Kiseleva M S and Tyutyunnik O A 2014 Microwave synthesis, properties and analytical possibilities of magnetite-based nanoscale sorption materials *J. Anal. Chem.* **69** 336–46

[28] Obodo K O, Ouma C N M and Bessarabov D 2022 First principles evaluation of the OER properties of TM–X (TM = Cr, Mn, Fe, Mo, Ru, W and Os, and X = F and S) doped IrO_2 (110) surface *Electrochim. Acta* **403** 139562
[29] Aigbe U O, Onyancha R B, Kingsley B, Ukhurebor E and Obodo K O 2020 Removal of fluoride ions using a polypyrrole magnetic nanocomposite influenced by a rotating magnetic field *RSC Adv.* **10** 595–609
[30] Prasad C, Krishna Murthy P, Krishna R H H, Rao R S, Suneetha V and Venkateswarlu P 2017 Bio-inspired green synthesis of RGO/Fe_3O_4 magnetic nanoparticles using murraya-koenigii leaves extract and its application for removal of Pb(II) from aqueous solution *J. Environ. Chem. Eng.* **5** 4374–80
[31] Cai H, Feng J, Wang S, Shu T, Luo Z and Liu S 2019 Tannic acid directed synthesis of Fe_3O_4 @TA@P(NVP-co-NIPAM) magnetic microspheres for polyphenol extraction *Food Chem.* **283** 530–8
[32] Vállez-Gomis V, Grau J, Benedé J L and Chisvert A 2024 Magnetic sorbents: synthetic pathways and application in dispersive (micro)extraction techniques for bioanalysis *TrAC, Trends Anal. Chem.* **171** 117486
[33] Pryazhnikov D V and Kubrakova I V 2021 Surface-modified magnetic nanoscale materials: preparation and study of their structure, composition, and properties *J. Anal. Chem.* **76** 685–706
[34] Effenberger F B *et al* 2017 Economically attractive route for the preparation of high quality magnetic nanoparticles by the thermal decomposition of iron(III) acetylacetonate *Nanotechnology* **28** 115603
[35] Esaifan M *et al* 2019 Synthesis of hydroxy-sodalite/cancrinite zeolites from calcite-bearing kaolin for the removal of heavy metal ions in aqueous media *Minerals* **9** 484
[36] Chen L F, Liang H W, Lu Y, Cui C H and Yu S H 2011 Synthesis of an attapulgite clay@carbon nanocomposite adsorbent by a hydrothermal carbonization process and their application in the removal of toxic metal ions from water *Langmuir* **27** 8998–9004
[37] Zhang T *et al* 2021 Removal of heavy metals and dyes by clay-based adsorbents: from natural clays to 1D and 2D nano-composites *Chem. Eng. J.* **420** 127574
[38] Nolan N T, Serry M K and Pillai S C 2011 Crystallization and phase-transition characteristics of sol-gel- synthesized zinc titanates scopus—document details *Chem. Mater.* **353** 737–40
[39] Li Z *et al* 2023 An overview of synthesis and structural regulation of magnetic nanomaterials prepared by chemical coprecipitation *Metals (Basel)* **13** 152(1–19)
[40] Das A, Natarajan K, Tiwari S and Ganguli A K 2020 Nanostructures synthesized by the reverse microemulsion method and their magnetic properties *Mater. Res. Express* **7** 104001
[41] Ali A *et al* 2021 Review on recent progress in magnetic nanoparticles: synthesis, characterization, and diverse applications *Front. Chem.* **9** 629054
[42] Wang B, Wei Q and Qu S 2013 Synthesis and characterization of uniform and crystalline magnetite nanoparticles via oxidation-precipitation and modified co-precipitation methods *Int. J. Electrochem. Sci.* **8** 3786–93
[43] Iida H, Takayanagi K, Nakanishi T and Osaka T 2007 Synthesis of Fe3O4 nanoparticles with various sizes and magnetic properties by controlled hydrolysis *J. Colloid Interf. Sci.* **314** 274–80
[44] Ansari S A M K *et al* 2019 Magnetic iron oxide nanoparticles: synthesis, characterization and functionalization for biomedical applications in the central nervous system *Materials (Basel)* **12**

[45] Osman A I *et al* 2023 *Methods to Prepare Biosorbents and Magnetic Sorbents for Water Treatment: a Review* 21 (Cham: Springer International Publishing)
[46] Liu G *et al* 2018 Adsorption and removal of organophosphorus pesticides from environmental water and soil samples by using magnetic multi-walled carbon nanotubes @ organic framework ZIF-8 *J. Mater. Sci.* **53** 10772–83
[47] Feng Yin X, Yu Wang Q, Zheng Ren F, Fang Pang G, Xu Zhang X and Xuan Li Y 2022 Efficiency and mechanism of C18-functionalized magnetic nanoparticles for extracting weakly polar pesticides from human serum determined by UHPLC-QTOF-MS and molecular dynamics simulations *Environ. Pollut.* **293** 118489
[48] Jiang M, Yang W, Zhang Z, Yang Z and Wang Y 2015 Adsorption of three pharmaceuticals on two magnetic ion-exchange resins *J. Environ. Sci.* **31** 226–34
[49] Tararchuk T, Soltys L and Macyk W 2023 Magnetic adsorbents for removal of pharmaceuticals: A review of adsorption properties *J. Mol. Liq.* **384** 122174
[50] Lei X *et al* 2023 A review of PFAS adsorption from aqueous solutions: current approaches, engineering applications, challenges, and opportunities *Environ. Pollut.* **321** 121138
[51] Sörengård M, Östblom E, Köhler S and Ahrens L 2020 Adsorption behavior of per- and polyfluoralkyl substances (PFASs) to 44 inorganic and organic sorbents and use of dyes as proxies for PFAS sorption *J. Environ. Chem. Eng.* **8** 103744
[52] Ahrens L, Taniyasu S, Yeung L W Y, Yamashita N, Lam P K S and Ebinghaus R 2010 Distribution of polyfluoroalkyl compounds in water, suspended particulate matter and sediment from Tokyo Bay, Japan *Chemosphere* **79** 266–72
[53] Benjelloun M, Miyah Y, Akdemir Evrendilek G, Zerrouq F and Lairini S 2021 Recent advances in adsorption kinetic models: their application to dye types *Arab. J. Chem.* **14** 103031
[54] Querol X, Alastuey A S, Pez-Soler A L and Plana F 1997 A fast method for recycling fly ash: microwave-assisted zeolite synthesis *Environ. Sci. Technol.* **31** 2527–33
[55] Siqueira Oliveira A M, Paris E C and Giraldi T R 2021 GIS zeolite obtained by the microwave-hydrothermal method: synthesis and evaluation of its adsorptive capacity *Mater. Chem. Phys.* **260** 124142

IOP Publishing

Environmental Applications of Magnetic Sorbents

Kingsley Eghonghon Ukhurebor and Uyiosa Osagie Aigbe

Chapter 7

The application of magnetic sorbents in soil decontamination

Onyedikachi Ubani, Sekomeng Johannes Modise, Harrison Ifeanyichukwu Atagana and Kevin Frank Mearns

The challenge of effectively removing toxic industrial waste pollutants has led to the popularity of soil remediation through sorbent amendment. To perform quantitative retrieval from soil, magnetically retrievable sorbents have been employed in soil remediation to provide effective sorption sites for pollutants. This eliminates the gradual leaching of contaminants and enhances remediation efficiency by removing hazardous substances. Magnetic sorbents, particularly magnetic nanoparticles (MNPs), significantly impact the speed, enrichment factor, anti-interference capability, selectivity, and reproducibility of magnetic solid-phase extraction (MSPE)-based methodologies. This chapter explores the synthesis and potential applications of MNP sorbents in the removal of pollutants from soil.

7.1 Introduction

Soil pollution, a consequence of human activities such as mining, pesticide use, and waste disposal, leads to environmental impacts such as soil quality degradation, which affects agriculture and ecological balance. Persistent pollutants, especially heavy metals (HMs), pose enduring challenges [1]. Soil, a vital ecosystem for human sustenance, is a significant reservoir of potentially toxic elements (PTEs), which can compromise human well-being and environmental safety. Global environmental concerns have heightened because of industrialization, urbanization, and increased chemical usage, which have led to an influx of pollutants such as HMs and antibiotics. Even at low concentrations, these pollutants can harm human health, cause diseases, and compromise the immune and reproductive systems [2–4] (figure 7.1). Research shows that HMs, prevalent in smelting sites, can disperse into the soil through leaching, causing deterioration and hindering various land uses [2]. Dissolved organic matter (DOM) in soil, a dynamic organic component, is crucial for maintaining soil

doi:10.1088/978-0-7503-5909-2ch7

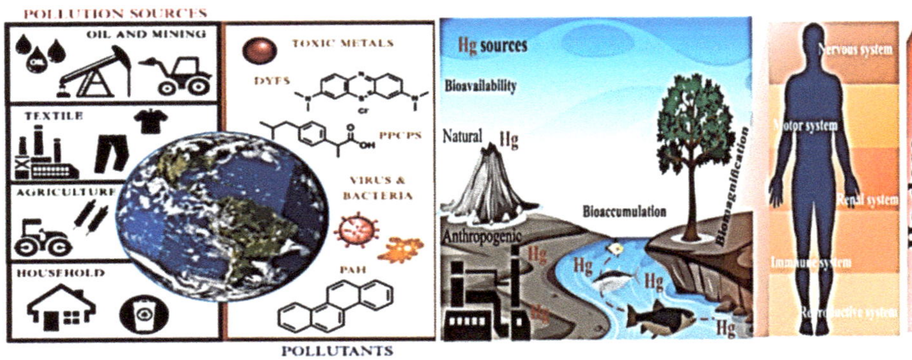

Figure 7.1. Pollutants' source, fate, and effect on human health (e.g. metals such as Hg). Reprinted from [4], Copyright (2022), with permission from Elsevier.

fertility and buffering PTEs. Mechanisms such as hydrogen bonding and complexation regulate the fate of PTEs, and DOM acts as an electron donor and acceptor in redox reactions. However, the specific interactions between soil DOM and PTEs remain incompletely understood [2]. Addressing environmental pollution requires urgent and practical solutions. Thus, magnetic sorbents (MSs) with porous structures are emerging as a promising technology because of their diverse origins, exceptional porosity, surface functional groups, and cost-effectiveness [3].

7.2 Magnetic sorbents

MSs, including nanoparticles (NPs) and nanocomposites, are crucial. These sorbents integrate magnetic particles into effective sorbents such as metal oxides (MeOs), silica-based materials, and diverse polymers, offering advantages such as high surface-area-to-volume ratios and increased active sites [5]. Micro and nanomaterials (NMs) enhance the effectiveness of MSPE methods by improving parameters such as separation speed, selectivity, and reproducibility. Various materials, such as silica-based materials, carbon-based materials, and metal/MeOs, have been incorporated into these sorbents to broaden their applicability [5]. On the other hand, activated carbon is commonly used for soil and sediment purification, and powdered activated carbon exhibits superior sorption kinetics. However, separating it from the cleaned matrices poses challenges. Magnetic activated carbon, or magnetic biochar (MBC), presents a promising solution by combining magnetic materials with biochar, allowing efficient reclamation from cleaned matrices [6]. MBC retains the qualities of biochar and offers the benefits of magnetic separation. Its use in soil remediation involves complex mechanisms impacting pollutant binding and soil fertility, necessitating a comprehensive understanding of magnetic modification trade-offs [6]. Researchers have explored the magnetization of sorbents to overcome biochar recovery limitations, addressing challenges in biochar applicability and potential secondary pollution during recovery from environmental matrices [6].

Soil remediation typically involves bringing a soil suspension into contact with a magnetic sorbent, using magnetic forces for separation, recovering the soil, and reprocessing the used sorbent [7]. In addressing the challenges posed by potential secondary pollution during recovery from environmental matrices, magnetic particles have proven effective for sorption and selective separation. Nano-to-micro-sized particles, particularly sorbents, are relevant because of their selective interaction and easy separation. Ideal sorbents should exhibit stability, a large surface area, functional groups, and cost-effectiveness. MSs, which offer swift collection and recyclability, stand out for their efficiency and unique properties, making them excellent candidates for diverse applications [8].

7.3 Types of magnetic sorbent

The classification and significance of MSs encompass diverse criteria, including synthesis techniques and applications. Various synthesis methods yield efficient sorbents, such as micro-emulsions, self-assembly, emulsion evaporation and hydrothermal, coprecipitation, sonochemical, and ultrasonic methods. In the hydrothermal method, which is simple and cost-effective, the reaction rate, particle size, and surface structure are influenced by temperature. The sol–gel technique is favored for its low calcination temperature and small particle size, whereas sonochemistry excels in its adaptability to mild conditions [9]. Increased sorption capacity, material stability, and pollutant selectivity is achieved by blending MNPs with other materials through functionalization, addressing limitations caused by the restricted functional groups of MNPs impacted by aggregation. Magnetic composite (MC) materials such as graphene oxide (GO), carbon nanotubes (CNTs), aptamers, metal–organic frameworks (MOFs), and layered double hydroxides effectively serve as sorbents. Unlike nonmagnetic materials, MCs offer advantages such as adaptable remediation [9].

Fe_3O_4-based MNPs, prepared through various techniques, are vital and have easily functionalized surfaces for extracting targeted pollutants. Magnetic molecularly imprinted polymers (MMIPs) require a complex preparation process that can lead to effective pollutant removal. They should be selective, stable, and have an elevated sorption ability for efficient MSPE [9]. As such, ongoing developments are focused on functionalizing MNPs with molecularly imprinted polymers (MIPs), covalent organic frameworks (COFs), and restricted-access materials (RAMs) to enhance sorption and separation efficiencies, expanding their applications. These materials are crucial in extracting various contaminants from complex matrices, showing potential in sample preparation [9].

While MOFs, COFs, MIPs, and RAMs have exhibited the successful extraction of diverse pollutants from various sources, specific sorbents, such as aptamers, functionalized dysprosium–Fe_3O_4 NPs, and magnetic copper (Cu) and lanthanide oxides, exhibit selectivity in extracting specific compounds from different biological samples. Efforts have been concentrated on evolving sorbents such as RAM–MIP for pollutant removal via MSPE [9]. These materials treat pollutants effectively because of their stability and functional group availability. Research primarily revolves around the surface modification of sorbents to enhance hydrophilic wettability via discrete functional groups. The development and utilization of various sorbents, such as Fe–Al/GO, also show excellent arsenic (As) absorption capacities. Magnetic foams, sponges, and hydrogels modified for hydrophilic or hydrophobic wettability, efficiently remove specific contaminants from diverse samples [9]. Ongoing efforts aim to enhance the recycling efficiency of magnetic materials by increasing their reuse cycles, especially in applications such as hydrocarbon remediation and oil–water separation.

The application of MSs, particularly in MSPE, stands out for its ability to separate even minute concentrations of target contaminants. Techniques such as solid-phase extraction (SPE), MSPE, and magnetic solid-phase microextraction (MSPμE) are crucial for isolating pollutants from complex systems. MSPE, an evolution of SPE, employs magnetic materials as sorbents, allowing for easy recovery [9].

Moreover, expended MSs are reusable, promoting environmental friendliness and cost savings. MSPE, compared to SPE, relies on magnetic materials as sorbents for targeted pollutant removal from complex mixtures, offering advantages such as rapid separation, high efficiency, sorbent reusability, and ease of operation. Advanced MCs with enhanced sorption and removal proficiency have been emphasized. Thus, different Fe_3O_4-based functionalized MNPs and MCs have been designed to enhance pollutant selectivity and effective extraction in MSPE. The use of suitable MSs significantly enhances MSPE performance compared to the performance of traditional SPE [9, 10].

7.4 Methods for magnetic sorbent production

The synthesis of MS involves various methods, each of which influences the physicochemical properties and sorption capacity of the resulting MBC. This overview examines biomass and prominent techniques (figure 7.2), such as coprecipitation, impregnation pyrolysis, reductive co-deposition, hydrothermal carbonization, and other emerging techniques [3, 11, 12].

7.4.1 Impregnation pyrolysis

In this one-step process, biomass impregnated with transition-metal salts undergoes pyrolysis in a restricted-oxygen or passive atmosphere. Operating parameters such as the pyrolysis temperature, the inert gas used, and the pyrolysis time tightly control the physicochemical properties of the resulting MBC, ensuring versatility and efficiency [3, 11, 12].

7.4.2 The pyrolysis temperature

The pyrolysis temperature significantly affects the production and quality of MBC. Diverse temperatures influence its superficial section, aperture, functional groups, and magnetization. Generally, the magnetic properties of MBC improve with increasing pyrolysis temperatures, but excessively high temperatures may diminish the number of functional groups. The raw material choice directly influences the pyrolysis temperature, impacting the material composition and pollutant sorption efficiency [12].

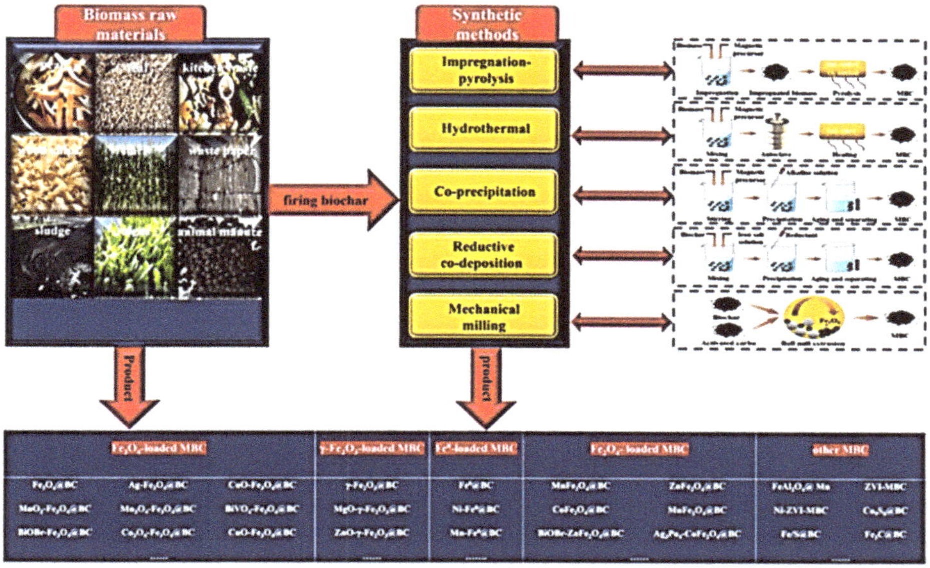

Figure 7.2. The synthesis of MBC by different methods. Reprinted from [3], Copyright (2022), with permission from Elsevier.

7.4.3 Coprecipitation

This procedure involves dissolving biochar in a modified metal solution, adding sodium or ammonium hydroxide, and stirring at a specific pH. Although more complex than impregnation pyrolysis, coprecipitation provides better control, ensuring stable adhesion of the magnetic medium to the biochar matrix. The choice between coprecipitation and impregnation pyrolysis depends on the target pollutant characteristics [3, 11, 12].

7.4.4 Reductive co-deposition

This is similar to coprecipitation but with added reduction by sodium or potassium borohydride agents; this method yields MBC with zero-valent metals, enhancing pollutant removal. However, safety risks are associated with the use of harmful reductants and hydrogen production during the reduction process, especially at scale [3, 11, 12].

7.4.5 Hydrothermal carbonization

This method involves reacting the biomass with a metal ion solution under milder conditions (100 °C–300 °C) and pressure generated by the reaction. Hydrothermal carbonization offers a more attractive and sustainable alternative to traditional methods because of its lower temperature requirements and the elimination of alkalis or strong reductants [3, 11, 12].

7.4.6 Other preparation methods

Recent advancements include ball milling, the direct pyrolysis of biomass/metal salts, cross-linking, and microwave-assisted pyrolysis. While coprecipitation, impregnation pyrolysis, and hydrothermal carbonization remain popular, the choice of method depends on the kind of raw material, the pollutant's physicochemical properties, and operability considerations [3, 11, 12]. Selecting an appropriate synthetic approach for MBC involves careful consideration of the raw material, pollutant characteristics, and method operability. The diverse methods discussed offer flexibility in tailoring MBC properties for effective pollutant removal in environmental applications [3, 11, 12].

7.5 Innovations in magnetic nanomaterials for soil remediation

NMs exhibit superparamagnetism; they are also small, recyclable, eco-friendly, and stable; they have low cytotoxicity and robust sorption capabilities and hold immense promise for remediating contaminated soil. By leveraging their exceptional magnetic attributes, nanostructured magnetic NPs and nanocomposites have proven highly effective in soil pollution remediation. Specifically, superparamagnetic iron oxide NPs (SPIONPs) excel in extracting pollutants from soil because of their nanoscale dimensions, enabling the efficient sorption of contaminants [13]. The magnetic behavior of pollutant–nanomagnetite aggregates allows accessible attraction by an external magnetic stimulus. SPIONPs are often modified or functionalized with chemical and biological entities to enhance their stability and properties. The combination of particle size, stability, the superparamagnetic characteristics of Fe-NPs, and their prompt reaction to magnetic fields makes SPIONs highly effective materials for soil treatment [10, 13].

Nanotechnology, operating at the 1–100 nm scale, has sparked interest in various applications, particularly the synthesis of magnetic NMs (MNMs). MNMs, exemplified by SPIONPs, possess unique properties such as superparamagnetism, lower coercivity, and extraordinary magnetic susceptibility compared to mass materials. This transition to superparamagnetism occurs because of the nanoscale magnetic domain size, which modifies internal structures and causes magnetic moment fluctuations in response to thermal energy [13]. MNMs, including Co, Fe, Ni, Mn, maghemite, magnetite, and mixed ferrites, can be synthesized in various forms, such as core–shell configurations, ferrofluids, and magnetic nanocomposites. These materials respond to external magnetic fields, offering versatility in terms of characteristics and applications. Their selection criteria include hydrophobicity, osmophilicity, crystallinity, dispersity, and magnetic recoverability [13].

Various synthesis methods for MS have been explored, including coprecipitation, microemulsion, hydrothermal processes, sol–gel methods, and the thermal decomposition of organometallic precursors. SPIONPs, particularly magnetite and maghemite, are widely used because of their superparamagnetic nature, ease of synthesis, and biocompatibility. Their integration with various chemical and biological components further enhances their properties [13]. The synthesis of SPIONPs involves the slow addition of NH_4OH to a ferrous oxide mixture. Maghemite NPs are obtained by oxidizing SPIONPs with $Fe(NO_3)_3$. These SPIONPs are the foundation for magnetic nanocomposites, which combine nanosized magnetic materials with host matrices such as organic polymers, metals, ceramics, and liposomes. These multiphase substances exhibit large saturation magnetization, magnetic moments, susceptibility, and anisotropy [13]. The chemical equations representing these processes are shown in equations (7.1) and (7.2):

$$Fe^{3+} + Fe^{2+} + 8OH^- \rightarrow Fe(OH)_2 + Fe(OH)_3 \rightarrow Fe_3O_4 + 4H_2O \quad (7.1)$$

$$Fe_3O_4 + O_2 \rightarrow 6Fe_2O_3 \quad (7.2)$$

Magnetic CNTs offer a versatile platform for attaching MNPs and enhancing their surface functionality. The decoration of NPs on CNTs is achieved through various techniques, making them versatile in electrical, thermal, and mechanical applications. Iron oxide NPs and nanocomposites, known for their magnetic properties and osmophilicity, have extensive applications in modern research, particularly in soil remediation [13].

7.6 Advanced sorbent materials for efficient microextraction and pollution remediation

As sorbents, micro- and NMs offer an extraordinary surface-area-to-volume ratio, unique size characteristics, and increased active sites for rapid equilibrium attainment. MSPE using MNPs such as Co, Ni, and Fe and metal oxides such as maghemite (γ-Fe_3O_4) and magnetite (Fe_3O_4) serves as an efficient technique. The functionalization of MNPs is crucial for analyte interaction, preventing oxidation and aggregation, and preserving their magnetic properties. Various materials such as graphene, phenyl, C8, C18, and CNTs have been employed to functionalize MNPs or magnetic particles [5].

7.7 Innovative MSPE for pollution remediation

MSPE is an innovative approach to pollution separation challenges; it involves the sorption of elements on a magnetic sorbent followed by phase separation using a permanent magnet. Recent developments include nanocomposites such as MNP@CNT, which exhibit a synergistic combination of the following characteristics: extraordinary sorption capacity, rapid mass-exchange kinetics, comprehensive sorption and desorption, various regeneration potentials, and supermagnetism. Due to its simple synthesis, cost-effectiveness, and accessibility, MNP@CNT has emerged as an appealing MC for pollution remediation [14].

The development of scalable sorbent materials with robust capacity for metal ions, organic matter, and unwanted entities is imperative to address contaminant removal challenges. Materials such as MOFs, metal-oxide NPs, and layered double hydroxides (LDHs) stand out because of their stability, low leaching, reusability, and versatile functional groups [8]. Current research is focused on enhancing hydrophilicity and sorption capacity through diverse functional group modifications. Examples include the synthesis of Fe–Al-layered multihydroxide/reduced GO intercalated with $NaC_6H_7O_6$ nanocomposites, which exhibit exceptional elimination efficiency (>98%) and high maximum sorption capacity (190.84 mg g^{-1}) for As (V). Hydrogels, beads, sponges, and foams modified for hydrophilic or hydrophobic wettability efficiently remove specific contaminants from diverse samples. Superhydrophobic foams, for instance, demonstrate remarkable efficiency in oil/organic solvent removal and have notable recyclability [8].

7.8 Magnetism-based sorbent classification: strategies for enhanced performance

MNPs with a magnetic core often face challenges such as aggregation, autoxidation, and limited target selectivity under ambient conditions. Solving these issues involves modifying the magnetic core by depositing protective chemical compounds in the shell. This shell can be organic (polymers and surfactants) or inorganic (silica, carbon, and noble metals); its purpose is to enhance stability and sorption capacity, preventing aggregation and oxidation. These chemical compounds can also be coupled with chelating agents, making the modified MNPs useful for various environmental applications [15].

7.8.1 Surface modification and functionalization

An alternative approach involves the *in situ* encapsulation of MNPs with diverse substrate materials, such as porous carbonaceous materials or inorganic clay, to create magnetically retrievable composites. These composites exhibit exceptional pollutant removal performance that is unachievable by the individual substrate components. Another strategy includes the preparation of magnetic polymer beads with uniformly distributed MNPs. This approach broadens the scope of magnetism-based composites, using carbonaceous materials, polymers, biomaterials, and inert oxides to enhance pollutant removal efficiency [15].

7.8.2 Classification criteria

MSs can be categorized based on the type of shell and porous substrate used. This classification system accounts for the different strategies employed in modifying MNPs, thereby enhancing their performance for environmental applications [15].

7.9 Surface-coated magnetism-based sorbents for enhanced contaminant removal

Surface-coated magnetism-based sorbents play a pivotal role in addressing the challenges associated with the stability and selectivity of MNPs. This section explores various inorganic and organic coating methods, including silica, carbon, polymers, carbonaceous substrates, and inorganic clay, used to enhance pollutant removal efficiency [15].

7.9.1 Inorganic coating methods

Silica coating: a widely employed method involves silica coating through the Stober process, which offers advantages such as chemical stability, low cost, and controlled porosity. Silica-coated MNPs exhibit enhanced sorption capability, but their susceptibility to degradation under typical conditions poses challenges. Various studies have demonstrated the efficacy of silica-coated MNPs in pollutant removal [15].

7.9.2 Carbon coating

Carbon-coated MNPs synthesized through chemical vapor deposition (CVD) and pyrolysis have better stability against oxidative degradation than silica-coated MNPs. Hollow Fe_3O_4 MNPs coated with mesoporous carbon (h-Fe_3O_4@mC) exhibited efficient pollutant sorption due to its mesoporous structure and carboxyl-functionalized surface [15].

7.9.3 Organic coating methods

Polymeric coatings: polymers with functional groups such as COOH, SO_4, and PO_4 can be chemically linked or physically sorbed onto magnetite particles, forming protective layers. Chitosan, a cost-effective and biocompatible biopolymer, was used to create ethylenediamine-modified magnetic chitosan (EMMC) complexes for HM removal. In addition, magnetic particles coated in ion-imprinted polymer (IIP) with embedded g-methacryloxypropyltrimethoxysilane (g-MPS) show promise for selective uranyl extraction [15].

7.9.4 Carbonaceous substrates

Carbonaceous substrates, including activated carbon, multiwalled CNTs (MWCNTs), mesoporous carbons, and GO, have garnered attention for their high specific surface areas, chemical inertness, and thermal stability. GO-based magnetite composites with anchored technology exhibit enhanced sorption efficiency for contaminants, especially cesium (Cs) [15].

7.9.5 Inorganic clay

The layered morphology and surface charge of clay minerals provide stable sites for magnetite nanoparticle loading. Various clay minerals, such as kaolinite, bentonite, and montmorillonite, have been used to prepare MCs for effective water contaminant sorption. The surface modification of magnetic Mg–Al layered double hydroxide with citrate acid enhances uranium (U(VI)) removal efficiency [15]. Surface-coated magnetism-based sorbents offer diverse strategies for enhancing pollutant removal, each with unique advantages and applications. The choice of coating method depends on factors such as cost, stability, and the specific pollutants targeted in environmental remediation efforts [15].

7.10 Deciphering soil contamination: unveiling mechanisms and widespread contaminants

Soil contamination is a critical aspect of soil degradation and has far-reaching consequences for water quality, food safety, and human health. The sources of contamination are diverse, arising from natural and human-induced activities, including industrial processes, agriculture, and mining. Their impacts extend beyond the soil, affecting groundwater, surface water, oceans, and the atmosphere [1–3].

7.10.1 Common contaminants and their environmental impacts

Soil contaminants encompass a range of pollutants such as HMs, radioactive elements, organic compounds, inorganic salts, and emerging pollutants, which pose significant environmental and human health risks. The soil's role as a contaminant repository contributes to issues such as pesticide contamination of groundwater, nitrate and phosphate pollution of surface water, microplastic pollution in oceans, and the emission of greenhouse gases into the atmosphere [16–19].

7.10.1.1 Soil contamination by various pollutants

Specific contaminants, such as HMs (Cu, Cr, As), organic compounds (pentachlorophenol, dioxins, furans), and radioactive elements, present unique challenges. Soil rich in nitrates poses a carcinogenic risk, and the growth of nuclear power raises concerns about radioactive soil contamination [16–19].

7.10.1.2 The mechanisms and pathways of soil contamination

Contaminants undergo natural biogeochemical and human-induced processes as they cycle through different environmental spheres. Physical, chemical, and biological conditions influence the movement of contaminants. Geological origins, biowaste disposal, and agrochemical use are identified as significant pathways for soil contamination [1–3, 16].

7.10.1.3 Soil's role in contaminant sequestration

Soil properties, including those of minerals and organic matter, are vital in retaining organic and inorganic contaminants. Sorption and immobilization mechanisms involve the retention of contaminants by soil particles and are influenced by factors such as pH, surface charge, and specific surface area. Chemical mobilization, speciation, and transformation are governed by pH and redox potential [1–3, 16].

7.11 The significance of soil decontamination

Soil plays a dual role as a contributor and a recipient of pollutants originating from natural processes and human activities. Natural sources yield inorganic pollutants, such as As, from weathering, whereas human actions introduce a mix of organic and inorganic contaminants, such as pesticides, HMs, and explosive residues. Soil acts as a mediator, transforming and distributing pollutants across environmental compartments, affecting their mobility and ecological impact [16]. Various contaminants enter soils through intentional use, improper disposal, accidents, and leaks. Inorganic compounds such as nitrates, phosphates, and perchlorates stem from manufacturing residues, while hydrocarbons and herbicides originate from spills and industrial activities. HMs, a common soil pollutant, often result from industrial processes, mining, and improper waste handling, posing varying toxicity levels [20, 21]. Rising industrial output continues to inundate the environment, necessitating efforts to reuse contaminated sites while enforcing stricter pollution control. However, effective, existing remediation methods such as excavation and chemical treatments are hindered by high costs, especially in developing nations. Therefore, there is a growing need for more cost-efficient remediation approaches [21].

Comprehensive research into contaminant behaviors under real-world conditions has taken place with the aim of enhancing soil remediation. This involves studying soil properties and mechanisms such as hyperaccumulation and hypertolerance and identifying more effective remediation strategies by thoroughly analyzing soil conditions and contaminants. Combining different remediation methods could offer a more potent decontamination process, particularly in sites with multiple contaminants [22]. Identifying contamination sources is crucial for selecting optimal soil remediation methods based on efficiency, biodiversity impact, and cost. MSs, which are versatile in application, immobilize inorganic pollutants and aid in mobilizing organic pollutants, offering sustainable remediation. While bioremediation is practical for various pollutants, it struggles with per- and polyfluoroalkyl substance (PFAS)-contaminated soil, necessitating the exploration of better methods. Combining techniques, especially NPs, in bioremediation shows promise but requires thorough environmental risk assessment. Studies of MSs have conducted sorption experiments for different pollutant classes, aiming to understand their mechanisms, potential modifications, and effectiveness in adsorbing hazardous pollutants, considering various influencing factors [16, 20–22].

7.12 Magnetic sorbents for heavy metal removal: opportunities and challenges

HM ions severely threaten human health because of potential biomagnification, leading to organ damage. Effective HM contamination mitigation is crucial for addressing ecological risks. Various methods, including electrokinetic remediation, sorption, and immobilization, have been explored. Sorption has gained prominence, particularly by employing cost-effective, highly efficient, and easily recoverable sorbents. Natural minerals such as clay, zeolite, and metal oxyhydroxides have been favored due to their economic feasibility. Magnetite and other materials have been explored for creating magnetically separable sorbents, addressing HM cation and oxyanion sorption challenges. Synthesized phosphate-modified magnetite@ferrihydrite sorbents show promise for efficient HM removal, as demonstrated by the sequestration of Cd(II) [23].

7.12.1 The advantages of magnetic sorbents

Recent advancements in MNMs offer environmentally friendly mechanisms for efficient soil contaminant removal. MSs that have low surface energy, improved surface properties, biodegradability, non-sinking behavior, easy recovery, reusability, and high sorption capacity are ideal sorbents. Fe_2O_3@C NPs exhibit unique properties, achieving reusability and recyclability through the use of an external magnetic field. Chitosan-capped MNPs demonstrate exceptional recyclability, maintaining effectiveness through multiple cycles. Even after functionalization, MNPs retain their superparamagnetic character, thereby emphasizing their benefits.

Choosing a suitable sorbent is crucial for pollutant separation, and MSs enable immediate separation. Hydroxyapatite (HAP), derived from calcium-rich animal waste such as seashells and cuttlefish bones, is effective due to various sorption mechanisms. A magnetic hydroxyapatite composite (MHAP) synthesized from snail shell powder using the hydrothermal method displayed the potential for sustainable pollutant separation [24, 25].

7.13 The applications of magnetic sorbents for heavy metal removal

7.13.1 The mechanisms of heavy metal sorption

Iron oxide NPs with superparamagnetic properties are effective sorbents for HMs because of their substantial surface area, superparamagnetic characteristics, prevention of agglomeration, and easy removal using an external magnetic field. Small-molecule affinity ligands can enhance the specificity of sorbents for HMs [26].

7.13.2 Advances in magnetic nanoparticles for analyte separation

Recent developments in MNPs have focused on enhancing their sorption capacity, selectivity, and stability in acidic environments. Various synthesis methods have been explored, including hydrothermal, sol–gel, and sonochemical methods. The functionalization and blending of MNPs with materials such as surfactants, CNTs, and MOFs have addressed challenges such as agglomeration and reduced sorption capacity, creating diverse magnetic nanocomposites [8].

7.13.3 Selective sorption materials for contaminant removal

MSs, including hydrogels, sponges, beads, and foams, are crucial in extracting, preconcentrating, and removing diverse analytes. Materials such as aptamers, MIPs, and RAMs perform selective preconcentration tasks. Graphene, CNTs, MOFs, and MeO excel in preconcentration and removal roles. Unique selective sorbents, including modified dysprosium-based NPs and magnetic lanthanide oxides, highlight the versatility of MSs [8].

7.13.4 Metal ion separation using modified chitosan

Chitosan, a natural polysaccharide derived from crustacean shells, is a cost-effective biosorbent with the potential for metal ion sorption. Cross-linking chitosan with chemical agents such as epichlorohydrin and glutaraldehyde enhances acid resistance but may reduce sorption ability. Modifying chitosan with grafted functional groups improves its sorption capacity. In a study of Cr(VI) removal, chemically modified chitosan from shrimp shell waste, crosslinked with Schiff's base and augmented with Fe_3O_4, demonstrated enhanced efficacy in magnetic phase separation [27].

7.14 Application challenges in different soil environments

Comprehensive evidence suggests that adding BC can impact soil hydrophobicity and that its hydrophilic or hydrophobic nature is contingent upon distinct surface functional groups. For instance, the hydroxyl (OH) and alkyl aliphatic (CH) groups within BC can induce either hydrophilic or hydrophobic behavior [1].

7.14.1 Factors affecting soil hydrophobicity

The sorption or degradation of hydrophobic organic compounds facilitated by BC contributes to the modification of soil hydrophilicity. This, in turn, diminishes soil water repellency, amplifies water content, and augments aeration, ultimately enhancing soil fertility. The alkaline nature of BC also plays a pivotal role by elevating soil pH, which is particularly beneficial for correcting the acidity of infertile soils such as those found in mining areas [1].

7.14.2 The effects of biochar on soil's physical properties

A review of BC application reveals noteworthy improvements in various soil physical properties. The addition of BC correlates with enhanced soil bulk density (an average increase of 7.5%), increased aggregate stability (an average rise of 8%), augmented available water content (AWC) (by approximately 15%), and elevated soil conductivity (by an average of 25%). Moreover, the size of BC particles influences the extent of their impact on soil properties, and significant benefits are observed in terms of moisture storage capacity [1].

7.14.3 Biochar's composition and functional groups

The chemical composition of BC, including its functional groups, significantly contributes to soil modification. Biochar addition increases the soil pH, cation exchange capacity (CEC), and nutrient content. This adjustment proves invaluable in combating soil depletion resulting from intensified agricultural practices. For instance, the use of straw-derived BC in loamy sand soils demonstrated substantial enhancements in bulk density, porosity, and AWC, and optimal results were obtained at a 4% addition rate [1].

7.14.4 Optimal ratios and sustainability

Despite the positive impacts, caution is warranted in determining the optimal BC-to-soil ratio. Adverse effects on soil respiration rates were noted with an 8% BC addition, underscoring the need for precise investigations into sustainability criteria.

The ratio's efficacy depends on BC properties such as carbon content, minerals, and HMs. Surface modifications, such as acid or alkaline treatments, offer novel properties for specific applications, improving BC performance in pollutant sorption and immobilization [1].

7.14.5 Mineral element incorporation

Incorporating essential chemical elements such as Ca, Fe, and Al into BC structures intended for transfer into the soil is crucial for effective soil amendment. Experiments involving incubation with representative soil minerals demonstrated the formation of organometallic compounds on or within BC, highlighting its potential as an effective sorbent. The type and concentration of mineral elements in BC depend on the feedstock, necessitating a tailored approach for optimal results [1]. To harness the full potential of BC for large-scale applications, it is imperative to understand the specific soil properties requiring modification. Experiments are essential to determine the ideal BC-to-soil ratio, ensuring optimal results while mitigating potential ecological consequences. Striking a balance is crucial, because excessive BC-to-soil ratios, particularly with metal-rich BCs, may pose risks such as releasing HMs into the environment [1].

7.15 Advances in magnetic sorbent design for environmental applications

Recent sorbent design and synthesis developments have emphasized the need for MSs that meet specific criteria. These include superhydrophobicity, osmophilicity, high sorption capacity, sustained retention, ease of application and recovery, biodegradability, cost-effectiveness, and convenient storage [13].

These are addressed in the following sections.

7.15.1 Nanotechnology-based magnetic sorbents

A novel nanotechnology-based approach that focuses on magnetic separation has been introduced. This technology allows for dispersion onto the soil via waterborne or aerial methods. From an environmental perspective, MNMs exhibit eco-friendly properties, are easily recoverable, are predominantly hydrophobic and oleophilic, and cause minimal environmental harm. The cost-effectiveness of MNMs, especially on a larger scale, further supports their viability [13].

7.15.2 Challenges and considerations

Despite the promising attributes of MSs, challenges persist in manufacturing high-quality, economically feasible, and stable sorbents while maintaining crucial characteristics such as high performance and nontoxicity. It is imperative to address practical issues, such as the required mass of sorbents for pollutant removal and the development of effective recovery techniques. Furthermore, extensive research is required to determine the optimal strength of the externally applied magnetic field [13]. Advancements in magnetic sorbent technology offer a promising avenue for addressing environmental concerns. However, the practical implementation of these sorbents will require the solution of the challenges related to mass requirements, recovery techniques, and determining the optimal magnetic field strength. The ongoing research in this field is critical for unlocking the full potential of MSs for environmental applications [13].

7.16 Techniques for enhancing magnetic biochar sorption performance

The efficacy of MBC sorption is intricately linked to its physical and chemical attributes. The magnetic medium usually hinders the pores of BC, resulting in low surface area and small porous structure, which makes unmodified MBC have relatively poor adsorption performance for contaminants. To address this, various modification methods have been employed to enhance the surface properties of MBC, including caustic and basic modification, oxidation modification, superficial efficient modification, nanoparticle loading, and chemical doping modification [3, 28].

7.16.1 Modification methods

The purpose of acid and alkaline modification is to eliminate pollutants such as HM ions and create caustic-containing functional groups (amines and carboxyls). It is implemented by improving MBC affinity for HMs via ion donation and complexation.

7.16.2 Oxidation modification

The purpose of oxidation modification is to enhance pore structures and increase functional groups on the MBC surface. It is implemented using Fenton and electrochemical oxidation. It is regarded as a novel strategy for creating highly adsorptive MBCs with improved pollutant removal capacity.

7.16.3 Surface functional modification

Surface functional modification influences the permeability and sorption execution of MBC. Creating functional groups with heteroatoms of N, O, or S enhances HM removal efficiency. It is crucial for improving MBC's sorption capacity for pollutant removal.

7.16.4 Nanoparticle loading

NP loading is carried out to improve the physical and chemical properties of MBC. Nanomaterial-loaded MBC exhibits exceptional sorption, especially for low concentrations of phosphorous (P).

7.16.5 Element doping modification

The elements used for element doping modification are Ca, Ce, La, Mn, N, and F. It is carried out via hydrothermal and coprecipitation methods. It substantially changes surface properties and enhances the sorption performance of MBC.

7.16.6 Biological modification: approach

Biological modification uses biological pretreatment with microorganisms (e.g. bacteria). It enhances MBC's sorption capacity in an environmentally friendly and cost-effective manner. However, concerns and considerations revolve around its technical difficulty, whereas surface modifications involve several operational steps, and nanoparticle modifications require careful handling in multiple steps. From commercial and viability perspectives, these modifications are cost-effective and have significant relevance, whereas nanoparticle modifications may face limitations due to metal clusters or nanoparticle aggregation. Thus, in terms of environmental friendliness, biological modification is environmentally preferred, whereas acid–alkaline and oxidation modifications may pose environmental risks due to residual toxicity and corrosiveness. The diverse modification methods offer tailored strategies for enhancing the sorption performance of MBCs, providing flexibility for various environmental applications. Consideration of the technical, economic, and environmental aspects guides the selection of the most suitable modification approach [3, 28].

7.17 Tailoring sorbent properties for targeted contaminant binding

The strategy outlined here introduces a valuable approach in which MNP sorbents are designed to include an organic ligand with a proven affinity for binding toxic HMs. This adaptable ligand allows for easy modification and offers a versatile foundation for HM sorption. The chosen ligands facilitate precise tuning of the affinity and specificity for HMs within this novel nanoparticle-based sorbent category. This process uses a common precursor to generate NPs with varied surface functionalities.

An iron oxide core was synthesized through high-temperature decomposition, resulting in monodisperse, superparamagnetic NPs passivated with lauric acid (LA). These LA-stabilized NPs possessed qualities such as high purity, crystallinity, superparamagnetism, and small size (approximately 8 nm in diameter), contributing to an exceptionally high active surface area (>100 m^2 g^{-1}). These attributes positioned the NPs as ideal candidates for diverse sorbent applications, including batch capture experiments and high-volume separation under flow conditions.

The creation of hydrophilic magnetic nanoparticle sorbents involves a flexible ligand exchange reaction, allowing the integration of various functional groups into the ligand shell without compromising the desirable characteristics of the precursor nanoparticle. Carboxylate-containing ligands, known for their weak surface binding, serve as anchor groups, enabling the development of magnetic nanoparticle sorbents with peripheral functional groups such as thiols, amines, polyethylene glycol (PEG), and carboxylates (COO). Carboxylate-for-carboxylate ligand exchange, conducted under varied conditions, offers flexibility, influencing the material's dispersibility and stability—critical factors for environmental applications focused on HM remediation and decontamination [26].

7.17.1 The physicochemical properties of magnetic biochar

The essential component of MBC is intricately linked to its source biomass. Diverse biomass origins yield varying total iron contents in MBC, enabling it to adsorb iron, oxygen, and distinct compounds. Plant-derived MBC, for instance, exhibits an affinity for alkaline metals, alkaline earth metals, Si, P, N, H, and S. Modification techniques, such as metal doping and organic adjustments, are commonly employed to enhance the performance, thereby affecting the elemental composition of MBC. In impregnation pyrolysis synthesis, carbon and iron in MBC exhibit inverse and direct relationships to temperature, respectively [29].

7.17.2 The point of zero charge (pHpzc) of magnetic biochar

pH, a crucial factor influencing physicochemical properties, plays a role in determining the surface charge and pHpzc of MBC. Measuring the zeta potential at various pH values and conducting reactions in NaCl solutions enable the determination of the pHpzc. Variations in pHpzc arise from the different synthesis methods, raw materials, and physicochemical properties of MBC [29].

7.17.3 The superficial area of magnetic biochar

The superficial area of MBC, a crucial factor in assessing sorbent quality, is closely tied to its porosity and varies based on the magnetic species' nature, synthesis method, time, and raw materials [29].

7.17.4 Morphology and magnetic medium species

The surface morphology of magnetic biochar, which is influenced by raw materials and synthesis methods, often exhibits a plane of superficial structures. Chemically bonded FeO, primarily α-Fe_2O_3, γ-Fe_2O_3, Fe_3O_4, FeO, and Fe0, enhance the material's environmental remediation capabilities. Incorporating magnetic materials into biochar addresses recovery concerns and improves pollutant removal efficiency [29].

7.17.5 The saturation magnetization of magnetic biochar

Saturation magnetization, a critical parameter indicating the largest magnetic induction that can be achieved, is essential for evaluating the salvaging capacity of MBC. Variations in saturation magnetisation are ascribed to several natural resources, auxiliary materials, synthesis techniques, and associated parameters [29].

7.17.6 The functional groups of magnetic biochar

The efficacy of MBC in pollutant sorption is intricately linked to its surface functional groups. Understanding the nature and mechanisms of these groups is pivotal for enhancing biochar sorption performance. Quantitative analysis is essential for identifying the critical functional groups responsible for pollutant removal and thus facilitating the development of optimized functionalization methods. Notably, persistent free radicals within MBC contribute significantly to pollutant removal. In addition, the composition of surface functional groups is influenced by synthesis/modification methods and raw materials [13, 29].

7.18 The applications of magnetic sorbents in soil decontamination

7.18.1 Multipurpose pollutant removal

MSs have extensive applications in tackling various pollutants, including HMs, inorganic anions, antibiotics, pesticides, organic dyes, and nuclear materials. Their versatility is exemplified by their efficacy in addressing pollutants, showcasing their broad utility in environmentally friendly remediation. Magnetic sorption and properties that support the catalytic degradation of organic pollutants underscore their multifaceted environmental applications [29].

7.19 Sorption

7.19.1 Heavy metal sorption

The sorption of HMs by MBC encompasses various mechanisms, including anionic (Cr (VI) and As) and cationic (cadmium (Cd) (II), lead (Pb) (II), Cu (II), nickel (Ni) (II), and mercury (Hg) (II)) types. The mechanisms involved are electrostatic sorption, reduction, ion exchange, complexation, and coprecipitation. MBC effectively eliminates multiple HMs, but competition for sorption sites can influence overall sorption behavior [29].

7.19.2 Inorganic anion sorption

MBC has been shown to have an extraordinary ability to remove inorganic anions such as phosphate, nitrate, and fluoride, and phosphate has been extensively studied. Modifying MBC can enhance phosphate sorption. The removal mechanisms include coprecipitation, electrostatic sorption, superficial, inner-sphere complexation, and ligand exchange. In addition to pollution remediation, MBC shows promise as a slow-release fertilizer, improving soil's elemental composition and improving fertility [29].

7.19.3 Combined pollutant sorption

Given the common co-occurrence of HMs, inorganic anions, and organic pollutants, MBC's outstanding sorption performance is harnessed for multicompound removal. While research into these combined systems is relatively limited, the interactive nature and competitive sorption between inorganic and organic pollutants underscore the need for further investigation into the nuanced sorption mechanisms of MBC in these scenarios [29].

7.20 Critical factors in pollutant sorption by magnetic sorbents

7.20.1 Contact time

The sorption process evolves over time; an initial rapid phase is followed by a slower stage until equilibrium is reached. The dynamic equilibrium state signifies a saturation point at which further sorption ceases. Specific studies, such as the remediation of ethylenediamine tetraacetic acid using MBC, have revealed varying sorption rates during the first 30 min, indicating abundant active sites. The sorption performance near equilibrium is essential for understanding the temporal dynamics of contaminant removal [3].

7.20.2 The influence of sorbent dosage

Contaminant removal efficiency correlates with sorbent dosage, reaching stability at higher doses. The increased number of binding sites enhances removal rates, but efficiency plateaus because of agglomeration, limiting sorption capacity beyond an optimal dosage. Understanding the dosage-dependent saturation of sorption sites is crucial for optimizing pollutant removal with different MBCs [3].

7.20.3 The influence of the initial pollutant concentration

Sorption efficiency generally increases with higher initial pollutant concentrations up to a certain point. For instance, the removal rate of norfloxacin demonstrated a direct relationship with sodium dodecyl sulfate concentration. However, there is a critical concentration beyond which efficiency decreases, highlighting the nuanced relationship between initial pollutant concentration and sorption performance [3].

7.20.4 The influence of pH

pH significantly influences sorption, impacting superficial charge, metal-type models, and the protonation of functional groups. The point of zero charge (pHpzc) is pivotal in determining the surface charge characteristics. For example, below the pHpzc, positively charged sorbents repel divalent metal ions, whereas above the pHpzc, negatively charged adsorbents attract positively charged HM species. The optimal pH is crucial for maximizing pollutant removal [3].

7.20.5 The influence of interfering ions

The complexity of water bodies makes it necessary to consider interfering ions. Various anions can affect the sorption efficiency; for example, sulfate substantially degrades Cr(VI) sorption. The presence of coexisting anions may slightly reduce the sorption efficiency. Significantly, the ions in the solution can inhibit sorption sites and alter the superficial properties of sorbents, emphasizing the impact of interfering ions on pollutant removal [3].

7.21 Mechanisms of heavy metal elimination from soil by magnetic sorbents

The elimination of HMs from soil using MSs has been extensively investigated, revealing multiple concurrent mechanisms, as indicated in figure 7.3 by Bing Xiao *et al* [12].

7.21.1 Sorption

Strong heating during the magnetization process enhances the specific surface area and permeability of MBC, facilitating the immobilization of HM ions through surface physisorption. At an optimal particle size, granular MBC (gMBC) exhibits effective recovery from wet and dry soil. The dynamic balance between sorption and the soluble forms of HMs indicates a resorption phenomenon.

7.21.2 Surface coprecipitation

HM cations combine with the mineral components of MBC, forming precipitates that contribute to immobilization. For instance, sulfide MBC effectively immobilizes HMs by sequestering them via ion exchange, sorption, and precipitation.

7.21.3 Electrostatic attraction

A static interface produced by an MBC's superficial charge and HM ions immobilizes metals. The pH of the ambient medium relative to the zero-charge point of the MBC governs the static impact, influencing the superficial charge of the MBC and subsequent interactions.

7.21.4 Double-pi interactions

Contacts involving HM multiplexes with –OH and aromatic complexes in MBC, termed π–π interactions, contribute to the removal process. These interactions

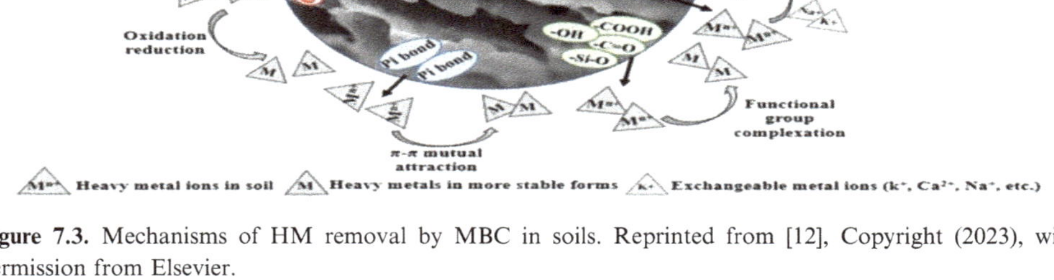

Figure 7.3. Mechanisms of HM removal by MBC in soils. Reprinted from [12], Copyright (2023), with permission from Elsevier.

enhance the sorption capacity and transport potential, particularly in biochar-functionalized nanocomposites. Understanding these mechanisms provides insights into the diverse and effective strategies for HM removal from soil by MSs, offering valuable implications for environmental remediation practices [12].

7.22 Heavy metal removal using modified magnetic biochar

Magnetite, an iron oxide displaying superparamagnetic behavior at nanoscale dimensions, has drawn attention because of its potential for use in HM removal. To address the prevalent health issue of HM pollution, researchers have explored the use of modified MBCs, focusing on the removal of Cd (II), Cr (VI), Pb (II), Cu (II), and Hg (II). The sorption mechanisms of MBC involve static magnetism, ion exchange, redox contact, superficial complexation, and H bonding (figure 7.4). Different MBC types, derived from various sources, such as steel pickling waste liquor and agricultural residues, have been developed for effective Cr (VI) elimination. A reduction of Cr (VI) to Cr (III) was monitored, and variations in reduction capacity were observed between MBC types. Characterization results indicated that electrostatic attraction and complexation with C– –O groups are crucial Cr (VI) reduction mechanisms [4, 12].

Intriguing permeable water-hyacinth-derived biochar (MPBCMW3) synthesized through microwave-assisted procedures demonstrated reduction reactions, as confirmed by x-ray photoelectron spectroscopy (XPS) peaks associated with Cr(VI) and Cr(III). Complexation and ion exchange processes were evident, exemplifying the versatility of MBCs in HM remediation. When employed in soil remediation for Pb, MBCs showed efficient recovery using regenerants such as HNO_3, HCl, and EDTA-2Na. Magnetized wheat-straw-derived MBCs exhibited varying sorption capacities and recovery efficiencies, emphasizing the role of saturation magnetization. MBCs have emerged as promising sorbents for soil and water remediation, employing different processes for HMs with distinct valency states. While divalent metals require processes such as complexation, ion exchange, and static contact, multivalent metals such as Cr engage in redox interplay [4, 12].

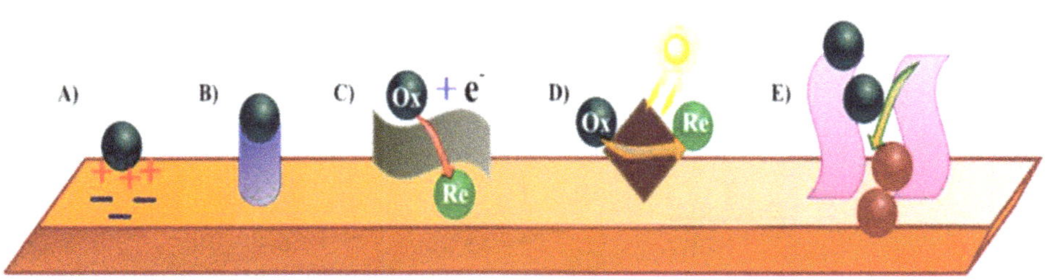

Figure 7.4. The mechanisms of common interactions between ions and materials: (A) electrostatic interaction, (B) complexation, (C) redox reaction, (D) photoreduction, and (E) ionic exchange. Reprinted from [4], Copyright (2022), with permission from Elsevier.

7.23 The elimination of persistent organic pollutants

In addition to HMs, MBC effectively removes persistent organic contaminants, including phenols and pesticides. For pesticides such as chlorpyrifos and hexaconazole, magnetic sporopollenin reinforced with cyanocalixarene (MSP-CyCalix) nanocomposites exhibit sorption mechanisms such as double-pi interaction, H bonding, and static interaction.

Phenol removal using a biochar-based Fe_2O_3 complex material (FeYBC) mainly involves physical sorption via π–π interaction and electron donor–acceptor complexes. Similarly, metolachlor residue in the soil is effectively adsorbed by MBC augmented with illite and $FeCl_3$ (IMBC), using mechanisms such as H bonding, double-pi interaction, crater filling, and static interaction.

MBC is a versatile and efficient modification for diverse pollutants because of its outstanding sorption, easy reusability, and conservational benefits. The use of magnetic carbon materials in soil and their subsequent recovery significantly reduces the ecological risks associated with organic pollutants [12].

7.24 The influence of various factors on heavy metal removal by magnetic sorbates

The effectiveness of HM removal using MBC is influenced by intrinsic HM properties, soil characteristics, the MBC feedstock, pyrolysis temperature, and applied MBC quantity [12].

7.24.1 The intrinsic properties of heavy metals

Environmental media contain various HM types that can be categorized into cationic and anionic species. The cationic species include Cd (II), lead (II), Ni (II), and Hg (II), whereas the anionic species consist of Cr (VI), As (III), and As (V). Research is currently focused on finding optimal sorbents for HM removal. Upon the addition of MBC, the soil pH significantly rises, facilitating the deprotonation of acidic functional groups. This enhances the negative charge of the soil surface colloids, attracting metal cations. Conversely, higher soil pH prompts stronger repulsion among metal anions. The surface charge of MBC, which is linked to pH, contributes to its high electronegativity, attracting metal ions via electrostatic forces. The soil characteristics—pH, organic matter (OM), and CEC—impact HM morphology and efficacy.

7.24.2 Magnetic biochar dosage

The dosage of MBC used to remediate HM-contaminated soil is pivotal, directly influencing remediation effectiveness. Unnecessary overdosage must be averted to avoid resultant toxic waste. For instance, iron-augmented biochar (NBC–Fe) administered at varying concentrations to As-contaminated soil showed increased crystalline aqueous oxide-bound residues at higher NBC–Fe doses. The optimal efficacy in reducing the amount of bound As was observed at a dose of 0.4% NBC–Fe.

7.24.3 Heavy metal forms

HM forms significantly affect mobility, biodegradability, and biotoxicity. Researchers alter the HM percentages to reduce bioavailability. Sequential extraction methods, such as the Tessier and BCR (Community Reference Bureau, now the EU Standard Measurement and Testing Scheme) methods, are used to study the HM geochemical fractions in soils after MBC addition. Different HM forms in soil influence the natural efficacy and migration capability. Adding MBC reduces accessible metals and increases residual-state metals. MBC is classed as a form of passivation remediation for soil-borne HMs; it encapsulates HMs to reduce risk. Studies using reed-straw-based MBCs revealed enhanced HM passivation, indicating effective metal encapsulation. In Cr (VI) passivation tests, calcium-based MBC (Ca-Fe-B) transformed Cr into reduced and residual Cr, demonstrating passivation efficiency. Understanding these influencing factors provides insights into optimizing MBC applications for effective HM removal and passivation in diverse environmental scenarios [12].

7.25 The applications of magnetic sorbents in various fields

MS are highly versatile materials with inherent properties that make them attractive for diverse applications. Their utility is particularly prominent in sorbent-based techniques because of their superior extraction capacity and selectivity. The choice of application depends on the physicochemical properties of both the analyte and the stationary phases, which influence the extraction mechanism or principle employed [28].

7.25.1 Solid-phase extraction

One of the most prevalent sorbent-based practices is SPE. CNTs are extensively used in this process because of their adequate sorption capacity for both organic and inorganic analytes [28].

7.25.2 Conventional solid-phase extraction

In conventional SPE, CNTs serve as sorbents for the extraction of both inorganic and organic analytes. Whether used in pristine form, functionalized, coated, as composites, or immobilized on solid-supported CNTs, these materials exhibit broad applicability, particularly in analyzing environmental matrices with varying complexities. Notably, their superior surface area, smaller particle size, and enhanced partitioning contribute to more effective mass transfer, overcoming the challenges encountered at low concentrations [28].

7.25.3 Enhancements for sorbent efficiency

To boost sorbent efficiency, polar groups such as COOH, CO, and OH are introduced through oxidation, acting as binding sites for specific compounds to enhance extraction selectivity. Amino compounds are effective chelating agents for metal extraction because of the free pair of electrons on the nitrogen atom. In addition, the incorporation of compounds such as 1-(2-pyridylazo)-2-naphthol

(PAN), diphenylcarbazide (DPC), and various amino acids enhance hydrophilicity, selectivity, and binding affinity to certain transition metals [28].

7.25.4 Innovative sorbent combinations

Studies have revealed that combining the high sorption ability of MWCNTs with the high porosity of cryogels yields cost-effective sorbents that have excellent sensitivity and recovery rates ranging between 89% and 98%. This innovation represents a promising high-performance approach for pollutant extraction [28].

7.26 Environmental impacts: the fate and transport of magnetic nanoparticles

7.26.1 Magnetic nanoparticle dispersion stability

Perfluorinated ligand-passivated MNPs prepared through biphasic ligand exchange exhibited enhanced dispersion stability. Nitrogen bubbling to reduce the O_2 content in the fluorous phase resulted in a stable MNP dispersion, maintaining a uniform single-particle size distribution. MNPs, which are susceptible to oxidation and corrosion, are protected by core–shell structures, which isolate the core with materials such as organic compounds or inorganic substances. Surface modification with long-chain carboxylic acids enables the solubilization of magnetic inorganic particles in organic solvents. Mussel-inspired multidentate block copolymers stabilized superparamagnetic MNPs in potential magnetic resonance imaging (MRI) contrast agent applications [11].

7.26.2 The environmental implications of ageing

MNPs applied to soil environments have enduring effects, potentially lasting hundreds to thousands of years. Studies have revealed that aged MNPs influence the physicochemical properties of black soils, affecting pH, carbon, P concentration, N content, the C/N ratio, and soil bulk density even after three years. The decomposition rates of MNPs and their stability, which are influenced by pyrolysis parameters, feedstock types, and ageing conditions, are crucial factors determining their persistence in soil. Surface charge alterations that take place due to ageing highlight the significance of surface functional groups, feedstock, and pyrolysis temperature [1].

7.26.3 The toxic effects of magnetic nanoparticles

The toxicity of MNPs is closely tied to their pyrolytic conditions; higher temperatures potentially reduce toxic effects. *In vitro* studies of nano-tobacco stem-pyrolyzed MNPs indicate cytotoxic and genotoxic effects, emphasizing the importance of mitigating toxicity in MNP production. Assessments of genotoxicity based on particle size and surface coatings reveal that the mutagenic potential of MNPs is influenced by the coating type. Efforts to reduce toxicity include modifying pyrolysis processes to reduce fluoride and HM leachability. Continued research aims to comprehend the potentially toxic effects of MNPs on production facility workers and farmers who may come into contact with MNPs [1].

7.26.4 Biocompatibility and toxicity studies

Various studies have demonstrated the biocompatibility of specific MNPs, including negligible toxicity in cell viability assays and minimal impact on organism-level health. *In vivo* studies of zebrafish showed efficient elimination of accumulated NPs when the fish were moved to NP-free water. Surface modifications, antioxidant coatings, and specific coatings such as PEG help to reduce the toxicity of MNPs. Studies of bare MNPs indicate the potential disruption of autophagic processes in human cells, emphasizing the need for further research to understand the degree to which MNPs are toxic to humans and other organisms. Understanding the fate, transport, ageing, and toxicity of MNPs is critical for responsible environmental application and to ensure minimal adverse effects on ecosystems and human health. Ongoing research is essential to address potential concerns and develop sustainable MNP production and application practices [11].

7.27 Future directions and required research

7.27.1 Challenges in biochar soil applications

Biochar has shown promise in soil amendment, contributing to reduced greenhouse gas emissions by sequestering carbon in soil structures. Despite progress in laboratory and pilot-scale studies, larger-scale applications face challenges [1]. Greenhouse gas emissions, mainly CO_2 and N_2O, from loess soils (accumulated wind-blown dust) require careful consideration. Although BC effectively decreases nitrogen emissions, field studies remain limited. Future research should bridge the gap between lab and field applications by correlating BC production conditions and feedstock types with soil effects.

7.27.2 Sustainability in the pyrolysis process

Sustainability in biochar production demands that attention be paid to the pyrolysis process; the energy generation required by this process can contribute to greenhouse gas emissions. To power biomass-to-biochar conversion, urgent measures are needed to develop green energy solutions, such as concentrated solar towers. Despite biochar's potential benefits, concerns persist about its environmental behavior. It is crucial to study biochar's fate, long-term effects, and mitigation of environmental impacts. Standardized analytical methods for environmental fate studies are urgently required.

7.27.3 Human health and economic considerations

Thorough investigations into biochar's potential human health impacts are necessary. Optimizing production conditions to produce less toxic biochar is vital for worker and farmer protection. Considering various feedstocks and pyrolytic conditions, economic evaluations of biochar are lacking. Economic assessments in biochar reports can aid decision-makers. Social studies of public perception and overcoming barriers to biochar's broader applications in soil amendment are recommended.

7.27.4 Future directions for magnetic biochar

Functionalized MBCs in environmental remediation face challenges, including the potential leaching of HMs. The HM salt content increase may impact water quality, affecting aquatic life. To address this issue, metal loading should be strengthened. The MBC preparation process, which mainly uses waste biomass, requires careful handling to prevent the production of toxic substances. Future research areas include enhancing MBC surface stability, exploring green and efficient preparation technologies, and conducting dynamic sorption tests for practical applications.

7.27.5 Addressing the challenges associated with magnetic biochar

Despite advancements, MBC research must address critical issues. The formation mechanism and regulation of crucial MBC components, environmental toxicity, and transformation in environmental remediation applications should be investigated. Preventing the release of adsorbed pollutants, understanding the effects of preparation parameters, improving selectivity, and conducting practical wastewater remediation are essential areas for future exploration. In summary, future research should focus on bridging the gap between lab and field applications for biochar, developing sustainable energy solutions for biochar production, investigating biochar's human health impacts and economic considerations, and addressing challenges in the formation, toxicity, and application of magnetic biochar. These directions will enhance our understanding and application of biochar and MBC in soil amendment and environmental remediation.

7.28 Conclusions

Today, advanced techniques are employed for the nanoscale synthesis of MC sorbents to address pollutant issues through adsorptive remediation. This chapter outlined the synthesis methods, modification strategies, and applications of modified MBC for removing hazardous pollutants from soil. Modified MBC exhibits excellent magnetic recovery features, reducing subsequent filtration costs and avoiding secondary pollution. This chapter highlighted the potential of modified MBCs to yield material benefits, efficiency, and harmful pollutant elimination mechanisms. Future investigations should concentrate on enhancing the superficial strength of metal-comprising MBCs to decrease metal seepage and investigating eco-friendly batch preparation technologies.

The use of MBC in environmentally friendly remedies is increasing, overcoming biochar limitations. However, research gaps persist. To address these issues, more investigations are needed. Magnetic materials, such as MIPs, have proven efficient for selective extraction, offering dense combinations without the need for pretreatment. Advances in careful removal procedures such as MSPE, magnetic solid-phase microextraction (MSPME), and in-tube solid-phase microextraction (IT-SPME) must enhance novelty, simplicity, and efficiency, emphasizing automation. MSs simplify sample pretreatment, ease recycling and reuse, and have applications in preconcentration and removal. Novel magnetic hybrid sorbents employing advanced nanostructured materials show promise for efficient and selective extraction. The field's future trends include the construction of MC sorbents with diverse functional groups, huge magnetic domains, and large pore sizes and surface areas to enhance active selectivity and efficacy. Developing novel MSs with high sorption ability to simultaneously remove contaminants is a significant trend that contributes to materials science, environmental chemistry, and engineering.

References

[1] Mohammadreza Kamali R D, Sweygers N, Al-Salem S, Appels L and Aminabhavi T M 2022 Biochar for soil applications-sustainability aspects, challenges and future prospects *Chem. Eng. J.* **428** 131189

[2] Ke W *et al* 2023 Remediation potential of magnetic biochar in lead smelting sites: insight from the complexation of dissolved organic matter with potentially toxic elements *J. Environ. Manage.* **344** 118556

[3] Jianhua Qu Y Z *et al* 2022 Applications of functionalized magnetic biochar in environmental remediation: a review *J. Hazard. Mater.* **434** 128841

[4] Yeisy G A O, Lopez C and Reguera E 2022 Hazardous ions decontamination: from the element to the material *Chem. Eng. J. Adv.* **11** 100297

[5] Yadeghari A and Farajzadeh M A 2021 Synthesis of a magnetic sorbent and its application in extraction of different pesticides from water, fruit, and vegetable samples prior to their determination by gas chromatography-tandem mass spectrometry *J. Chromatogr.* A **1635** 461718

[6] Han Z *et al* 2015 A critical evaluation of magnetic activated carbon's potential for the remediation of sediment impacted by polycyclic aromatic hydrocarbons *J. Hazard. Mater.* **286** 41–7

[7] Dulanská S, Macášek F and Mičeková P 2001 Decontamination of soil by a magnetic sorbent *9th Int. Conf. SIS'01 Abstracts of the 9th International Conference SIS'01* 102–3 https://www.osti.gov/etdeweb/servlets/purl/20180033

[8] Faraji M, Shirani M and Rashidi-Nodeh H 2021 The recent advances in magnetic sorbents and their applications *TrAC—Trends Anal. Chem.* **141** 116302

[9] Zhang S *et al* 2022 Analytical perspective and environmental remediation potentials of magnetic composite nanosorbents *Chemosphere* **304** 135312

[10] Maqbool A, Wang H, Saeed M and Hafeez A 2021 Magnetic nanocomposite-system for the remediation of lead-contaminated urban surface *E3S Web Conf* **266** 08007

[11] Su C 2017 Environmental implications and applications of engineered nanoscale magnetite and its hybrid nanocomposites: a review of recent literature *J. Hazard. Mater.* **322** 48–84

[12] Bing Xiao J J, Y K, Wang W, Zhang B, Ming H, Ma S and Zhao M 2023 A review on magnetic biochar for the removal of heavy metals from contaminated soils: preparation, application, and microbial response *J. Hazard. Mater. Adv.* **10** 100254

[13] Singh H, Bhardwaj N, Arya S K and Khatri M 2020 Environmental impacts of oil spills and their remediation by magnetic nanomaterials *Environ. Nanotechnol. Monit. Manag.* **14** 100305

[14] Grazhulene S S, Zolotareva N I, Red'kin A N, Shilkina N N, Mitina A A and Kolesnikova A M 2018 Magnetic sorbent based on magnetite and modified carbon nanotubes for extraction of some toxic elements *Russ. J. Appl. Chem.* **91** 1849–55

[15] Husnain S M, Um W, Woojin-Lee and Chang Y S 2018 Magnetite-based adsorbents for sequestration of radionuclides: a review *RSC Adv.* **8** 2521–40

[16] Sarkar B, Mukhopadhyay R, Ramanayaka S, Bolan N and Ok Y S 2021 The role of soils in the disposition, sequestration and decontamination of environmental contaminants *Philos. Trans. R. Soc.* B*376 20200177*

[17] Valdiviezo Gonzales L G, Castañeda-Olivera C A, Cabello-Torres R J, García Ávila F F, Munive Cerrón R V and Alfaro Paredes E A 2023 Scientometric study of treatment technologies of soil pollution: present and future challenges *Appl. Soil Ecol.* **182** 104695

[18] Yeung A T and Gu Y Y 2011 A review on techniques to enhance electrochemical remediation of contaminated soils *J. Hazard. Mater.* **195** 11–29

[19] Dongdong Wen Q L and Fu R 2021 Removal of inorganic contaminants in soil by electrokinetic remediation technologies: a review *J. Hazard. Mater.* **401** 123345

[20] Kavamura V N and Esposito E 2010 Biotechnological strategies applied to the decontamination of soils polluted with heavy metals *Biotechnol. Adv.* **28** 61–9

[21] Ubani O 2021 Development of an active bacterial formulation for degradation of complex crude oil wastes *PhD* University of South Africa https://hdl.handle.net/10500/28221

[22] Cluis C 2004 Junk-greedy greens: phytoremediation as a new option for soil decontamination *BioTeach J.* **2** 61–7 https://bioteach.ubc.ca/Journal/V02I01/phytoremediation.pdf

[23] Fu H, He H, Zhu R, Ling L, Zhang W and Chen Q 2021 Phosphate modified magnetite@ferrihydrite as an magnetic adsorbent for Cd(II) removal from water, soil, and sediment *Sci. Total Environ.* **764** 142846

[24] Sghaier R B *et al* 2023 Green magnetic snail shell hydroxyapatite sorbent for reliable solid-phase extraction of pesticides from water samples *J. Sep. Sci.* **46** e2300290

[25] Jebali S *et al* 2023 The potential of three different adsorbents in solid-phase extraction of antihistaminic and antimigraine drugs from water samples using ultra-high-performance liquid chromatography-ultraviolet analysis *Sep. Sci. Plus* **7** 2300194

[26] Warner C L *et al* 2010 High-performance, superparamagnetic, nanoparticle-based heavy metal sorbents for removal of contaminants from natural waters *ChemSusChem.* **3** 749–57

[27] Elwakeel K Z, Elgarahy A M and Mohammad S H 2017 Magnetic Schiff's base sorbent based on shrimp peels wastes for consummate sorption of chromate *Water Sci. Technol.* **76** 35–48

[28] Bárbara Socas-Rodríguez M A-R, Herrera-Herrera A V and Hernández-Borges J 2014 Recent applications of carbon nanotube sorbents in analytical chemistry *J. Chromatogr.* A **1357** 110–46

[29] Yi Y *et al* 2020 Magnetic biochar for environmental remediation: a review *Bioresour. Technol.* **298** 122468

IOP Publishing

Environmental Applications of Magnetic Sorbents

Kingsley Eghonghon Ukhurebor and Uyiosa Osagie Aigbe

Chapter 8

The application of magnetic sorbents in seed germination

Kingsley Eghonghon Ukhurebor, Uyiosa Osagie Aigbe, Joseph Onyeka Emegha, Lucky Evbuomwan, Bamikole Olaleye Akinsehinde, Olusoji Anthony Ayeleso, Rout George Kerry, Benedict Okundaye, Atala Bihari Jena, Francis Jesmar P Montalbo, Grace Jokthan, Aizebeoje Balogun Vincent and Ahmed El Nemr

Soil polluted with dyes, heavy metals, metalloids, and other potentially detrimental domestic and industrial materials and contaminants poses a significant threat to the colonization of several plant species. The removal of these potentially detrimental materials and other contaminants that have accumulated in anthropogenic soils in several substantially altered places is a critical issue. Hence, this chapter explores and reviews some recent publications dedicated to the application of sorbents (particularly magnetic sorbents) to plants, specifically for seed germination.

8.1 Introduction

The environment contains naturally occurring, potentially detrimental materials (PDMs) as well as other contaminants [1–4]. But only trace amounts are produced by weathering, volcanic eruptions, and other natural processes [5]. Anthropogenic activities, including industrial production, the rise of urbanization, and mining and smelting, are the main causes of elevated concentrations of PDMs and other contaminants, particularly in soil [6–9].

PDMs can spread to nearby uncontaminated soil because PDM-contaminated soils have intermittent to non-existent plant cover and are subject to water erosion (wearing away of the soil), wind, radiation from the Sun (solar radiation), and solifluction [10–13]. Since PDMs are harmful to living things and have a negative impact on environmental quality, rising PDM concentrations in the soil now rank among the most critical environmental issues [14–16]. Remedial techniques must be used to lower their content in the soil and eradicate their detrimental effects [17].

doi:10.1088/978-0-7503-5909-2ch8

Conventional techniques, including landfilling, soil replenishment, and washing, have been employed in the past. These methods' drawbacks include their impracticality, time commitment, and material needs [18–20].

Today, there is a tendency to employ appropriate remediation methods from a social as well as ecological and economic perspectives [19]. The use of sorbents, particularly magnetic and naturally occurring sorbents, seems to be one of these tendencies. Solid materials with the ability to adsorb or absorb another material are known as sorbents [21]. Sorbents such as bentonite, perlite, charcoal, $CaCO_3$, manure, and some nanoparticles (NPs) are often utilized [22].

One of the most prevalent minerals in soil is bentonite [23]. It is a kind of clay rock that is mostly made up of montmorillonite; other minerals that are present include quartz impurities, minerals that contain kaolinite and illite, and minerals from the smectite clay group [24]. These have been demonstrated to be useful sorbents of some heavy metals (HMs) such as Cd, Cr, Cu, Hg, Mn Ni, Pb, and Zn from marine settings [25–28]. They also increase soil moisture, enhance water utilization efficiency, and boost the accessibility of water, potassium, and carbon-based organic materials in the soil, all of which have a favorable impact on plant development and yield [29–31].

When organic-based matter with a high carbon-based material content and high aromaticity is thermally decomposed at a temperature of around 9.0×10^2 °C in an oxygen-deficient environment, the resultant solid carbonized product is called biochar [11, 32]. Manure is composed of beneficial nutrients and is used as fertilizer, but plant-based biochar is often low in beneficial nutrients [33, 34]. In addition to boosting soil nutrient availability, cation exchange capacity, specific surface area (SSA), total nitrogen (N), carbon-based materials, and water retention [35–37], this sorbent aids in carbon storage and greenhouse gas reduction [38]. It may raise the pH of the soil, provide a liming action that effectively lowers the movement of As, Cd, Cu, Ni, Pb, and Zn, and immobilize HMs by absorbing them on its surface [39, 40]. In soils polluted with HMs, it has a beneficial impact on microbial activity and plant development [33, 41]. Pb, Zn, Cd, and Cu concentrations can be considerably lowered using biochar [42–44].

Organo-zeolitic substrates are made up of an alkaline zeolite combined with a combination of sorbents of biological (organic) origin (such as manure, charcoal, compost/peat, etc.) as well as an additional sorbent of abiotic origin (such as $CaCO_3$) if desired. Each of these elements: perlite (used to sorb Pb, Cu, Ni, and Cd) [45], chicken manure (waste) (used to sorb Cd and Ni) [46, 47], and $CaCO_3$ (used to sorb Cd, Cu, and Pb) [47, 48] has had a beneficial impact in lowering the levels of various PDMs. Zeolites are beneficial to soil fertility, sorption characteristics, and phytomass production [49, 50]. They function as fully soluble fertilizers, enhancing the sorption and the equilibrium of the water (water balance), particularly in light, sandy soils, leading to increased yield and improved quality. Reactive N (NO_3) generation is reduced and N release is regulated in soil that has been fertilized or enhanced with organic zeolite combinations [51].

The amount of organic matter and clay in the soil and its pH all affect the availability of Cu and other HMs [52]. HM bioavailability may generally be decreased by a large percentage of organic materials [53]. HM immobilization is increased when clay-based minerals have a significant SSA and the pH of the soil is elevated, particularly when the HMs are cations [54–57]. It is crucial to get rid of PDMs that have collected in anthropogenic soils in various locations, as they pose a significant threat to the colonization of several plant species. Consequently, this chapter examines and evaluates some recent studies that have focus on the use of sorbents (particularly magnetic sorbents (MSs)) in plants, particularly in relation to seed germination.

8.2 Techniques for producing magnetic sorbents

Due to the exceptional qualities of MSs, much research has been conducted to develop techniques for their production. Notably, the design, morphology, magnetic characteristics, and particle size of MSs are primarily determined by their production process [56, 58, 59]. As illustrated in figure 8.1 (according to Osman *et al* [59]),

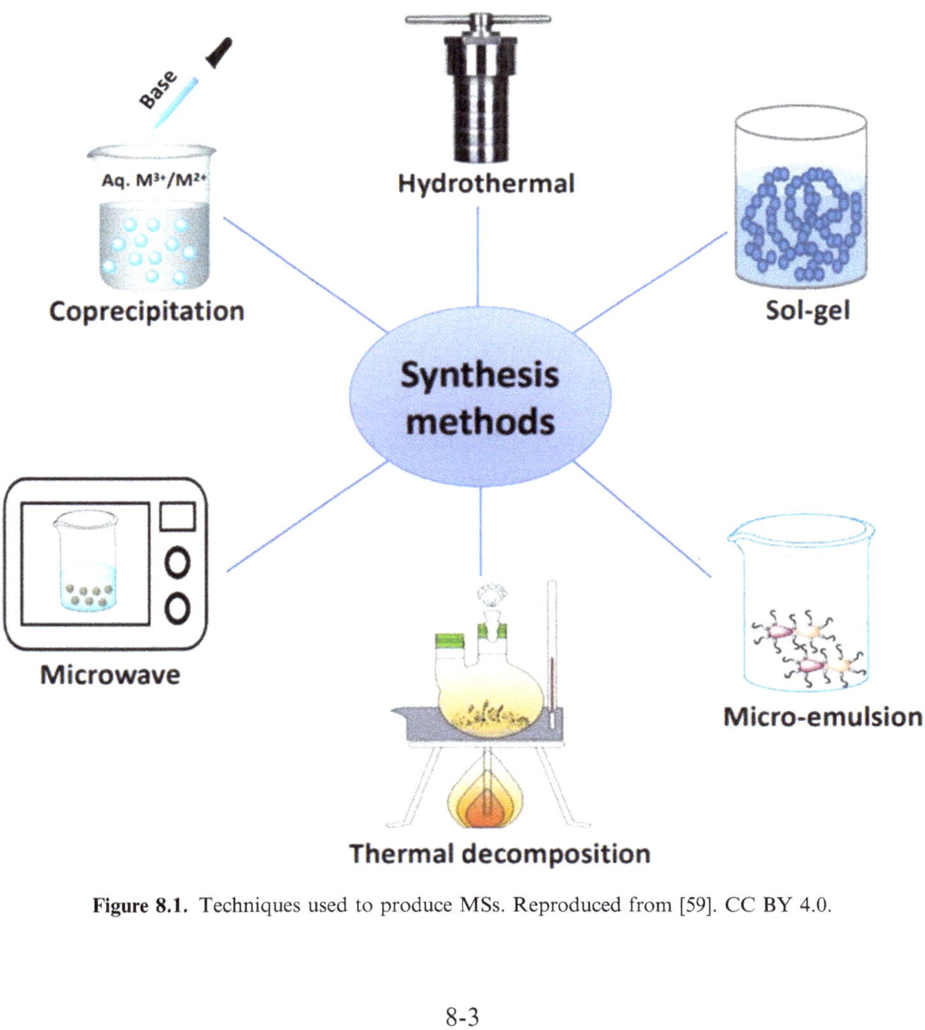

Figure 8.1. Techniques used to produce MSs. Reproduced from [59]. CC BY 4.0.

the techniques/methods that are typically used for producing MSs include the coprecipitation, hydrothermal, thermal decomposition, polyol, microwave, sol–gel, and micro-emulsion methods.

A precipitating agent may be used in coprecipitation to produce MSs. Furthermore, making MSs is a commonly involves the sol–gel technique [59]. MSs can be made via direct thermal decomposition, the microwave-assisted method, or via heating in a reactor (the hydrothermal process) [59]. MSs may also be produced by other techniques, such as the micro-emulsion method. The terms M^{2+} and M^{3+} denote metal valences [59]. Osman *et al* [59] contains more details of the techniques (methods) that are generally used for producing MSs.

8.3 The influence of magnetic nanoparticles on plant germination and growth

A growing number of researchers are interested in studying NPs because of their special physicochemical characteristics [60, 61]. Numerous opportunities exist for their use in agriculture, industry, technology, and medicine [58, 60, 62, 63]. Applications for agriculture include sensors that track the condition of the soil and plants, as well as fertilizers and insecticides that combat diseases and pests [3, 62]. To improve plant development (growth) and protection, nanocarriers containing nutrients or nanolayers of macro- and micronutrients are also employed [64].

The plant species as well as the composition, concentration, size, and physical characteristics of NPs all influence how they affect plants. Siddiqui and Al-Whaibi [65] used SiO_2 nanocrystals on tomatoes and saw favorable impacts on the growth and development of plants, yield, (beneficial) mineral nutrition, and photosynthesis. Rizwan *et al* [66] investigated the impact of silver (Ag) NPs (Ag NPs) on grown plants. Depending on the concentration of NPs, Ag NPs exerted growth-inhibiting effects on the mitotic index and root development of sorghum (*Sorghum bicolor*) and *Vicia faba* seedlings. When TiO_2 NPs were applied to canola seeds, fennel seeds, and peppermint seeds (*Mentha piperita*), Mahmoodzadeh *et al* [67], Feizi *et al* [68], and Samadi *et al* [69] respectively reported an upsurge in the height of the root and a favorable impact on the photosynthetic pigments' concentration.

Zinc oxide (ZnO) NPs exhibited the greatest phytotoxicity compared to aluminum oxide (Al_2O_3), silicon dioxide (SiO_2), magnetite (Fe_3O_4), and ZnO NPs on *Arabidopsis thaliana* [70]. Research was performed to examine the effects of NPs (aluminum, alumina, multilayer carbon nanotubes, Zn, and ZnO) on plant root development and seed germination in cucumber, lettuce, maize, radish, rapeseed, and ryegrass [71]. While NPs had no influence on seed germination, the investigated plant species' roots grew longer as a result of the application of nano-Zn and nano-ZnO solutions.

Zhu *et al* [72] discovered that Fe_3O_4 NPs may absorb, transfer, and accumulate NPs in tissues during their investigation of the impact of Fe_3O_4 NPs on pumpkins (*Cucurbita maxima*).

As reported by Ghafariyan *et al* [71], soybean leaves treated with superparamagnetic Fe oxide NPs (SPIONPs) had higher amounts of chlorophyll. In soft wheat, spherical FeO NPs and Fe_3O_4 NPs inhibited germination and leaf expansion (*Triticum vulgare* Vill.) When wheat seedlings were exposed to dryness and excessive salt, Fe NPs had a good impact on their germination and growth; this was demonstrated by an upsurge in the height and weight of the seedlings, shoots/ sprouts, and roots [73].

Spinach seeds were treated with an aqueous solution of non-paramagnetic iron pyrite ($FeS_{2+}H_2O$) NPs by Shrivastava *et al* [69] as well as Das *et al* [74]. The aqueous solution of nano-Fe pyrite considerably increased the amount of spinach produced in both situations. In the first experiment, seedlings were planted in the field after being bathed in a solution of aqueous NPs. Water-treated seeds served as the control group. Shrivastava *et al* compared the crop yields from seeds treated with iron pyrite NPs and control seeds. In addition to having considerably wider leaf shapes, more leaves overall, and higher biomass, the plants grown from NP-treated seeds also had higher leaf concentrations of Ca, Mn, and Zn than the control. A nano-ceria oxide (CeO_2) was combined with a suspension of aqueous FeS_2 NPs to lower the concentration of H_2O_2 produced during the seed treatment. It was shown that applying FeS_2 produced leaves with greater levels of glucose and chlorophyll compared to the leaves of plants treated with water (the control), nano-CeO_2, or $FeS_{2+}CeO_2$. It was discovered that nano-CeO_2 had the opposite effect to nano-FeS_2, which accelerated germination and served as a seed vigor booster. The data gained can be utilized for various agro-technical treatments, such as controlling the population of weeds, storing seeds under critical circumstances, and accelerating or delaying germination.

As a result of their exceptional qualities—such as their tiny dimension, high SSA-to-volume ratio, surface modifiability, amplified magnetic characteristics, and exceptional biocompatibility—ZnO NPs have drawn a lot of interest for use in environmental protection technology [75]. Boparai *et al* [76] and Celebi *et al* [77] employed zero-valent Fe NPs (nZVI) to remove/confiscate Cd^{2+} ions and Ba^{2+} ions, respectively. Fe_3O_4-FeB, or amorphous FeB alloy-modified magnetite NPs, were employed to eliminate Cr^{6+}-related water contamination. Ramandi and Shemirani [78] used a selective ionic liquid ferrofluid in dispersive solid-phase extraction to remove Pb^{2+} and Cd^{2+} residues from blood plasma and human urine as well as milk. Fe NPs may be employed to sequester Ni^{2+} from water by acting as a sorbent and reductant, as demonstrated by Li and Zhang [79].

Synthesized (produced) kaolin-supported nanoscale zero-valent Fe (K-nZVI) was utilized for the removal of Pb^{2+} from an aqueous solution [80, 81]. It was also successful in the removal of Cd^{2+}, Ni^{2+}, Pb^{2+} (98.80%), and Cr (99.80%) from galvanic/voltaic sewage.

Ge *et al* [82] removed HM ions (Cd^{2+}, Cu^{2+}, Pb^{2+}, and Zn^{2+}) from an aqueous solution using magnetic NPs modified with Fe_3O_4 and a polymer.

Surface-active magnetic Fe3O4 NPs were shown to have superparamagnetic properties and excellent efficacy in eliminating germs and harmful HM ions (As^{3+}, Cd^{2+}, Co^{2+}, Cr^{3+}, Cu^{2+}, Ni^{2+}, and Pb^{2+}) from polluted water [83]. To extract Cu^{2+} ions from an aqueous solution, Badruddoza *et al* [84] synthesized carboxymethyl-β-cyclodextrin-modified Fe_3O_4 NPs (CMCD-MNPs).

Compared to monometallic Fe NPs, bimetallic iron NPs have notably superior physical and chemical characteristics (such as magnetism) and the capacity to decrease other metals [85]. Fe NPs coated with ferromagnetic carbon and coated with a certain quantity of carboxylic functional groups were able to physically adsorb other metals onto their coating [80, 81], removing over 95% of Cr^{6+} from wastewater. Fe_3O_4@SiO_2, a unique magnetic NP (core–shell) designed to extract HM ions from aqueous environments, was developed by Wang *et al* [86].

The procedure of adsorption is solution pH dependent, according to some research into the utilization of Fe_2O_3 NPs for the selective confiscation/sequestration of HMs (especially the hazardous ones such as Cr^{6+}, Cu^{2+}, and Ni^{2+}) from industrial effluent [87]. Fe3O4 NPs coated with humic acid (Fe_3O_4/HA) were utilized by Liu *et al* [88] to extract over 95% of Cu^{2+} and Cd^{2+} and over 99% of Hg^{2+} and Pb^{2+} from water. NP composites with Fe NPs implanted in the walls (surroundings) of a macro-permeable polymer were produced by Savina *et al* [89]. They performed well in confiscating As^{3+} traces from a solution. To confiscate Cr^{6+} (97%), Tang *et al* [90] produced/developed Fe-doped ordered mesoporous/meso-permeable carbon (Fe/CMK-3).

To cure plant diseases, Corredor *et al* [91] employed NPs to preferentially transfer nourishing chemicals to the affected areas. It was shown that the cells of pumpkin plants (*Cucurbita pepo*) transport Fe and Cu NPs.

To stimulate seeds before sowing, constant and variable electric fields (EFs) and magnetic fields (MFs) (electromagnetic fields, ELMFs), ionizing radiation, microwave and laser radiation [92, 93], and magnetized water have often been employed. It has been shown that treating/preserving seeds using ELMF fields has a positive impact on germination and increases agricultural yield [64]. According to Binhi [94], when the physiological window and the electromagnetic window align, the magnetobiological effects in a live cell can be replicated. Depending on the field's strength, frequency, and amplitude and the duration of exposure, either positive or negative impacts may be seen.

Reina *et al* [95] proposed that the MF interacts with ionic electric currents in the cell of the embryo membrane and controls the water penetration into seeds through variations in osmotic pressure and ion concentration. The transport of chemicals into cells can also be impacted by MFs, since they can alter the porosity of the ion channels inside the membrane. MFs also have the ability to increase the production of free radicals, alter hormone and enzyme concentrations, and alter the synthesis or transfer of DNA. The water molecules and the dissolved ions' outer electron shells may become polarized due to the action of an MF. This may lead to modifications in the ion hydration conditions, which may act as nuclei for crystallization. When an MF interacts with water in plant material, the electrically conductive properties of the water can increase, while its surface tension can decrease. The Lorentz force, which causes the charged electrons, particles, or molecules in water (ionization) to temporarily polarize in accordance with an external MF, may be the origin of this phenomenon [96, 97].

It is uncommon to conduct research in which seeds are subjected to MFs from the time of germination through the development of seedlings [98]. Mroczek-Zdyrska *et al* [99] subjected lupine seeds and developing sprouts/seedlings to a continuous MF of 130.0 mT during a 14-day vegetative phase. Researchers also looked at the effects of a changing MFs on the seeds of narrow-leafed lupines (*Lupinus angustifolius L.*), using a field of 0.20 mT and frequencies of 16.0 Hz and 50.0 Hz. MFs at both frequencies decreased the number of carotenoids and chlorophyll in leaves [100].

A review of the research literature has revealed that few studies have examined the effects of ferromagnetic particles on plant development and seed germination in the presence of stable MFs. There are few reports of the amounts of certain components in seeds and seedlings exposed to MFs. Hence, it is understandable that Kornarzyński *et al* [64] want to find out how an aqueous solution of iron NPs affects the growth and development of sprouts/seedlings, the concentration of certain elements in sprouts/seedlings, and the germination of sunflower (*Helianthus annuus L.*) seeds. Investigations were conducted to investigate the effects of stable MFs on the aforementioned processes in systems with and without NPs. The study's objectives were to find out how Fe_3O_4 NPs (Fe NPs) affected sunflower seed germination, seedling growth in the early stages, and the concentrations of certain components in seedlings. Constant MF effects were studied in arrangements with and without Fe NPs. In the presence of a solution comprising 0, 50, or 500 ppm Fe NPs, seeds were exposed to germination and growth experiments under constant (unchanging or stable) MFs (0 (as the control), 5, 25, and 120 mT) for seven days. It was not shown that the MF or Fe NPs had any discernible effects on seed germination or seedling growth. A drop in the germination rate was noted in most cases. The bulk of the samples showed a relative drop in elemental concentrations, and those devoid of Fe NPs showed the largest decline. Interestingly, it was found that the quantities of trivalent elements, such as Fe, and hazardous elements were significantly lower in the Fe-NP-containing samples than in the control samples. According to the authors, in this instance, Fe NPs bound (adsorbed) these elements in the sunflower roots and seeds. This explains why the seedlings had less iron than the seeds before they were sown. These results may point to a synergistic effect—a strengthening of the MF in conjunction with Fe NPs. The results obtained suggest that further iron nanoparticle magnetization might enhance the particles' capacity to adsorb certain components from the seeds and roots of sunflower plants. It is necessary to do more research on the makeup of sunflower seedlings/sprouts and roots that have been exposed to solutions containing varying concentrations of Fe NPs. Finding the lowest concentrations (limits) at which the impacts or influences occur would be ideal.

When organic materials with high carbon content and high aromaticity are thermally decomposed at a temperature below 900 °C in an oxygen-deficient environment, the resultant solid carbonized product is known as biochar [11, 32, 101–105]. Manure contains beneficial nutrients and is used as a fertilizer, but plant-based biochar is often low in nutrients [33, 34]. In addition to boosting soil nutrient availability, the SSA, the cation exchange capacity, total N, carbon-based materials, and water retention [35–37], this sorbent aids in carbon storage and greenhouse gas reduction [38]. It may raise the pH of the soil, provide a liming action that considerably lowers the mobility of As, Cd, Cu, Ni, Pb, and Zn, and immobilize HMs by absorbing them on its surface [39, 40]. In soils polluted with HMs, it has a beneficial impact on microbial activities and plant development and growth [33, 41]. Pb, Zn, Cd, and Cu concentrations can be considerably lowered using biochar [42–44].

Organo-zeolitic substrates are made up of an alkaline zeolite combined with a mixture or combination of sorbents of biological (organic) origin (such as manure, charcoal, peat, etc.) and an additional sorbent of abiotic origin (such as $CaCO_3$) if desired. Each of these elements: perlite (used to sorb Pb, Cu, Ni, and Cd) [45], chicken manure (waste) (used to sorb Cd and Ni) [46, 106], and CaCO3 (used to sorb Cd, Cu, and Pb) [47, 48] has had a beneficial impact in lowering the level of various PDMs. Zeolites are beneficial to soil fertility, sorption characteristics, and phytomass production [49, 50]. They function as fully soluble fertilizers, enhancing the sorption and the equilibrium of the water (water balance), particularly in light, sandy soils, leading to increased yield and improved quality [107].

Reactive N (NO_3) generation is reduced and N release is regulated in soil that has been fertilized or enhanced with organic zeolite combinations [51]. The amount of organic material, clay, and pH in the soil all affect the availability of Cu and other HMs [52]. HM bioavailability may generally be decreased by a large percentage of organic materials [53]. HM immobilization is increased when clay-based minerals have a significant SSA and the pH of the soil is elevated, particularly when the HMs are cations [54, 55].

Hence, a study by Mozdzen *et al* [107] which is a continuation of a study by Turisová *et al* [108] aimed to find out how adding five different sorbents to soil that had been polluted with Cu and other possible PDMs affected the germination and development of Eurasian common grasses. The experiment demonstrated that the plant response to the addition of a sorbent to Cu-contaminated soil varies with respect to the species and developmental stage of the plant. According to the findings, there was no single universal sorbent among the four that were chosen for this analysis that would benefit all the species examined, both during germination and growth. Grass seedlings grew more readily in Cu-contaminated soil than in soil that had any sorbent added to it. The data obtained here, however, also demonstrated that the inclusion of an organo-zeolitic substrate could, but not always, have a distinctly favorable influence throughout the growth and germination processes. Furthermore, seeds cultivated with added chicken manure had by far the smallest seed emergence in the investigated grasses; yet, in the final phases, this sorbent had a good influence on the development of seedlings/sprouts. Thus, it follows that different sorbents should presumably be employed for the treatment and management of seed with respect to those used to encourage plant development. Not only should sorbents designed for direct application throughout plant growth be made available for purchase but also seeds should be prepared with sorbents to speed up germination in soils with specific contaminants.

8.4 Conclusions

Several studies' reports have demonstrated that a plant's reaction to sorbents (especially MSs) added to soil polluted with HMs and other PDMs varies depending on the species of plant and stage of development. There are hardly any universal sorbents, however, that benefit different species at different phases of germination and growth. It has also been noted that adding an organo-zeolitic substrate may, though not always, have a favorable influence on the germination and growth phases.

Reportedly, the seeds treated with added poultry manure had the lowest seed emergence by a substantial margin; yet, in the later phases, these sorbents improved the development of the seedlings. Thus, it follows that different MSs should presumably be employed for seed treatment with respect to those used to encourage plant development. Not only should sorbents designed for direct application during plant growth be made available for purchase but also seeds should be prepared with sorbents to speed up germination in soils with specific contaminants. This might play a significant role in hastening the process of reclaiming land affected by polluting industrial activities. However, since there are still many unanswered questions, more research into this aspect of sorbents (especially MSs) needs to be carried out.

References

[1] El-Nemr M, Aigbe U, Hassaan M, Ukhurebor K, Ragab S, Onyancha R, Osibote O and El Nemr A 2022 The use of biochar-NH_2 produced from watermelon peels as a natural adsorbent for the removal of Cu(II) ion from water *Biomass Conv. Bioref.* **14** 1975–91

[2] Ukhurebor K, Hossain I, Pal K, Jokthan G, Osang F, Ebrima F and Katal D 2023 Applications and contemporary issues with adsorption for water monitoring and remediation: a facile review *Top. Catal.* **67** 140–55

[3] Ukhurebor K, Aigbe U, Onyancha R, Balogun V, Anani O, Adama K, Pal K, Kusuma H and Darmokoesoemo H 2023 An overview of magnetic nanomaterials *Magnetic Nanomaterials* ed U Aigbe, K Ukhurebor and R Onyancha (Cham: Springer Nature) 1–120

[4] Eleryan A, Aigbe U, Ukhurebor K, Onyancha R, Hassaan M, Elkatory M, Ragab S, Osibote O, Kusuma H and El Nemr A 2023 Adsorption of direct blue 106 dye using zinc oxide nanoparticles prepared via green synthesis technique *Environ. Sci. Pollut.* **30** 69666–82

[5] Anani A, Adama K, Ukhurebor K, Habib A, Abanihi V and Pal K 2023 Application of nanofibrous protein for the purification of contaminated water as a next generational sorption technology: a review *Nanotechnology* **34** 1–18

[6] Aidonojie P, Ukhurebor K, Oaihimire I, Ngonso B, Egielewa P, Akinsehinde B, Heri S and Darmokoesoemo H 2023 Bioenergy revamping and complimenting the global environmental legal framework on the reduction of waste materials: a facile review *Heliyon* **9** E12860

[7] Li Y, Padoan E and Ajmone-Marsan F 2021 Soil particle size fraction and potentially toxic elements bioaccessibility: a review *Ecotoxicol. Environ. Saf.* **209** 1–9

[8] Tchounwou B P, Yedjou G C, Patlolla K A and Sutton J D 2012 Heavy metals toxicity and the environment *Mol. Clin. Environ. Toxicol., Exp. Suppl.* **2012** 133–64

[9] Kabata-Pendias A 2011 *Trace Elements in Soils and Plants* (Boca Raton, FL: CRC Press)

[10] Neolaka Y, Riwu A, Aigbe U, Ukhurebor K, Onyancha R, Darmokoesoemo H and Kusuma H 2023 Potential of activated carbon from various sources as a low-cost adsorbent to remove heavy metals and synthetic dyes *Results Chem.* **5** 100711
[11] 2009 *Biochar for Environmental Management Science, Technology and Implementation* ed J Lehmann and S Joseph (London: Earthscan)
[12] Turisová I, Sabo P, Štrba T, Korópy S, Andráš P and Širka P 2016 Analyses of floristic composition of the abandoned Cu-dump field piesky (staré hory mountains, slovakia) *Web. Ecol.* **16** 97–111
[13] Järup L 2003 Hazards of heavy metal contamination *Br. Med. Bull.* **68** 167–82
[14] Khalid S, Shahid M, Niazi K, Murtaza B, Bibi I and Dumat C 2017 A comparison of technologies for remediation of heavy metal contaminated soils *J. Geochem. Explor.* **187** 247–68
[15] Vareda P, Valente J and Durães L 2016 Heavy metals in Iberian soils: removal by current adsorbents/ amendments and prospective for aerogels *Adv. Colloid Interface Sci.* **237** 28–42
[16] Houben D, Evrard L and Sonnet P 2013 Beneficial effects of biochar application to contaminated soils on the bioavailability of Cd, Pb and Zn and the biomass production of rapeseed (Brassica napus L.) *Biomass Bioener.* **57** 196–204
[17] Aigbe U, Ukhurebor K, Onyancha R, Okundaye B, Pal K, Osibote O, Esiekpe E, Kusuma H and Darmokoesoemo H 2022 A facile review on the sorption of heavy metals and dyes using bionanocomposites *Adsorpt. Sci. Technol.* **8030175** 1–36
[18] Rizwan S, Imtiaz M, Zhu J, Yousaf B, Hussain M, Ali L, Ditta A, Ihsan Z, Huang G, Ashraf M *et al* 2021 Immobilization of Pb and Cu by organic and inorganic amendments in contaminated soil *Geoderma* **385** 114803
[19] Sun W, Ji B, Khoso A, Tang H, Liu R, Wang L and Hu Y 2018 An extensive review on restoration technologies for mining tailings *Environ. Sci. Poll. Res.* **25** 33911–25
[20] Abumaizar R and Smith E 1999 Heavy metal contaminants removal by soil washing *J. Hazard. Mater.* **70** 71–86
[21] Deng S 2006 Sorbent technology *Encyclopedia of Chemical Processing* ed S Lee (Boca Raton, FL: CRC Press) 1st edn 2825–45
[22] Gong Y, Zhao D and Wang Q 2018 An overview of field-scale studies on remediation of soil contaminated with heavy metals and metalloids: technical progress over the last decade *Water Res.* **147** 440–60
[23] Ling W, Shen Q, Gao Y, Gu X and Yang Z 2007 Use of bentonite to control the release of copper from contaminated soils *Aust. J. Soil Res.* **455** 618–23
[24] Haydn M 2006 Bentonite applications *Dev. Clay Sci.* **2** 111–30
[25] Wahab N, Saeed M, Ibrahim M, Munir A, Saleem M, Zahra M and Waseem A 2019 Synthesis, characterization, and applications of silk/bentonite clay composite for heavy metal removal from aqueous solution *Front. Chem.* **7** 654
[26] Sun Y, Li Y, Xu Y, Liang X and Wang L 2015 In situ stabilization remediation of cadmium (Cd) and lead (Pb)co-contaminated paddy soil using bentonite *Appl. Clay Sci.* **105–6** 200–6
[27] Kumararaja P, Manjaiah K, Datta C and Shabeer A 2014 Potential of bentonite clay for heavy metal immobilization of soil *Clay Res.* **33** 83–96
[28] Hamidpour M, Kalbasi M, Afyuni M, Shariatmadari H, Holm P and Hansen G 2010 Sorption hysteresis of Cd(II) and Pb(II) on natural zeolite and bentonite *J. Hazard. Mater.* **181** 686–91

[29] Mi J, Gregorich E, Xu S, McLaughlin N, Ma B and Liu J 2017 Effect of bentonite amendment on soil hydraulic parameters and millet crop performance in a semi-arid region *J. Field Crop. Res.* **212** 107–14
[30] De Castro M, Abad M, Sumalinog D, Abarca R, Paoprasert P and de Luna M 2018 Adsorption of methylene blue dye and Cu(II) ions on EDTA-modified bentonite: Isotherm, kinetic and thermodynamic studies *J. Sustain. Environ. Res.* **28** 197–205
[31] Kusuma H, Aigbe U, Ukhurebor K, Onyancha R, Okundaye B, Ama O, Darmokoesoemo H, Widyaningrum B, Osibote O and Balogun V 2023 Biosorption of methylene blue using clove leaves waste modified with sodium hydroxide *Results Chem.* **5** 1–15
[32] Ahmad M, Rajapaksha A, Lim E, Zhang M, Bolan N, Mohan D, Vithanage M, Lee S and Ok S 2014 Biochar as a sorbent for contaminant management in soil and water: a review *Chemosphere* **99** 19–33
[33] Lu H, Li Z, Fu S, Méndez A, Gascó G and Paz-Ferreiro J 2015 Combining phytoextraction and biochar addition improves soil biochemical properties in a soil contaminated with Cd *Chemosphere* **119** 209–16
[34] Uchimiya M, Lima I, Thomas Klasson K, Chang S, Wartelle L and Rodgers J 2010 Immobilization of heavy metal ions (CuII, CdII, NiII, and PbII) by broiler litter-derived biochars in water and soil *J. Agr. Food Chem.* **58** 5538–44
[35] Lee S, Shah H, Awad Y, Kumar S and Ok Y 2015 Synergy effects of biochar and polyacrylamide on plants growth and soil erosion control *Environ. Earth Sci.* **74** 2463–73
[36] Laird D, Brown R, Amonette J and Lehmann J 2009 Review of the pyrolysis platform for coproducing bio-oil and biochar *Biofuels Bioprod. Biorefin.* **3** 547–62
[37] Major J, Lehmann J, Rondon M and Goodale C 2010 Fate of soil-applied black carbon: downward migration, leaching and soil respiration *Glob. Chang. Biol.* **16** 1366–79
[38] Wang D, Jiang P, Zhang H and Yuan W 2020 Biochar production and applications in agro and forestry systems: a review *Sci. Total Environ.* **723** 137775
[39] Rajapaksha A, Ahmad M and Vithanage M 2015 The role of biochar, natural iron oxides, and nanomaterials as soil amendments for immobilizing metals in shooting range soil *Environ. Geochem. Health* **37** 931–42
[40] Lucchini P, Quilliam R, DeLuca T, Vameruli T and Jones D 2014 Does biochar application alter heavy metal dynamics in agricultural soil? *Agric. Ecosyst. Environ.* **184** 149–57
[41] Park J, Choppala G, Bolan N, Chung J and Chuasavathi T Biochar reduces the bioavailability and phytotoxicity of heavy metals *Plant Soil* **348** 439–51
[42] Nie C, Yang X, Niazi N, Xu X, Wen Y, Rinklebe J, Ok Y, Xu S and Wang H 2018 Impact of sugarcane bagasse-derived biochar on heavy metal availability and microbial activity: a field study *Chemosphere* **200** 274–81
[43] Puga P, Abreu C, Melo L and Beesley L 2015 Biochar application to a contaminated soil reduces the availability and plant uptake of zinc, lead and cadmium *J. Environ. Manag.* **159** 86–93
[44] Bian R, Joseph S, Cui L, Pan G, Li L, Liu X, Zhang A, Rutlidge H, Wong S, Chia C *et al* 2014 A three-year experiment confirms continuous immobilization of cadmium and lead in contaminated paddy field with biochar amendment *J. Hazard. Mater.* **272** 121–8
[45] Vijayaraghavan K and Raja D Experimental characterisation and evaluation of perlite as a sorbent for heavy metal ions in single and quaternary solutions *J. Water Process Eng.* **4** 179–84

[46] Barakat M, Ismail S and Ehsan M 2016 Immobilization of Ni and Zn in soil by cow and chicken manure *Int. J. Waste Res.* **6** 1–7

[47] Huang G, Su X, Rizwan M, Zhu Y and Hu H 2016 Chemical immobilization of Pb, Cu, and Cd by phosphate materials and calcium carbonate in contaminated soils *Environ. Sci. Poll. Res. Int.* **23** 16845–56

[48] He G, Zhang Z, Wu X, Cui M, Zhang J and Huang X 2020 Adsorption of heavy metals on soil collected from lixisol of typical karst areas in the presence of $CaCO_3$ and soil clay and their competition behaviour *Sustainability* **12** 7315

[49] Damian G, Andráš P, Damian F, Turisová I and Iepure G 2018 The role of organo-zeolitic material in supporting phytoremediation of a copper mining waste dump *Int. J. Phytoremediat.* **20** 1307–16

[50] Damian F, Jelea S, Lăcătusu R and Mihali C 2019 The treatment of soil polluted with heavy metals using the sinapis alba l. and organo zeolitic amendment *Carpathian J. Earth Environ. Sci.* **14** 409–22

[51] Jakkula V and Wani S 2018 Zeolites: potential soil amendments for improving nutrient and water use efficiency and agriculture productivity *Sci. Revs. Chem. Commun.* **8** 1–15

[52] Zeng F, Ali S, Zhang H, Ouyang Y, Qiu B, Wu F and Zang G 2011 The influence of pH and organic matter content in paddy soil on heavy metal availability and their uptake by rice plants *Environ. Poll.* **159** 84–91

[53] Castaldi P, Alberti G, Merella R and Melis P 2005 Study of the organic matter evolution during municipal solid waste composting aimed at identifying suitable parameters for the evaluation of compost maturity *Waste Manag.* **25** 209–13

[54] Herath I, Kumarathilaka P, Navaratne A, Rajakaruna N and Vithanage M 2015 Immobilization and phytotoxicity reduction of heavy metals in serpentine soil using biochar *J. Soils Sediments* **15** 126–38

[55] Herath I, Iqbal M, Al-Wabel M, Abduljabbar A, Ahmad M, Usman A, Ok Y, M and Vithanage 2017 Bioenergy-derived waste biochar for reducing mobility, bioavailability, and phytotoxicity of chromium in anthropized tannery soil *J. Soils Sediments* **17** 731–40

[56] Usman A, Kuzyakov Y and Stahr K 2005 Effect of clay minerals on immobilization of heavy metals and microbial activity in a sewage sludge-contaminated soil (8pp) *J. Soils Sediments* **5** 245–52

[57] Prost R and Yaron B 2001 Use of modified clays for controlling soil environmental quality *Soil Sci.* **166** 880–95

[58] Onyancha R, Aigbe U, Ukhurebor K, Kusuma H, Darmokoesoemo H, Osibote O and Pal K 2022 Influence of magnetism-mediated potentialities of recyclable adsorbents of heavy metal ions from aqueous solution–an organized review *Results Chem.* **4** 100452

[59] Osman A *et al* 2023 Methods to prepare biosorbents and magnetic sorbents for water treat: a review *Environ. Chem. Lett.* **21** 2337–98

[60] Onyancha R, Oyomo B, Aigbe U and Ukhurebor K 2023 Application of magnetic nanomaterials as drug and gene delivery agent *Magnetic Nanomaterials* ed U Aigbe, K Ukhurebor and R Onyancha (Cham: Springer Nature) 201–16

[61] Aigbe U, Onyancha R, Ukhurebor K and Obodo K 2020 Removal of fluoride ions using polypyrrole magnetic nanocomposite influenced by rotating magnetic field *RSC Adv.* **10** 595–609

[62] 2023 *Magnetic Nanomaterials: Synthesis, Characterization and Applications* ed U O Aigbe, K E Ukhurebor and R B Onyancha (Cham: Springer Nature)

[63] Aigbe U *et al* 2023 Utility of magnetic nanomaterials for theranostic nanomedicine *Magnetic Nanomaterials* ed U O Aigbe, K E Ukhurebor and R B Onyancha (Cham: Springer) 47–86

[64] Kornarzyński K, Sujak A, Czernel G and Wiącek D 2020 Effect of Fe_3O_4 nanoparticles on germination of seeds and concentration of elements in Helianthus annuus L. under constant magnetic field *Sci. Rep.* **10** 8068

[65] Siddiqui M and Al-Whaibi M 2014 Role of nano-SiO_2 in germination of tomato (Lycopersicum esculentum seeds mill) *Saudi J. Biol. Sci.* **21** 13–7

[66] Rizwan M *et al* 2017 Effect of metal and metal oxide nanoparticles on growth and physiology of globally important food crops: a critical review *J Hazard.Mater.* **322** 2–16

[67] Mahmoodzadeh H, Nabavi M and Kashef H 2013 Effect of nanoscale titanium dioxide particles on the germination and growth of canola (Brassica napus) *J. Ornam. Hortic. Plants* **3** 25–32

[68] Feizi H, Kamali M, Jafari L and Moghaddam P 2013 Phytotoxicity and stimulatory impacts of nanosized and bulk titanium dioxide on fennel (Foeniculum vulgare Mill) *Chemosphere* **91** 506–11

[69] Srivastava G *et al* 2014 Seed treatment with iron pyrite (FeS_2) nanoparticles increases the production of spinach *RSC Adv.* **4** 58495–504

[70] Lee C *et al* 2010 Developmental phytotoxicity of metal oxide nanoparticles to Arabidopsis thaliana *Environ. Toxicol. Chem.* **29** 669–75

[71] Ghafariyan M, Malakouti M, Dadpour M, Stroeve P and Mahmoudi M 2013 Effects of magnetic nanoparticles on soybean chlorophyll *Env. Sci. Technol.* **47** 10645–52

[72] Zhu H, Han J, Xiao J and Jin Y 2008 Uptake, translocation, and accumulation of manufactured iron oxide nanoparticles by pumpkin plants *J. Env. Monitor.* **10** 713–7

[73] Yasmeen F, Raja N, Razzaq A and Komatsu S 2016 Gel-free/label-free proteomic analysis of wheat shoot in stress tolerant varieties under iron nanoparticles exposure *Biochim. Biophys. Acta* **1864** 1586–98

[74] Das C *et al* 2016 The seed stimulant effect of nano iron pyrite is compromised by nano cerium oxide: regulation by the trace ionic species generated in the aqueous suspension of iron pyrite *RSC Adv.* **6** 67029–38

[75] Xu P *et al* 2012 Use of iron oxide nanomaterials in wastewater treatment: a review *Sci. Total Environ.* **424** 1–10

[76] Boparai H K, Joseph M and O'Carroll D 2011 Kinetics and thermodynamics of cadmium ion removal by adsorption onto nano zerovalent iron particles *J. Hazard. Mater.* **186** 458–65

[77] Celebi O, Uzumb C, Shahwan T and Erten H 2007 A radiotracer study of the adsorption behaviour of aqueous Ba2+ ions on nanoparticles of zero-valent iron *J. Hazard. Mater.* **148** 761–7

[78] Ramandi N and Shemirani F 2015 Selective ionic liquid ferrofluid based dispersive-solid phase extraction for simultaneous preconcentration/separation of lead and cadmium in milk and biological samples *Talanta* **131** 404–11

[79] Li X and Zhang W 2006 Iron nanoparticles: the core–shell structure and unique properties for Ni(II) sequestration *Langmuir* **10** 4638–42

[80] Zhang X, Lin S, Lu X-Q and Chen Z 2010 Removal of Pb(II) from water using synthesized kaolin supported nanoscale zero-valent iron *Chem. Eng. J.* **163** 243–8

[81] Zhang D *et al* 2010 Carbon-stabilized iron nanoparticles for environmental remediation *Nanoscale* **2** 917–9

[82] Ge F, Li M-M, Ye H and Zhao B-X 2012 Effective removal of heavy metal ions Cd^{2+}, Zn^{2+}, Pb^{2+}, Cu^{2+} from aqueous solution by polymer-modified magnetic nanoparticles *J. Hazard. Mater.* **211–2** 366–72

[83] Singh S, Barick K and Bahadur D 2011 Surface engineered magnetic nanoparticles for removal of toxic metal ions and bacterial pathogens *J. Hazard. Mater.* **192** 1539–47

[84] Badruddoza A, Tay A, Tan P, Hidajat K and Uddin M 2011 Carboxymethyl-β-cyclodextrin conjugated magnetic nanoparticles as nano-adsorbents for removal of copper ions: synthesis and adsorption studies *J. Hazard. Mater.* **185** 1177–86

[85] Liu W-J, Qian T-T and Jiang H 2014 Bimetallic Fe nanoparticles: recent advances in synthesis and application in catalytic elimination of environmental pollutants *Chem. Eng. J.* **236** 448–63

[86] Wang J *et al* 2010 Amino-functionalized Fe_3O_4@SiO_2 core–shell magnetic nanomaterial as a novel adsorbent for aqueous heavy metals removal *J. Colloid Interface Sci.* **349** 293–9

[87] Hu J C G and Lo I M 2006 Selective removal of heavy metals from industrial wastewater using maghemite nanoparticle: performance and mechanisms *J. Env. Eng.* **132** 709–15

[88] Liu J, Zao Z and Jang G 2008 Coating Fe_3O_4 magnetic nanoparticles with humic acid for high efficient removal of heavy metals in water *Environ. Sci. Technol.* **42** 6949–54

[89] Savina I *et al* 2011 High efficiency removal of dissolved As(III) using iron nanoparticle-embedded macroporous polymer composites *J. Hazard. Mater.* **192** 31002–8

[90] Tang L *et al* 2014 Synergistic effect of iron doped ordered mesoporous carbon on adsorption-coupled reduction of hexavalent chromium and the relative mechanism study *Chem. Eng.* **239** 114–22

[91] Corredor E *et al* 2009 Nanoparticle penetration and transport in living pumpkin plants: in situ subcellular identification *BMC Plant Biol.* **9** 1–11

[92] Aladjadjiyan A 2010 Influence of stationary magnetic field on lentil seeds *Int. Agrophys.* **24** 321–4

[93] Galland P and Pazur A 2005 Magnetoreception in plants *J. Plant. Res.* **118** 371–89

[94] Binhi V 2002 *Magnetobiology Underlying Physical Problems* (New York: Academic)

[95] Garciá Reina F, Arza Pascual L and Almanza Fundora I 2001 Influence of a stationary magnetic field on water relations in lettuce seeds. Part II: experimental results *Bioelectromagnetics* **22** 596–602

[96] Goldsworthy A, Whitney H and Morris E 1999 Biological effects of physically conditioned water *Water Res.* **33** 1618–26

[97] Coey J and Cass S 2000 Magnetic water treatment *J. Magn. Magn. Mater.* **209** 71–4

[98] Pietruszewski S and Kornarzyński K 2001 Germination of wheat grain in an alternating magnetic field *Int. Agrophys.* **15** 269–71

[99] Mroczek-Zdyrska M, Kornarzyński K, Pietruszewski S and Gagos M 2016 Stimulation with a 130-mT magnetic field improves growth and biochemical parameters in lupin (Lupinus angustifolius L.) *Turkish. J Biol.* **40** 699–705

[100] Mroczek-Zdyrska M, Kornarzyński K, Pietruszewski S and Gagoś M 2017 The effects of low-frequency magnetic field exposure on the growth and biochemical parameters in lupin (Lupinus angustifolius L.) *Plant Biosyst.* **151** 504–11

[101] Eleryan A, Hassaan M, Aigbe U, Ukhurebor K, Onyancha R, Kusuma H, El-Nemr M, Ragab S and El Nemr A 2023 Biochar-C-TETA as a superior adsorbent to acid yellow 17 dye from water: isothermal and kinetic studies *J. Chem. Technol. Biotechnol.* **2023** 1–14

[102] Eleryan A, Hassaan M, Aigbe U, Ukhurebor K, Onyancha R, El Nemr M *et al* 2023 Kinetic and isotherm studies of acid orange 7 dye absorption using sulphonated mandarin biochar treated with teta *Biomass Convers. Biorefinery* **2023** 1–12

[103] Eldeeb T, Aigbe U, Ukhurebor K, Onyancha R, El-Nemr M, Hassaan M *et al* 2022 Adsorption of methylene blue (MB) dye on ozone, purified and sonicated sawdust biochars *Biomass Convers. Biorefinery* **14** 9361–83

[104] Eleryan A, . El Nemr A *et al* 2022 Copper (II) ion removal by chemically and physically modified sawdust biochar *Biomass Convers. Biorefinery* **2022** 1–38

[105] El-Nemr M, Aigbe U, Ukhurebor K, Onyancha R, El Nemr A, Ragab S, Osibote O and Hassaan M 2022 Adsorption of Cr^{6+} ion using activated Pisum sativum peels-triethylene-tetramine *Environ. Sci. Pollut. Res.* **2022** 1–25

[106] Huang Q, Wan Y, Luo Z, Qiao Y, Su D and Li H 2019 The effects of chicken manure on the immobilization and bioavailability of cadmium in the soil-rice system *Arch. Agr. Soil Sci.* **66** 1753–64

[107] Możdżeń K, Barabasz-Krasny B, Kviatková T, Zandi P and Turisová I 2021 Effect of sorbent additives to copper-contaminated soils on seed germination and early growth of grass seedlings *Molecules* **26** 5449

[108] Turisová I, Kviatková T, Możdżeń K and Barabasz-Krasny B 2020 Effects of natural sorbents on the germination and early growth of grasses on soils contaminated by potentially toxic elements *Plants* **9** 1591

Chapter 9

The use of magnetic sorbents in improving photovoltaic performance

David O Idisi and Edson L Meyer

The undeniable global energy crisis has been a topical issue, attracting researchers' attention. Photovoltaics, as a renewable energy source and a technical way of alleviating the energy crisis, has been a front-runner and a subject of recent/ongoing research. The chapter examines the use of magnetic sorbents in optimizing the performance of photovoltaic-based energy harvesting. It focuses on the incorporation of magnetic sorbents into nanomaterials for solar cells, the preparation of such nanomaterials, and their various solar energy harvesting properties. It also highlights insights into the recent trends and outlook of magnetic sorbent-based photovoltaics and provides a future perspective on the subject.

9.1 Introduction

The roadmap of the performance of the various photovoltaic technologies that have been used to reach the current 25% has been intriguing. The process has combined different technologies such as inorganic solar cells made from crystalline, polycrystalline, amorphous, and microcrystalline Si, III–V compounds and alloys, CdTe, chalcopyrite compounds, and copper indium gallium diselenide (CIGS). In addition, organic, dye-sensitized, and perovskite solar cells (PSCs) have been on the front line of emerging technologies for photovoltaics [1]. While strategies for improving the lifetime and performance of these technologies is still the subjects of ongoing research, the improvement of photovoltaic performance using magnetic sorbents is gaining attention from researchers.

Magnetic sorbents are a class of nanoparticles with physical, chemical, and magnetic properties that tend to interact with chemical mixtures [2]. The interactive tendency of these sorbents leads to an improvement in the physicochemical properties of the parent material, which can be brought to bear in applications of interest. Due to the remarkable physicochemical properties of magnetic sorbents,

doi:10.1088/978-0-7503-5909-2ch9

they have found use in water purification and heavy metal removal [3]. Recently, the electronic properties of magnetic sorbents such as electrochemical sensors [4] have been used to overcome some of their limitations. Hence, researchers have devoted attention to magnetic sorbents because of their potential use in optoelectronic applications.

In addition, the magnetic field emanating from the Coulomb exchange interaction caused by particular sorbents can be explored for its Hall-effect properties and channeled toward applications such as magnetic gas sensing [5]. Combining magnetic spin ordering and electrochemical properties forms the backbone of sorbent-based photovoltaics applications. The incorporation of magnetic spins from the sorbent as an additional parameter in materials for photovoltaic applications can extend their degree of freedom and improve their long-term performance. While sorbents have mainly been employed for pollutant remediation and wastewater treatment applications [6, 7], sorbents in the form of magnetic nanoparticles can be used to form magnetic hybrids and harnessed in photovoltaic applications.

This chapter examines organic solar cell-based photovoltaics and their synthesis and focuses on the impact of magnetic spin ordering on the performance of solar cells. It provides insights into some magnetic nanoparticle sorbent types and the photocurrent response results recently reported in the research literature.

9.2 The synthesis of magnetic sorbents for photovoltaic applications

It has been widely established that the nanomaterials synthesis methods offer the opportunity to control particle morphology and size, which inevitably influences the resulting properties of the nanomaterials [8]. The attainment of homogeneous sorbent nanoparticles is crucial for maintaining enhanced magnetic response and electrical conductivity [9, 10]. Synthesis routes are chosen that prioritize the monodisperse and homogeneous distribution of the magnetic sorbent on the surface of the parent material. As a broad overview, most magnetic-sorbent-based nanocomposite preparation uses a wet chemistry process, while other techniques are employed to modify the end product to produce the desired nanocomposite. The next section examines the sol–gel method as the predominant procedure; subsequently, other modification routes are presented.

9.2.1 The sol–gel method

The sol–gel synthesis method is the most widely used method of nanocomposite synthesis, due to the ease of control of each chemical precursor with respect to concentration, purification, and other parameters. The sol–gel method involves the mixture of chemical precursors in aqueous form to produce the desired slurry/precipitate. Due to its ease of implementation, this method has been employed to incorporate magnetic sorbents into parent nanomaterials [11–13]. A recent study by Hammad *et al* [12] used the sol–gel method to prepare a hybrid nanocomposite of Fe_2O_3:ZnO. In this study, Fe_2O_3 nanoparticles represented the magnetic sorbents incorporated into inorganic–organic ZnO nanomaterial. The procedure involved stirring an aqueous mixture of ferric nitrate ($Fe(NO_3)_3.9H_2O$) and zinc acetate

($(CH_3CO_2)_2Zn$) at room temperature for approximately 30 min. A polymeric capping agent was introduced via the same sol–gel technique to obtain another solution. Both solutions were subsequently mixed via stirring to obtain a capped Fe_2O_3:ZnO nanocomposite. Figures 9.1(a) and (b) show the microstructure and morphological properties of the polymer-capped Fe_2O_3:ZnO nanocomposite. As indicated in the x-ray diffraction (XRD) data, the peak for Fe_2O_3 is prominent in the nanocomposite, which implies that the nanocomposite nucleation process was successful.

The homogeneous distribution of the Fe_2O_3 sorbent on ZnO is evident in the scanning electron microscopy image shown in figure 9.1(b). As mentioned earlier, a homogeneous distribution is crucial to obtain the desired magnetic properties. The ease of the sol–gel process has been applied in the production of other magnetic sorbents, such as zinc ferrite nanoparticles. Ali and colleagues [15] varied the concentration of Zn in a Co–Zn–Fe nanocomposite using the sol–gel route. Their process was similar to that reported by Hammad except for the chemical precursors used. Citric acid was used as a stabilizing agent alongside nitric metal at a ratio of 1:2 to obtain the different nanocomposites. The final product was heated to evaporate excess solvent concentrates and complete the gel transformation. The gel was further calcinated at 400°C for 6 h to reduce the impact of impurities. Figures 9.2(a) and (b) show the microstructure and morphology of the $Co_{1-x}Zn_xFe_2O_4$ magnetic nanocomposite. A microstructural analysis performed using XRD shows Zn and Fe-related peaks, which are associated with the ZnO and Fe_3O_4 microstructure [16]. As indicated, the citric acid enabled the nanoparticles to disperse and display only weak agglomeration.

The sol–gel technique's advantages allow the synthesis process to be modified or combined with other techniques to obtain the desired morphology and microstructure for specific applications. The following section examines some of the techniques researchers have employed recently.

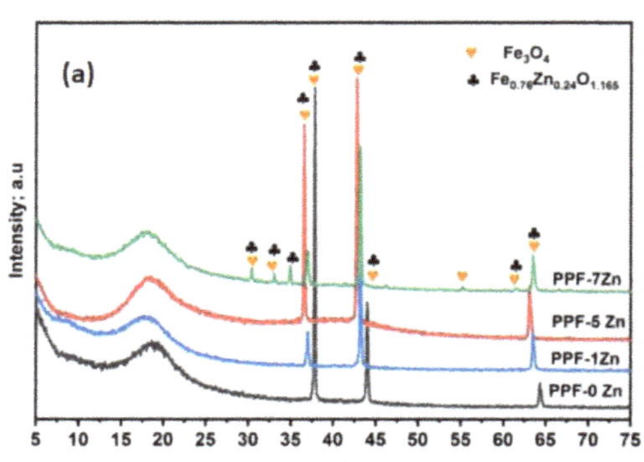

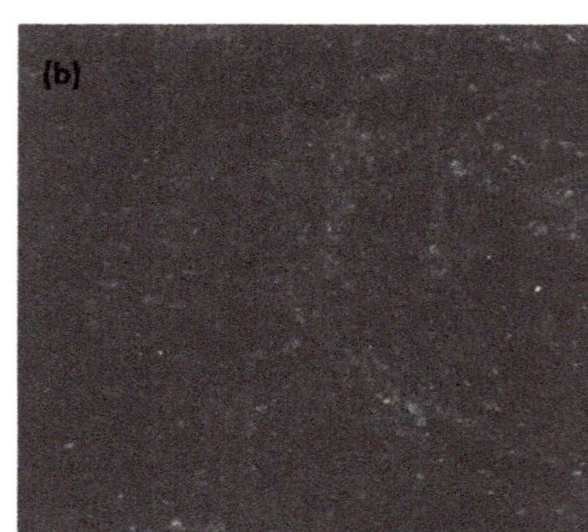

Figure 9.1. (a) The microstructure and (b) the morphology of the polymer-capped Fe_2O_3:ZnO nanocomposite. Image reproduced under open access licence from Hammad *et al* [14]. Copyright Springer Publishing, 2023.

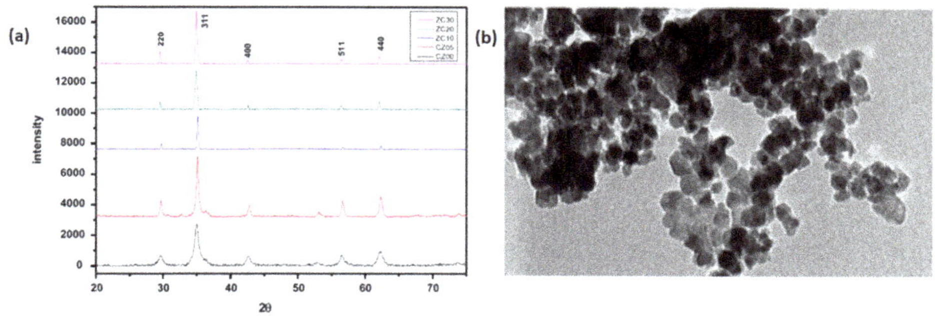

Figure 9.2. (a) The microstructure and (b) the morphology of the $Co_{1-x}Zn_xFe_2O_4$ magnetic nanocomposite. Image reproduced with permission from Ali *et al* [15]. Copyright Elsevier, 2016.

9.2.2 Coprecipitation

The coprecipitation method is a widely used synthesis route for producing ceramic-based nanoparticles. The ease of controlling the pH, reactant concentration, reaction temperature, and time has led to increased use of the coprecipitation method [17, 18]. Due to the potential to obtain high-yield magnetic nanocomposites and the ease of control of the synthesis process, the coprecipitation method has gained enormous attention from researchers. The process described in the early work of Gopalan and co-workers [19] established the effectiveness of coprecipitation in preparing a manganese ferrite nanostructure. The desired stoichiometric amounts of $Fe(NO_3)_3.9H_2O$ and $Mn(NO_3)_2.4H_2O$ were dissolved in deoxidized water. NaOH, a base precursor, was added to the dissolved mixture dropwise with concurrent magnetic stirring. The continuous addition of the drops and stirring at a constant temperature of 70°C resulted in a pH of 13. The resulting precipitate was washed several times and sintered at 600°C for 5h.

Figures 9.3(a) and (b) show the microstructure and morphological properties of the coprecipitate-prepared $MnFe_2O_4$ nanoparticles. The XRD peak positions represent a typical magnetite–spinel structural phase [10] with the appearance of an MnO_6 octahedral phase [20]. The transmission electron microscopy (TEM) image shows sparsely dispersed particles averaging 28nm in size. Moreover, the coprecipitation method has been extended to carbon-based sorbents such as iron–carbon nanocomposites. In their recent study, Chen *et al* [21] developed an iron–carbon sorbent using the coprecipitation route. Their procedure was like that of Gopalan *et al* with slight modifications to account for the carbon content. Activated carbon was used in conjunction with $FeSO_4.7H_2O$ and $FeCl_3.6H_2O$ reagents. Figure 9.4 shows similar features to the spinel phase of hematite, suggesting the effectiveness of the co-precipitation method.

In summary, the coprecipitation method can be extended to magnetism-based sorbents, as indicated by other studies [22]. However, significant distinctions are usually found in the chemical precursors, which must be modified to suit the desired application.

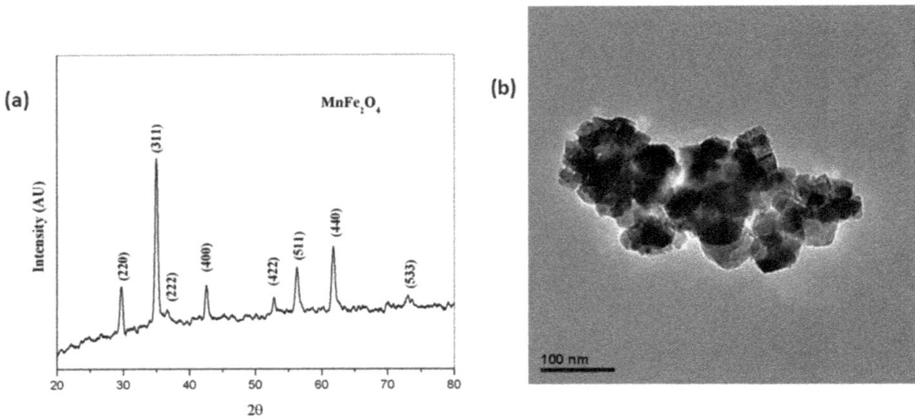

Figure 9.3. The microstructure and morphology of the $MnFe_2O_4$ nanoparticles. (a) XRD spectra; (b) TEM images. Image reproduced by permission from Gopalan *et al* [19], copyright, IOP Publishing, 2008.

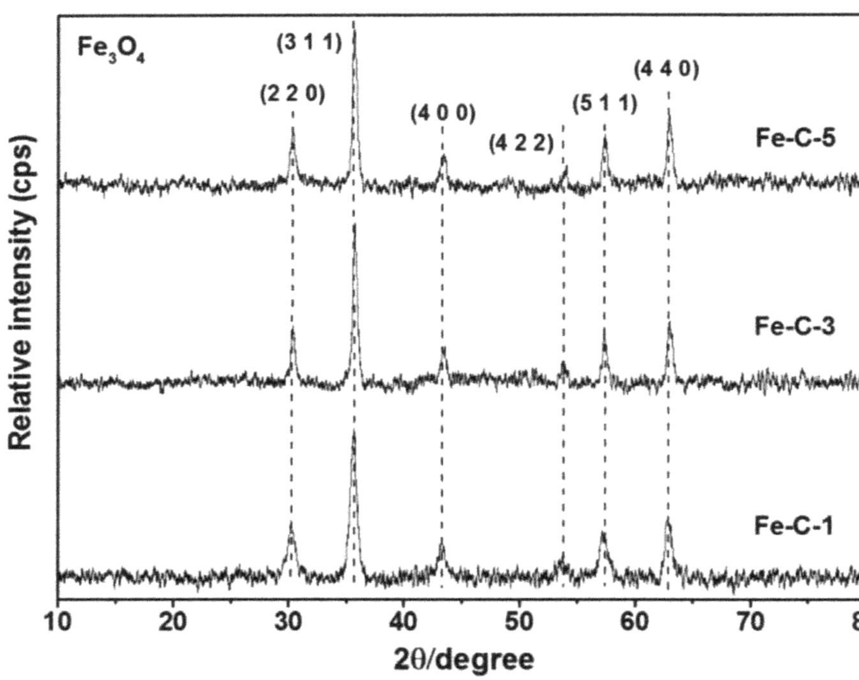

Figure 9.4. The microstructural properties of the iron–carbon sorbent that exhibited a spinel phase. Image adapted with permission from Chen *et al* [21], copyright, Elsevier, 2020.

9.2.3 The hydrothermal method

Due to heat's significance in enhancing the processes of nucleation and nano-composite formation, hydrothermal synthesis has been used for the preparation of magnetic sorbent-based nanoparticles. The advantage of the hydrothermal method is that it produces nanoparticles of homogeneous size, morphology, crystallinity, and phase [23]. This process involves heating a stoichiometric mixture of the desired precursors in a Teflon steel autoclave chamber at specific temperatures and for specific durations [24]. Among numerous studies of hydrothermally prepared magnetic sorbents [25–27], Pandi *et al* [28] synthesized an iron oxide nanoparticle-based sorbent using the hydrothermal route. In their approach, a desired stoichiometric aqueous mixture of $Ca(NO_3)_2.4H_2O$ and $Al(NO_3)_3.9H_2O$ was added to another dissolved mixture of Na_2CO_3 and NaOH. The entire slurry was subsequently added to a combined solution of $FeCl_2.4H_2O$ and $FeCl_3.6H_2O$. The basic reaction properties within the preceding mixtures produced a magnetic oxide precipitate with a pH of 10. The precipitate thus obtained was subsequently transferred to an autoclave in an oven at a temperature of 700°C for 12 h. The final product was magnetically separated and washed several times with deoxidized water to reduce the impact of trace impurities. Figures 9.5(a) and (b) show the morphology and molecular bonding properties of the iron oxide magnetic nanoparticle sorbent. The blackish spheres shown in the scanning electron microscopy image depict the iron nanoparticles, whereas the greyish flakes represent the chitosan molecules that were attached to the magnetic sorbent for the desired application. The peaks depicted in the Fourier transform infrared (FTIR) spectroscopy are consistent with typical hematite nanoparticles [29] and confirm the electronic bonding of the individual elemental components.

The versatility of the hydrothermal method allows it to be reproduced for other magnetic nanoparticle sorbents. Xiong *et al* [30] used the hydrothermal route to prepare $CuCrO_2$ nanocrystals. Their procedure utilized $Cr(NO_3)_3.9H_2O$, Cu $(NO_3)_2.3H_2O$, and NaOH as chemical precursors. A mixture of stoichiometric measures of $Cr(NO_3)_3.9H_2O$ and $Cu(NO_3)_2.3H_2O$ was dissolved in deoxidized water to obtain an aqueous solution and was stirred magnetically. The subsequent addition of a stoichiometric measure of NaOH was introduced into the dissolved mixture for mineralization.

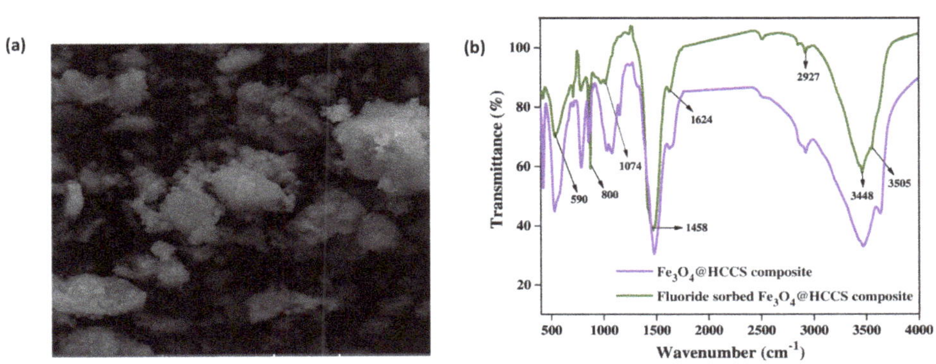

Figure 9.5. (a) A scanning electron microscopy image and (b) FTIR analysis of the iron oxide magnetic nanoparticle sorbents. Image reproduced with permission from Pandi *et al* [28], copyright, Elsevier, 2019.

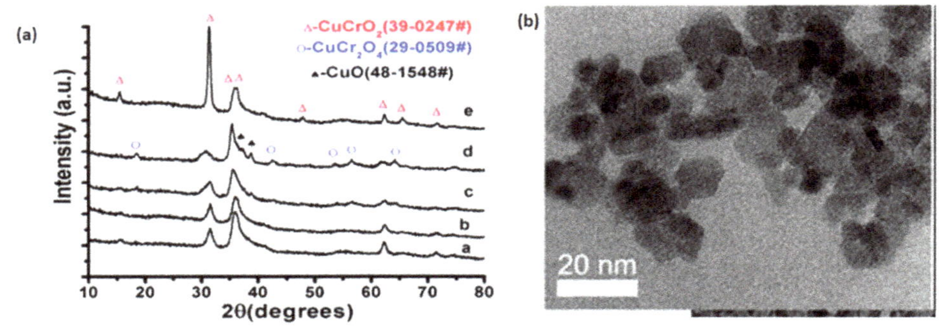

Figure 9.6. The microstructure and morphological properties of $CuCrO_2$. (a) XRD Spectra (b) TEM images. Image reproduced with permission from Xiong *et al* [30]. Copyright RSC Publishing, 2012.

The obtained mineralized mixture was transferred to a 100 ml autoclave and heated at 240 °C for 60 h. The final product was purified several times with diluted hydrochloric acid and ethanol. Figures 9.6(a) and (b) show the microstructure and morphology obtained from XRD and TEM measurements. The prominent peaks from the XRD measurement are consistent with the hexagonal R3m crystalline phase given in JCPDS card #39-0247. The TEM images indicate a homogeneous distribution of the nanocrystals with negligible aggregation, which could be important for energy harvesting applications. The ease of nucleation of the nanoparticles/clusters has led to the popularity of the hydrothermal synthesis method among researchers. Hence, a series of samples with varying parameters, such as heating times, can be used to tune the morphology of the magnetic sorbent-based nanoparticles/composites.

9.2.4 Ball milling

The ball milling technique is a widely used synthesis method that enables the production of ultrafine nanoparticles. The adaptability of ball milling synthesis has led to its use in wet and dry nanomaterial preparation. Due to the milling properties such as collision force, direction, and kinetic energies of the balls, the method can be adapted to small- and large-scale nanomaterial production [31].

The working principle of the ball milling technique is based on the geometrical shape of the balls, which are cylindrical and rotate horizontally along their axes. Due to the spaces between the milling materials, which can be iron, steel, flint (silica), or porcelain bearings, particle loads can be easily milled. The cascade repetition of the balls rolling along the cylindrical wall results in the crushing and refinement of the particles. The effectiveness of the milling method has resulted in the production of different types of nanoparticles and composites [32].

The studies of Carvalho and co-workers [33] utilized the ball milling technique to obtain Fe_3O_4 nanoparticles. The process involved using iron powder on a planetary miller with an angular velocity of 300 rpm. The ball-to-powder mass ratio was set to 20:1 with a corresponding 200 rpm rotation speed for a milling duration of 96 h and powder data were acquired at desired intervals. Figure 9.7 shows the XRD spectra

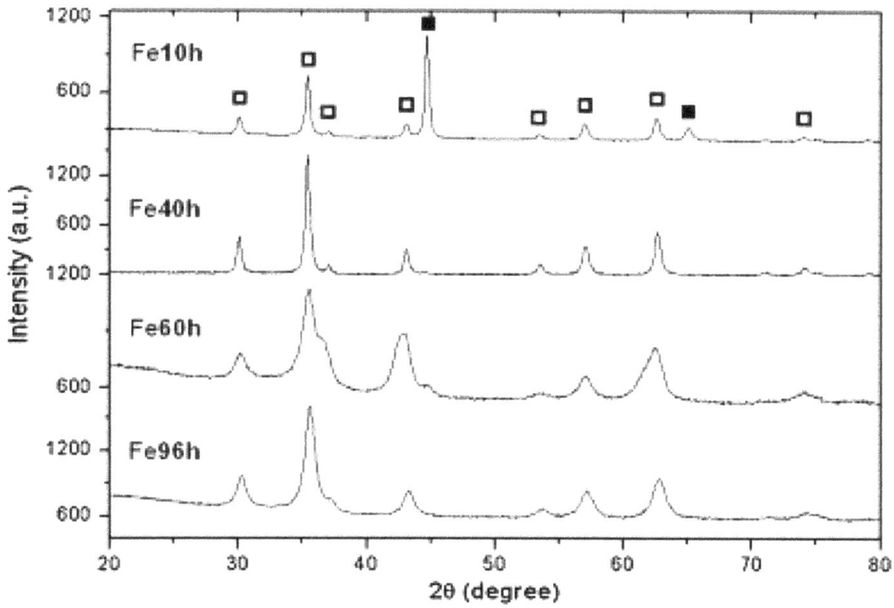

Figure 9.7. The XRD spectral data of Fe_3O_4 obtained using the ball milling technique. Image reproduced by permission from Carvalho *et al* [33], copyright, Elsevier, 2013.

of the obtained microstructure with corresponding decreasing particle sizes as the milling times increased.

The balling milling method can be extended to nanocomposites using a two-step process. First, ingots of the nanocomposite are prepared via another physical mixture method, such as arc melting or other heating routes. The alloyed ingots are then transferred to a ball miller for crushing and fine powder milling. As explored by Su's group [34], ball milling can be effectively employed in preparing nanocomposites. $SmCo_5$ ingots were initially prepared by a vacuum arc melting process in their study. Thereafter, the ingots were crushed and ball milled for 5 h at a powder ratio of 20 to 25:1.

As indicated in figure 9.8, the XRD data showed peaks for Sm–Co and Fe, which implies the formation of a Sm–Co:Fe nanocomposite. While the ball milling technique is popular for dry samples, its use can be extended to wet nanocomposites. The working principle for wet samples is similar to that used for dry samples; the main difference lies in the first step, which involves the mixture of the chemical precursors. The precipitate obtained from the chemical mixtures can be ball milled to obtain finer nanoparticles and characterized for the desired applications.

An instance of this procedure was utilized by Narayanaswamy *et al* [35], in which the ball milling technique was employed to obtain a GO:Fe_3O_4 nanocomposite following the initial use of a wet chemistry route. The GO was obtained using the Hummers method, a wet chemistry route. The obtained precipitate was then oven dried and mixed with a stoichiometric measure of Fe_3O_4 to formulate the composite.

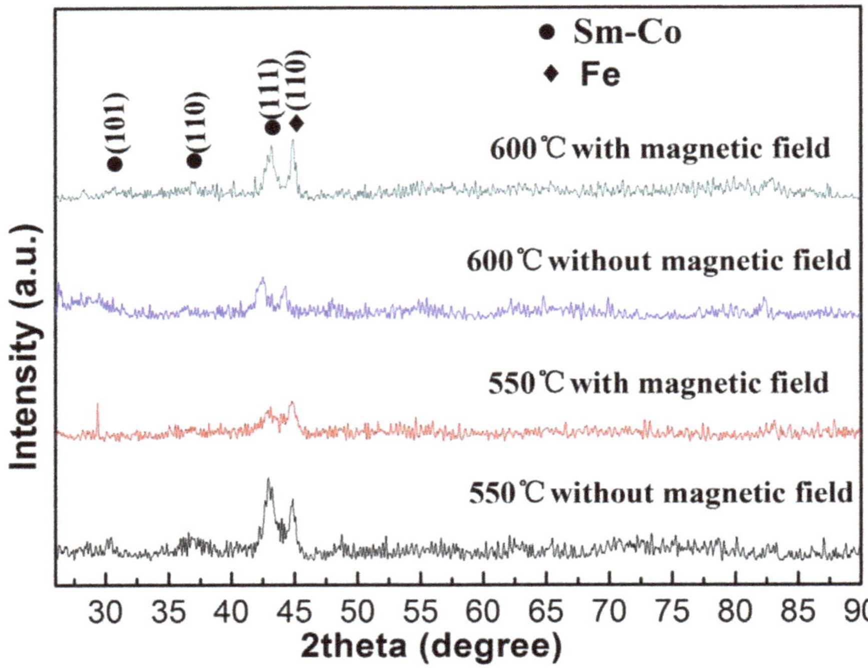

Figure 9.8. The microstructural properties of the Sm–Co:Fe nanocomposite produced using the ball milling technique. Image adapted with permission from Su *et al* [38]. Copyright Elsevier, 2015.

As described earlier, the mixture was transferred to a ball miller at a ball-to-powder ratio of 20:1 and milled for 45 h; samples were collected at 25, 35, 40, and 45 h.

Other synthesis routes, such as green-assisted synthesis, are currently being explored [36, 37]. In these processes, plant extracts are used with other chemicals to improve chemical reduction and bindings. The plant extracts are not used as standalone precursors mainly because they are low yield and require chemical reagents to boost their concentrations. It is pertinent to note that the described synthesis methods are not unique to the current focus of this chapter; they can be used in other applications. Table 9.1 summarizes some of the magnetism-based sorbent nanoparticles/composites that have been discussed in this section.

9.3 The application of magnetic spin in photovoltaics

The introduction of sorbent-based magnetic nanoparticles into organic solar cells has recently gained huge popularity. The inclusion of magnetic spin as an additional degree of freedom in organic solar cells has permitted improved electronic interactions within the molecular structure of the organic materials [39]. Depending on the nature of the photovoltaic device, magnetic spin interaction with the active material results in tuning the excitonic properties of the intrinsic photogenerated electron–hole (e–h) pairs. This section examines the effect of magnetic ordering on the photovoltaic performance of different organic solar cells.

Table 9.1. A summary of magnetism-based photovoltaic solar cells and their synthesis routes.

Magnetic sorbents	Synthesis route	Photovoltaic type	References
Fe_2O_3 γ-Fe_2O_3	Coprecipitation Sol–gel/hydrothermal	Organic solar cells Dye-sensitized solar cells	[44, 47–49]
Fe_3O_4	Coprecipitation Ball milling	Perovskite solar cells (PSCs)	[43, 61, 62]
$CoFe_2O_4$	Coprecipitation	Organic solar cells Perovskites	[45, 63, 64]
$ZnFe_2O_4$	Coprecipitation	Organic solar cells Dye-sensitized solar cells	[45, 65, 66]
$NiFe_2O_4$	Coprecipitation	Organic solar cells PSCs	[45, 67]
Co_3O_4	Coprecipitation	PSCs	[55, 58, 68]
Mn-NP	Sol–gel	PSCs	[56, 69, 70]

Recent results related to their performance are examined and summarized in table 9.1.

9.3.1 Magnetism-based organic solar cells

Organic solar cells (OSCs), also known as bulk heterojunction polymer solar cells, are a class of photovoltaics that have become popular due to their light weight (<100 nm) and economic viability in comparison with commercially available inorganic solar cells [40]. While numerous attempts have been made to improve the performance of OSCs, the incorporation of magnetic sorbents has offered a new opportunity to advance OSC-based photovoltaics. The active organic material for OSCs possesses photogenerated charge carriers, which interact by Coulomb interactions. The Coulomb interactions between the electrons lead to the formation of e–h pairs. Hence, including an external magnetic field in an OSC can increase the acceleration of charge carrier transport [41].

Due to the experimental realization of magnetic sorbent incorporation into OSCs [42], researchers have attempted to harvest the potential of the external magnetic fields in different types of organic materials for improved photovoltaic performance. Among the numerous attempts by different researchers, the early study of Zhang's group [43] incorporated Fe_3O_4 magnetic sorbent-based nanoparticles into a poly(3-hexylthiophene) (P3HT) and [6,6]-phenyl C61-butyric acid methyl ester (PCBM) blend. In this study, oleic acid was used as a capping agent for easy attachment of

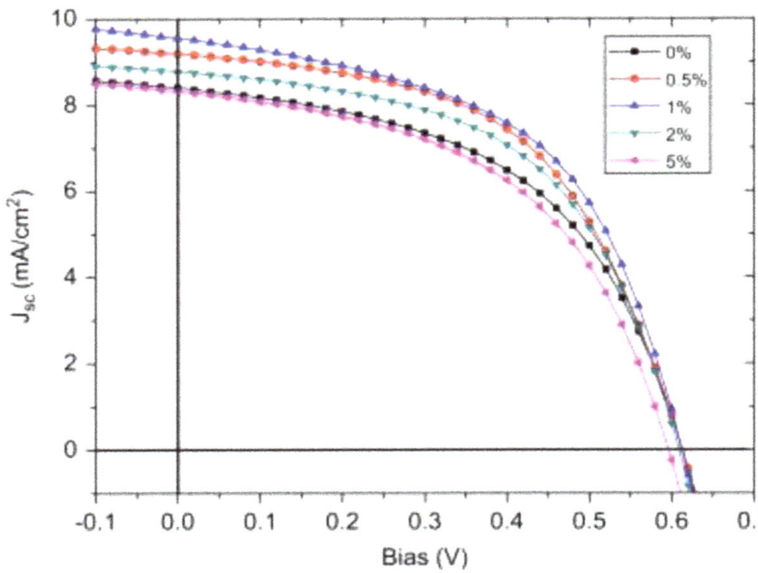

Figure 9.9. The *J–V* curve for the P3HT:PCBM blend for different concentrations of Fe_3O_4 magnetic sorbent-based nanoparticles. Image reproduced with permission from Zhang *et al* [43]. Copyright Elsevier, 2011.

the Fe_3O_4 nanoparticles (NP) onto the P3HT:PCBM blend. Figure 9.9 shows the *J–V* curve of Fe_3O_4 NPs incorporated P3HT:PCBM at different Fe concentrations. A calculated power conversion efficiency (PCE) of 3.1% for optimum Fe atomic incorporation was obtained compared to the value of 2.6% for pristine P3HT:PCBM. The increase in the PCE was attributed to the effect of the Fe_3O_4 NPs, and the group confirmed the influence by conducting several independent experiments to isolate other influences.

Meanwhile, incorporating nanocomposite into the polymer active layer of an OSC can improve the photovoltaic device's performance. The collaborative work of Pereira [44] established the possibility of including an Fe-based nanocomposite into P3HT:PCBM. Their study doped Fe onto SnO_2 and covalently attached the nanocomposite to P3HT:PCBM via the sol–gel route. As indicated in figure 9.10, the OSC devices possessing reinforced magnetic sorbents exhibited a slightly enhanced short-circuit current ($11.25 \rightarrow 12.09$ cm^{-2}). The slightly enhanced current results imply that including the magnetic sorbents improves the separation of photogenerated excitons into free charge carriers, which leads to increased short-circuit current density.

While the study of Wang and Pereira focused on the active polymer material, the effect of magnetic sorbents on OSC devices can be crucial in improving photovoltaic performance. The incorporation of magnetic nanoparticles as a transport layer was proposed by Kovalenko *et al* [45], who also reported its effect on the performance of the photovoltaic device. Their study explored the effect of ferro- and superparamagnetism-based sorbent nanoparticles on the OSC's performance. They

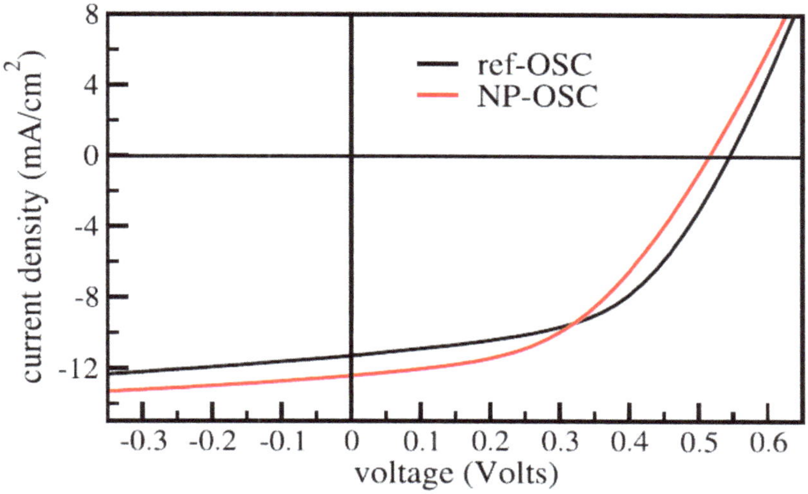

Figure 9.10. The current density curve for a reference OSC and Fe–SnO_2 NP-OSC cells. Image reproduced with permission from Pereira *et al* [44]. Copyright Elsevier, 2017.

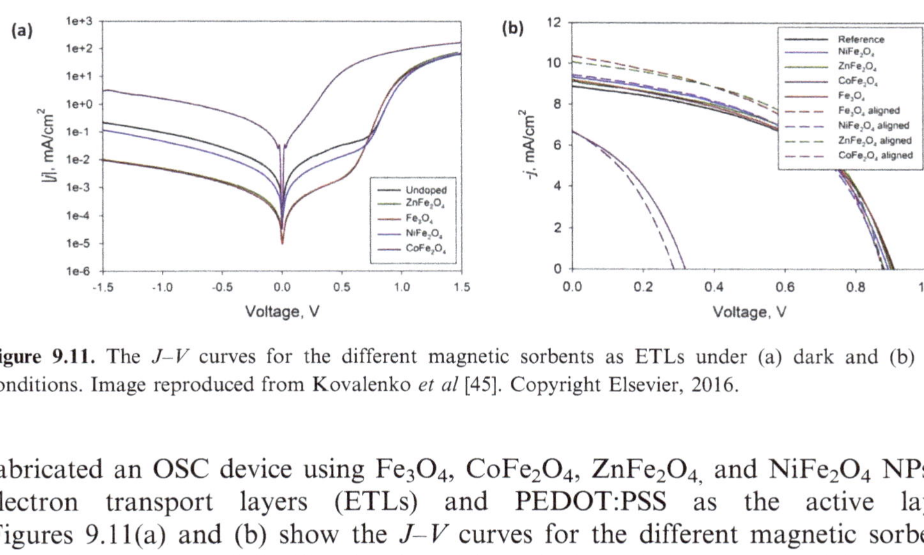

Figure 9.11. The *J–V* curves for the different magnetic sorbents as ETLs under (a) dark and (b) light conditions. Image reproduced from Kovalenko *et al* [45]. Copyright Elsevier, 2016.

fabricated an OSC device using Fe_3O_4, $CoFe_2O_4$, $ZnFe_2O_4$, and $NiFe_2O_4$ NPs as electron transport layers (ETLs) and PEDOT:PSS as the active layer. Figures 9.11(a) and (b) show the *J–V* curves for the different magnetic sorbents ETLs for dark and light illumination. The calculated PCEs improved for Fe_3O_4, $ZnFe_2O_4$, and $NiFe_2O_4$ NPs. The improvement was attributed to the magnetic moments. In addition, aligning the magnetic sorbents using the applied magnetic field at constant temperature resulted in an enhanced OSC PCE. The additional enhancement was attributed to improved charge collection under short-circuit conditions, leading to increased short-circuit current.

9.3.2 Magnetism-based dye-sensitized solar cells

Dye-sensitized solar cells (DSSCs) are another form of photovoltaics that harvest energy via light–matter interactions. The operating mechanism of DSSCs is similar to that of OSCs; however, it possesses unique components such as a conductive transparent glass anode, a mesoporous metal-oxide layer, a conductive dye layer, an organic conductive electrolyte, and a glass cathode [46]. Due to the global increased energy consumption and needs, there is always a demand for additional photovoltaic performance, which DSSCs are not exempted from. In addition, DSSCs have shown limitations with regard to the interfacial electronic transfer processes between the metal oxide, conductive dye, and electrolyte. Hence, ongoing research is on the rise to improve the interfacial electronic transport properties, which have a long-term impact on the durability and PCE of DSSCs. Among the different routes used by different researchers, the introduction of magnetic sorbents into DSSCs has received enormous attention, owing to the effect of magnetic fields in accelerating the intrinsic photogenerators of DSSCs [47].

Among the numerous studies that have focused on incorporating magnetic sorbents into DSSCs, Kilic *et al* [48] used Fe_2O_3 NP as their sorbent material. Their study used the hydrothermal method to prepare a TiO_2-Fe_2O_3 transport layer and fabricate DSSCs using N719 dye on a conductive fluorine-doped tin oxide (FTO) glass electrode. Figure 9.12 shows the *J–V* curve of the Fe_2O_3 NP-containing

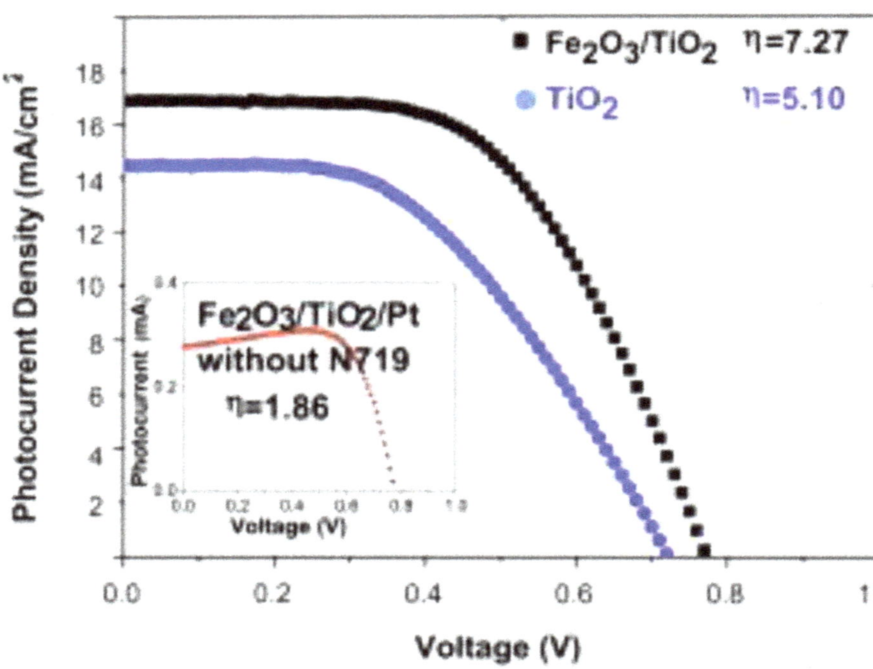

Figure 9.12. The *J–V* curve of an Fe_2O_3-modified TiO_2 electron transport layer for a DSSC. Image reproduced with permission from Kilic *et al* [48]. Copyright Elsevier, 2015.

TiO_2 transport layer. The effect of the magnetic field produced by the Fe_2O_3 NPs resulted in an increase of 7.27% compared to the value of 5.1% for pristine TiO_2. The study suggests that the applied magnetic field enables accelerated electron transport properties, improving the DSSC device's PCE.

A similar study was performed by Cai *et al* [49], who incorporated γ-Fe_2O_3 into TiO_2 alongside N719 dye. The magnetic hysteresis loop indicated in figure 9.13(a) suggests a ferromagnetic feature which could easily interact with the TiO_2, which is typically ferromagnetic [50]. The exchange interaction between Fe 3d and Ti 3d orbital states led to accelerated electronic transport with a corresponding PCE of 6.92%. In addition, the injection rate of the electrons depends on the intrinsic energy overlap between the excited states and the semiconductor acceptor states. Hence, the DSSCs combined with (N719) dye possess strong spin–orbit coupling originating from the ruthenium-rich atomic center that accelerates the intersystem crossing. The intersystem acceleration leads to the formation of additional injections through lower-energy-triplet excited-state process [49].

While the electron transport layer has the major influence on the performance of DSSCs, manipulating the electrolyte component may prove crucial in improving the device's performance. Yang *et al* [51] included Co_3O_4 in a polymer electrolyte component, using TiO_2 as the electron transport layer. The magnetic potential of Co_3O_4 can easily be harnessed; hence, the study used an applied magnetic field to explore the possibility of accelerating the electron transport properties within the DSSC device. The photovoltaic response results indicated an improved performance of the DSSC with Co_3O_4 included in the polymer electrolyte. The PCE increased (1.27% to 1.73%) for the pristine polymer and Co_3O_4-modified polymer electrolyte under the influence of an external magnetic field (see figures 9.14(a) and (b)). Therefore, magnetic sorbent contributions in DSSCs do not only affect the electron transport layer but can be extended to other layers of DSSC devices.

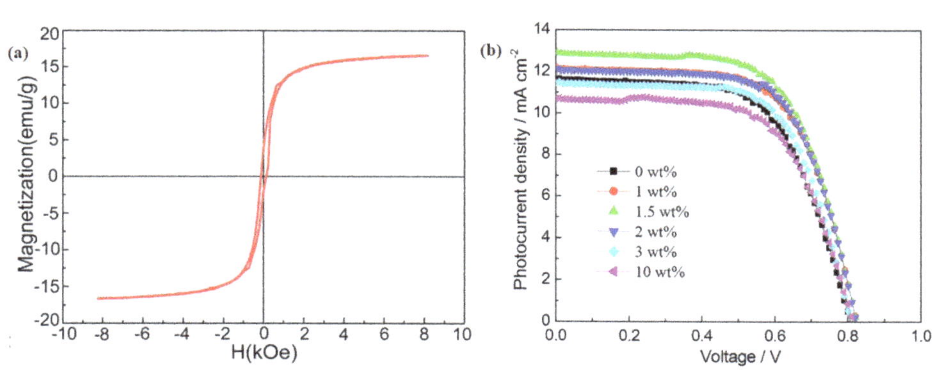

Figure 9.13. (a) The magnetic hysteresis loop of γ-Fe_2O_3 and (b) the *J–V* curve of the TiO_2:γ-Fe_2O_3 nanocomposite. Image reproduced under open access license from Cai *et al* [49]. Copyright RSC Publishing, 2015.

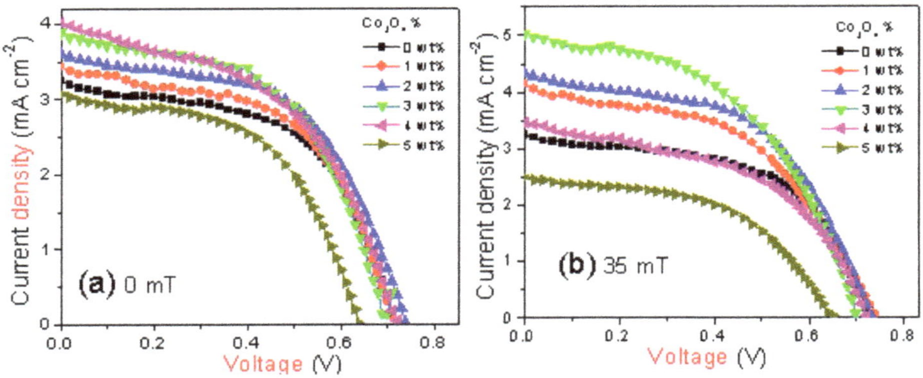

Figure 9.14. (a) and (b) The *J–V* curves of the CO_3O_4-modified polymer electrolyte layer under the influence of an external magnetic field. Image reproduced with permission from Yang *et al* [51], copyright, Elsevier, 2015.

9.3.3 Magnetism-based perovskite solar cells

Due to the increasing PCE growth in the photovoltaic industry, PSCs have become very popular among researchers. Within a decade, the PCE of PSCs has reached 25.7%, which is close to that of commercially available inorganic silicon solar cells [52]. Since the process of light energy harvesting in PCEs is the same as other that of organic-based solar cells, it is crucial to maintain high-quality individual layers to mitigate losses such as recombination (originating from defect-induced interfacial nonradiative and interfacial energy level mismatches) to retain the optimum efficiency of the PSC device. The process of mitigating these factors has created room for diverse research opportunities, in which the incorporation of magnetic sorbents has been used to accelerate intersystem transport properties. The inclusion of magnetic fields such as ferroelectric domain features has been suggested to improve the separation of electrons and holes, which overall improves the PSC device's bandgap and open-circuit voltage [53].

Among the numerous attempts to incorporate magnetic sorbent-based nanoparticles into PSCs [54, 55], Ren *et al* [56] proposed the doping of Mn into a $CH_3NH_3PbI_3$ PSC to study the effect of magnetic properties on their interlayer interactions. Figure 9.15 shows the magnetic and electrical conductivities of pristine and Mn-doped $MAPbI_3$ PSC devices. As indicated by the magnetic properties, a switch from diamagnetic to ferromagnetic behavior was observed when Mn was included in the PSC material, suggesting that the magnetic moments were mostly aligned in either the up or down spin direction. The *J–V* curve and the corresponding measured PCE indicated a decrease (15.1% ← 8.87%). The decreased PCE was attributed to reduced carrier mobility in the Mn-doped $MAPbI_3$ PSC. While their result is undesirable, the study describes the experimental realization of PSC doping and creates room for future research opportunities.

Although Mn doping results in a decreased PCE, other materials could prove effective, and Fe_3O_4 NPs yielded promising results. A study proposed by Xu *et al* [57] incorporated Fe_3O_4 NPs into the active layer, which formed a $CH_3NH_3PbI_3$:

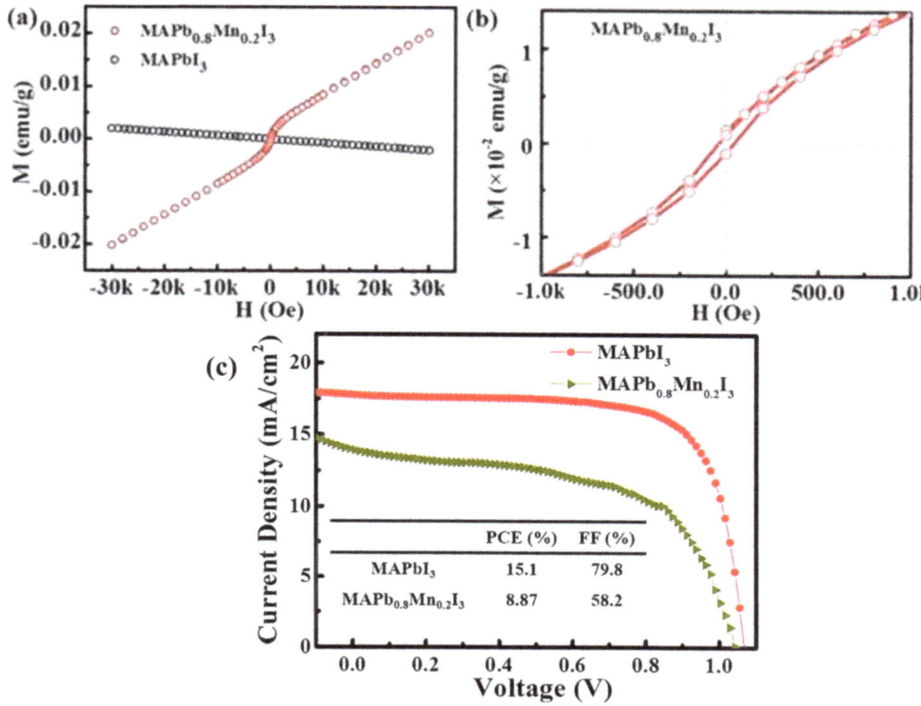

Figure 9.15. (a)–(b) The magnetic hysteresis and (c) the *J*–*V* curve of the Mn-doped $MAPbI_3$ PSC device. Image reproduced with permission from Ren *et al* [56]. Copyright ACS Publishing, 2020.

Fe_3O_4 composite. The impact of the exchange interaction of the intrinsic magnetic field of Fe_3O_4 resulted in a decrease in photocurrent hysteresis and a corresponding PCE of 20.2%. The enhanced PCE (relative PCE =2.5%) was attributed to increased suppression of charge carrier recombination with a corresponding shortened charge carrier extraction time.

Moreover, cobalt-based magnetic sorbents are intriguing materials that can improve PSC performance, as indicated in a study by Di *et al* [58]. Their recent study incorporated different concentrations of Co into $CH_3NH_3PbI_3$ PSC material. As indicated in figure 9.16(a), a ferromagnetic behavior was observed in the Co-doped $CH_3NH_3PbI_3$ composite, which suggests alignment of the magnetic spins in a single direction. The effect of the magnetic alignment can be observed in the *J*–*V* curve (see figure 9.16(b)), where the open-circuit voltage is significantly improved. The resulting electrical conductivity yielded a PCE of 10.5% in comparison to the value of 8.3% for pristine $CH_3NH_3PbI_3$. It can be inferred from the cited studies that the effect of magnetic interactions with the photogenerators results in improved electron–hole separation within PSC devices.

The electron and hole separation can be attributed to ferromagnetic contributions caused by polarization induced by the non-centrosymmetry of the crystal structure intrinsic between the magnetic sorbents and the PSC materials [59]. Table 9.1

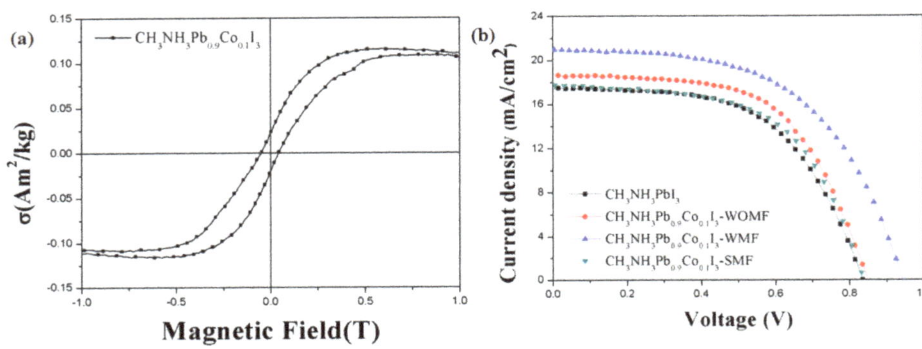

Figure 9.16. (a) The magnetic and (b) electrical conductivities of Co-doped $CH_3NH_3PbI_3$. Image reproduced with permission from Di *et al.* Reprinted from [58]. Copyright (2022), with permission from Elsevier.

summarizes the magnetism-based photovoltaic materials discussed in this chapter with respect to their solar cell type and synthesis. While few studies have been examined in this chapter, the detailed review outlined by Abboubi and San [60] gives insights into the recent magnetism-based photovoltaics trends.

9.4 Conclusions

This chapter focused on the incorporation of magnetic sorbent-based nanoparticles into organic photovoltaics, which has the potential to improve their performance. The studies considered in the current chapter suggest that incorporating magnetic sorbent into organic photovoltaics improves their performance. The interaction between their intrinsic magnetic fields and the atomic bonds leads to improved electronic transport, which results, in most cases, in improved photovoltaic performance. In addition, the examined reports suggest improved efficiencies for dye-sensitized organic solar cells and minimal PCEs for organic solar cells. Meanwhile, magnetic perovskites show the maximum PCEs, comparable to those of conventional organic PSCs. The trend toward magnetism-based photovoltaics is an ongoing research field that is still attracting interest, and the encouraging results can be improved for future solar energy harvesting.

References

[1] Todorov T, Gunawan O and Guha S 2016 A road towards 25% efficiency and beyond: perovskite tandem solar cells *Mol. Syst. Design Eng.* **1** 370–6

[2] Faraji G, Kim H S and Kashi H T 2018 Severe plastic deformation methods for sheets *Severe Plastic Deformation* (Amsterdam: Elsevier) 113–29 ch 3

[3] Vojoudi H, Badiei A, Bahar S, Mohammadi Ziarani G, Faridbod F and Ganjali M R 2017 A new nano-sorbent for fast and efficient removal of heavy metals from aqueous solutions based on modification of magnetic mesoporous silica nanospheres *J. Magn. Magn. Mater.* **441** 193–203

[4] Bojdi M K, Behbahani M, Hesam G and Mashhadizadeh M H 2016 Application of magnetic lamotrigine-imprinted polymer nanoparticles as an electrochemical sensor for trace determination of lamotrigine in biological samples *RSC Adv.* **6** 32374–80

[5] 2017 *Nanomagnetism: Applications and Perspectives* ed C Fermon and M Van de Voorde (New York: Wiley)

[6] He M, Chen Z, Xu C, Chen B and Hu B 2021 Magnetic nanomaterials as sorbents for trace elements analysis in environmental and biological samples *Talanta* **230** 122306

[7] Abdel Maksoud M, Fahim R A, Bedir A G, Osman A I, Abouelela M M, El-Sayyad G S *et al* 2021 Engineered magnetic oxides nanoparticles as efficient sorbents for wastewater remediation: a review *Environ. Chem. Lett.* **20** 519–62

[8] Wang L and Gao L 2009 Morphology-controlled synthesis and magnetic property of pseudocubic iron oxide nanoparticles *J. Phys. Chem.* C **113** 15914–20

[9] Idisi D O, Ali H, Oke J A, Sarma S, Moloi S J, Ray S C S C, Wang H T, Jana N R N R, Pong W F and Strydom A M A M 2019 Electronic, electrical and magnetic behaviours of reduced graphene-oxide functionalized with silica coated gold nanoparticles *Appl. Surf. Sci.* **483** 106–13

[10] Idisi D O, Oke J A, Sarma S, Moloi S J, Ray S C, Pong W F and Strydom A M 2019 Tuning of electronic and magnetic properties of multifunctional r-GO-ATA-Fe_2O_3-composites for magnetic resonance imaging (MRI) contrast agent *J. Appl. Phys.* **126** 35301

[11] Kayani Z N, Arshad S, Riaz S and Naseem S 2014 Synthesis of iron oxide nanoparticles by sol–gel technique and their characterization *IEEE Trans. Magn.* **50** 1–4

[12] Oke J A, Idisi D O, Sarma S, Moloi S J, Ray S C, Chen K H, Ghosh A, Shelke A and Pong W F 2019 Electronic, electrical, and magnetic behavioral change of SiO_2-NP-decorated MWCNTs *ACS Omega* **4** 14589–98

[13] Oke J A, Idisi D O, Sarma S, Moloi S J, Ray S C, Chen K H, Ghosh A, Shelke A, Hsieh S-H and Pong W F 2019 Tuning of electronic and electrical behaviour of MWCNTs-TiO_2 nanocomposites *Diam. Relat. Mater.* **100** 107570

[14] Abou Hammad A B, Mansour A, Elhelali T M and El Nahrawy A M 2023 Sol–gel/gel casting nanoarchitectonics of hybrid Fe_2O_3–ZnO/PS-PEG nanocomposites and their opto-magnetic properties *J. Inorg. Organomet. Polym. Mater.* **33** 544–54

[15] Ali M B, El Maalam K, El Moussaoui H, Mounkachi O, Hamedoun M, Masrour R, Hlil E and Benyoussef A 2016 Effect of zinc concentration on the structural and magnetic properties of mixed Co–Zn ferrites nanoparticles synthesized by sol/gel method *J. Magn. Magn. Mater.* **398** 20–5

[16] Abbas H, Krishnan A and Kotakonda M 2020 Antifungal and antiovarian cancer properties of α-Fe_2O_3 and α-Fe_2O_3/ZnO nanostructures synthesized by *Spirulina platensis* *IET Nanobiotechnol.* **14** 774–84

[17] Ahn T, Kim J H, Yang H-M, Lee J W and Kim J-D 2012 Formation pathways of magnetite nanoparticles by coprecipitation method *J. Phys. Chem.* C **116** 6069–76

[18] Idisi D O, Ahia C C, Meyer E L, Bodunrin J O and Benecha E M 2023 Graphene oxide: Fe_2O_3 nanocomposites for photodetector applications: experimental and *ab initio* density functional theory study *RSC Adv.* **13** 6038–50

[19] Gopalan E V, Malini K, Saravanan S, Kumar D S, Yoshida Y and Anantharaman M 2008 Evidence for polaron conduction in nanostructured manganese ferrite *J. Phys. D: Appl. Phys.* **41** 185005

[20] Saines P J, Melot B C, Seshadri R and Cheetham A K 2010 Synthesis, structure and magnetic phase transitions of the manganese succinate hybrid framework, Mn ($C_4H_4O_4$) *Chem.–Eur. J.* **16** 7579–85

[21] Chen Y, Guo X and Wu F 2020 Development and evaluation of magnetic iron-carbon sorbents for mercury removal in coal combustion flue gas *J. Energy Inst.* **93** 1615–23

[22] Nasir S and Anis-ur-Rehman M 2011 Structural, electrical and magnetic studies of nickel–zinc nanoferrites prepared by simplified sol–gel and co-precipitation methods *Phys. Scr.* **84** 025603

[23] Torres-Gómez N, Nava O, Argueta-Figueroa L, García-Contreras R, Baeza-Barrera A and Vilchis-Nestor A R 2019 Shape tuning of magnetite nanoparticles obtained by hydrothermal synthesis: effect of temperature *J. Nanomater.* **2019** 7921273

[24] Idisi D O, Aigbe U O, Ahia C C and Meyer E L 2023 Graphene oxide: Fe_2O_3 nanocomposite: synthesis, properties, and applications *Carbon Lett.* **33** 605–40

[25] Tadic M, Trpkov D, Kopanja L, Vojnovic S and Panjan M 2019 Hydrothermal synthesis of hematite (α-Fe_2O_3) nanoparticle forms: synthesis conditions, structure, particle shape analysis, cytotoxicity and magnetic properties *J. Alloys Compd.* **792** 599–609

[26] Tadic M, Kralj S and Kopanja L 2019 Synthesis, particle shape characterization, magnetic properties and surface modification of superparamagnetic iron oxide nanochains *Mater. Charact.* **148** 123–33

[27] Otari S V, Patel S K, Kalia V C and Lee J-K 2020 One-step hydrothermal synthesis of magnetic rice straw for effective lipase immobilization and its application in esterification reaction *Bioresour. Technol.* **302** 122887

[28] Pandi K, Viswanathan N and Meenakshi S 2019 Hydrothermal synthesis of magnetic iron oxide encrusted hydrocalumite-chitosan composite for defluoridation studies *Int. J. Biol. Macromol.* **132** 600–5

[29] Azizi A 2020 Green synthesis of Fe_3O_4 nanoparticles and its application in preparation of Fe_3O_4/cellulose magnetic nanocomposite: a suitable proposal for drug delivery systems *J. Inorg. Organomet. Polym. Mater.* **30** 3552–61

[30] Xiong D, Xu Z, Zeng X, Zhang W, Chen W, Xu X, Wang M and Cheng Y-B 2012 Hydrothermal synthesis of ultrasmall $CuCrO_2$ nanocrystal alternatives to NiO nanoparticles in efficient p-type dye-sensitized solar cells *J. Mater. Chem.* **22** 24760–8

[31] Neikov O D 2009 Mechanical crushing and grinding *Handbook of Non-Ferrous Metal Powders* ed O D Neikov, S S Naboychenko, I V Murashova, V G Gopienko, I V Frishberg and D V Lotsko (Oxford: Elsevier) 47–62 ch 2

[32] Chakka V M, Altuncevahir B, Jin Z Q, Li Y and Liu J P 2006 Magnetic nanoparticles produced by surfactant-assisted ball milling *J. Appl. Phys.* **99** 08E912

[33] De Carvalho J, De Medeiros S, Morales M, Dantas A and Carriço A 2013 Synthesis of magnetite nanoparticles by high energy ball milling *Appl. Surf. Sci.* **275** 84–7

[34] Su Y, Su H, Zhu Y, Wang F, Du J, Xia W, Yan A, Liu J P and Zhang J 2015 Effects of magnetic field heat treatment on Sm–Co/α-Fe nanocomposite permanent magnetic materials prepared by high energy ball milling *J. Alloys Compd.* **647** 375–9

[35] Narayanaswamy V, Obaidat I M, Kamzin A S, Latiyan S, Jain S, Kumar H, Srivastava C, Alaabed S and Issa B 2019 Synthesis of graphene oxide-Fe_3O_4 based nanocomposites using the mechanochemical method and *in vitro* magnetic hyperthermia *Int. J. Mol. Sci.* **20** 3368

[36] Huang L, He M, Chen B, Cheng Q and Hu B 2017 Facile green synthesis of magnetic porous organic polymers for rapid removal and separation of methylene blue *ACS Sustain. Chem. Eng.* **5** 4050–5

[37] Bagheri A R, Arabi M, Ghaedi M, Ostovan A, Wang X, Li J and Chen L 2019 Dummy molecularly imprinted polymers based on a green synthesis strategy for magnetic solid-phase extraction of acrylamide in food samples *Talanta* **195** 390–400
[38] Jia Y, Zeng Y, Li X and Meng L 2020 Effect of Sr substitution on the property and stability of $CH_3NH_3SnI_3$ perovskite: a first-principles investigation *Int. J. Energy Res.* **44** 5765–78
[39] Zhang C, Sun D, Sheng C, Zhai Y, Mielczarek K, Zakhidov A and Vardeny Z 2015 Magnetic field effects in hybrid perovskite devices *Nat. Phys.* **11** 427–34
[40] Pathakoti K, Manubolu M and Hwang H-M 2018 Nanotechnology applications for environmental industry ed C Mustansar Hussain *Handbook of Nanomaterials for Industrial Applications* (Amsterdam: Elsevier) ch 48 pp 894–907
[41] Wei M, Wang Z, Wen Z, Hao X and Qin W 2018 Utilizing magnetic field to study the impact of intramolecular charge transfers on the open-circuit voltage of organic solar cells *Appl. Phys. Lett.* **113** 093301
[42] Kannan U, Giribabu L and Jammalamadaka S N 2019 Demagnetization field driven charge transport in a TiO_2 based dye sensitized solar cell *Sol. Energy* **187** 281–9
[43] Zhang W, Xu Y, Wang H, Xu C and Yang S 2011 Fe_3O_4 nanoparticles induced magnetic field effect on efficiency enhancement of P3HT:PCBM bulk heterojunction polymer solar cells *Sol. Energy Mater. Sol. Cells* **95** 2880–5
[44] Pereira M S, Lima F A S, Ribeiro T S, da Silva M R, Almeida R Q, Barros E B and Vasconcelos I F 2017 Application of Fe-doped SnO_2 nanoparticles in organic solar cells with enhanced stability *Opt. Mater.* **64** 548–56
[45] Kovalenko A, Yadav R S, Pospisil J, Zmeskal O, Karashanova D, Heinrichová P, Vala M, Havlica J and Weiter M 2016 Towards improved efficiency of bulk-heterojunction solar cells using various spinel ferrite magnetic nanoparticles *Org. Electron.* **39** 118–26
[46] Gong J, Liang J and Sumathy K 2012 Review on dye-sensitized solar cells (DSSCs): Fundamental concepts and novel materials *Renew. Sustain. Energy Rev.* **16** 5848–60
[47] Klein M, Pankiewicz R, Zalas M and Stampor W 2016 Magnetic field effects in dye-sensitized solar cells controlled by different cell architecture *Sci. Rep.* **6** 30077
[48] Kılıç B, Gedik N, Mucur S P, Hergul A S and Gür E 2015 Band gap engineering and modifying surface of TiO_2 nanostructures by Fe_2O_3 for enhanced-performance of dye sensitized solar cell *Mater. Sci. Semicond. Process.* **31** 363–71
[49] Cai F, Zhang S and Yuan Z 2015 Effect of magnetic gamma-iron oxide nanoparticles on the efficiency of dye-sensitized solar cells *RSC Adv.* **5** 42869–74
[50] Kumar A, Kashyap M K, Sabharwal N, Kumar S, Kumar A, Kumar P and Asokan K 2017 Structural, optical and weak magnetic properties of Co and Mn co-doped TiO_2 nanoparticles *Solid State Sci.* **73** 19–26
[51] Yang Y, Gao J, Yi P, Cui J and Guo X 2015 The influence of Co_3O_4 concentration on quasi-solid state dye-sensitized solar cells with polymer electrolyte *Solid State Ionics* **279** 1–5
[52] Liu S, Biju V P, Qi Y, Chen W and Liu Z 2023 Recent progress in the development of high-efficiency inverted perovskite solar cells *NPG Asia Mater.* **15** 28
[53] Pecchia A, Gentilini D, Rossi D, Auf der Maur M and Di Carlo A 2016 Role of ferroelectric nanodomains in the transport properties of perovskite solar cells *Nano Lett.* **16** 988–92
[54] Hsiao Y, Wu T, Li M and Hu B 2015 Magneto-optical studies on spin-dependent charge recombination and dissociation in perovskite solar cells *Adv. Mater.* **27** 2899–906

[55] Wang H, Lei J, Gao F, Yang Z, Yang D, Jiang J, Li J, Hu X, Ren X and Liu B 2017 Magnetic field-assisted perovskite film preparation for enhanced performance of solar cells *ACS Appl. Mater. Interfaces* **9** 21756–62
[56] Ren L, Wang Y, Wang M, Wang S, Zhao Y, Cazorla C, Chen C, Wu T and Jin K 2020 Tuning magnetism and photocurrent in Mn-doped organic–inorganic perovskites *J. Phys. Chem. Lett.* **11** 2577–84
[57] Xu W, Zhu T, Yang Y, Zheng L, Liu L and Gong X 2020 Enhanced device performance of perovskite photovoltaics by magnetic field-aligned perovskites-magnetic nanoparticles composite thin film *Adv. Funct. Mater.* **30** 2002808
[58] Di J, Zhong M, Wang Y and Zhou J 2022 Effects of Co^{2+} doping and magnetic field actions on the stability and efficiency of perovskite solar cells and their mechanisms *J. Alloys Compd.* **891** 161910
[59] Seyfouri M M and Wang D 2021 Recent progress in bismuth ferrite-based thin films as a promising photovoltaic material *Crit. Rev. Solid State Mater. Sci.* **46** 83–108
[60] El Abboubi M and San S E 2023 Integration of spinel ferrite magnetic nanoparticles into organic solar cells: a review *Mater. Sci. Eng.* B **294** 116512
[61] Yang X, Qi Y, Wei P, Hu Q, Cheng J and Xie Y 2023 Enhanced efficiency and stability of carbon-based $CsPbI_2Br$ perovskite solar cells by introducing metal organic framework-derived Fe_3O_4@ NC interfacial layer *J. Power Sources* **566** 232927
[62] Ansari F, Salavati-Niasari M, Amiri O, Mir N, Abdollahi Nejand B and Ahmadi V 2020 Magnetite as inorganic hole transport material for lead halide perovskite-based solar cells with enhanced stability *Ind. Eng. Chem. Res.* **59** 743–50
[63] Li R, Liao Y, Dou Y, Wang D, Li G, Sun W, Wu J and Lan Z 2021 $CoFe_2O_4$ nanocrystals for interface engineering to enhance performance of perovskite solar cells *Sol. Energy* **220** 400–5
[64] Pereira M de S, Lima F A de S, Almeida R Q de, Martins J L da S, Bagnis D, Barros E B, Sombra A S B and Vasconcelos I F de 2020 Flexible, large-area organic solar cells with improved performance through incorporation of $CoFe_2O_4$ nanoparticles in the active layer *Mater. Res.* **22** e20190417
[65] Mahmoudi M, Alizadeh A, Roudgar-Amoli M and Shariatinia Z 2023 Rational modification of TiO_2 photoelectrodes with spinel $ZnFe_2O_4$ and Ag-doped $ZnFe_2O_4$ nanostructures highly enhanced the efficiencies of dye sensitized solar cells *Spectrochim. Acta, Part* A **289** 122214
[66] Hu J, Xie Y, Zhou X and Yang J 2016 Solid-state synthesis of ZnO and $ZnFe_2O_4$ to form p–n junction composite in the use of dye sensitized solar cells *J. Alloys Compd.* **676** 320–5
[67] Mao J, He B, Sui H, Cui L, Chen H, Duan Y, Yang P and Tang Q 2023 Interfacial modification of *in situ* polymerized AMPS/$NiFe_2O_4$ quantum dots for efficient and air-stable $CsPbBr_3$ perovskite solar cells *Chem. Eng. J.* **461** 141943
[68] Zhou Y, Zhang X, Lu X, Gao X, Gao J, Shui L, Wu S and Liu J 2019 Promoting the hole extraction with Co_3O_4 nanomaterials for efficient carbon-based $CsPbI_2Br$ perovskite solar cells *Sol. RRL* **3** 1800315
[69] Rajamanickam N, Chowdhury T H, Isogami S and Islam A 2021 Magnetic properties in $CH_3NH_3PbI_3$ perovskite thin films by Mn doping *J. Phys. Chem.* C **125** 20104–12
[70] Zhang K, Zhao J, Hu Q, Yang S, Zhu X, Zhang Y, Huang R, Ma Y, Wang Z and Ouyang Z 2021 Room-temperature magnetic field effect on excitonic photoluminescence in perovskite nanocrystals *Adv. Mater.* **33** 2008225

Chapter 10

The application of magnetic sorbents in microbial bioremediation

Joseph Onyeka Emegha, Timothy Imanobe Oliomogbe, Cyril Chinedu Otali, Udoka Bessie Igue, Odunayo Tope Ojo and Kingsley Eghonghon Ukhurebor

The world is currently grappling with environmental pollution. Waste materials can be biologically changed into a form that other creatures can use, which is the idea behind bioremediation. Microorganisms offer a promising solution due to their exceptional metabolic activity and ability to thrive in diverse environmental conditions. Their wide range of nutrients makes them suitable for the bioremediation of environmental pollutants. Microorganism-based magnetic sorbents are essential for the removal and breakdown of many kinds of physical and chemical waste. This chapter elaborates on the potential and practical applications of magnetic sorbents as environmentally friendly materials for addressing environmental contamination issues.

10.1 Introduction

The environment is now exposed to a variety of dangerous contaminants as a result of growth in urbanization and industrial development. These contaminants, stemming from a range of industrial activities, significantly contribute to water and soil pollution [1–4]. Industrial processes generate diverse heavy metals (HMs), which are subsequently released as effluents during further production stages [2]. As an example, wastewater discharged by dye manufacturers contains significant levels of HMs such as mercury, chromium, and antimony [5, 6]. The use of fertilizers, pesticides, and herbicides in agriculture introduces HM pollutants such as aluminum, nickel, zinc, lead, arsenic, and copper into the environment [7, 8]. In addition, untreated waste from agri-food industries that is disposed into waterways has harmful effects on the environment [9]. Crude oil also significantly contributes to environmental pollution through pipeline vandalism, transportation leaks, and accidental spills [10, 11]. Mining activities release toxic chemicals such as lead,

doi:10.1088/978-0-7503-5909-2ch10

arsenic, cadmium, and copper as well as other hazardous substances including cyanide and sulfuric acid [1, 4, 11].

Similarly, other waste products from industrial activities such as cement production release zinc, copper, and cadmium into the topsoil [1]. Pharmaceutical effluents contribute chromium and lead to water pollution, while plastics containing lead further contaminate water with silver, iron, manganese, and copper [4, 12–14]. Copper, arsenic, nickel, cadmium, and zinc are environmental contaminants produced by the coal industry. These HMs have harmful effects on both aquatic and terrestrial ecosystems along with their inhabitants [12, 15].

In humans, these HMs can cause serious health issues including disturbances to the central nervous system, especially in infants, through mercury exposure; cadmium toxicity, which leads to kidney disease; cardiovascular disorders; and reproductive and immune system malfunctions [1, 12, 16–20], as illustrated in figure 10.1.

Moreover, recent studies have linked cadmium to cancer and reproductive system failure [18, 19]. Waste materials containing high levels of metallic elements are

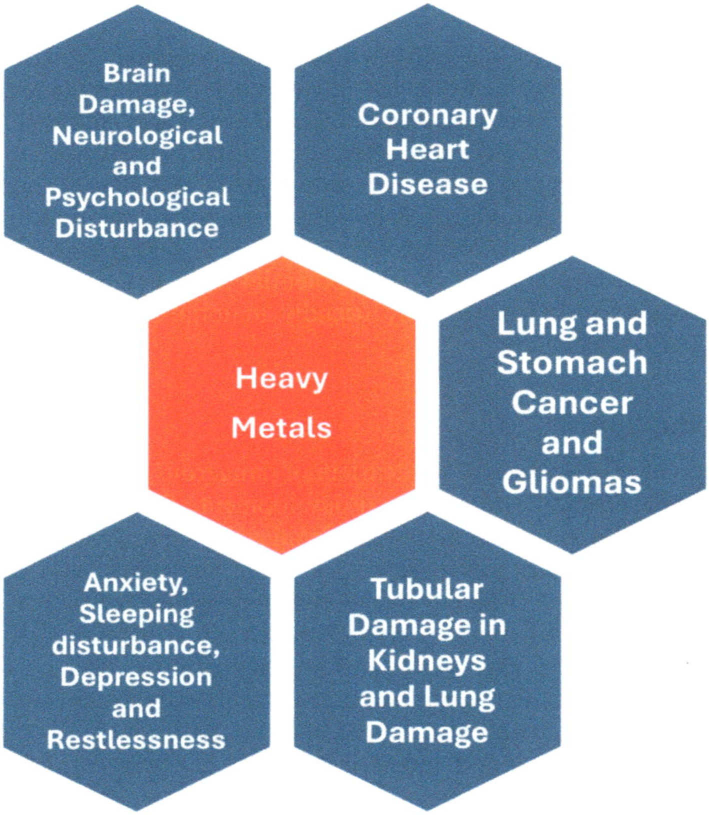

Figure 10.1. The impact of HMs on human health.

frequently inappropriately disposed of in the soil and bodies of water. When they are discarded into water, they can result in the death of aquatic organisms such as fish [1]. In addition, if not properly managed, these pollutants can become more concentrated within the food chain through a process known as biomagnification, leading to chronic illnesses in both humans and animals [20, 21]. Consequently, there is a necessity to address these contaminants through physical, chemical, or biological remediation [1], as illustrated in figure 10.2. Even though chemical and physical methods have been used for a long time, they have drawbacks. For example, chemical bioremediation procedures need specialized equipment, and physical bioremediation methods are quite expensive [22, 23]. This has led to the necessity for an improved option known as biological remediation. Bioremediation is a highly effective and environmentally friendly technology for altering contaminants [24–27]. It can involve the use of both plants and microorganisms; however, plants have a longer growth period and are not as easily controlled as microbes, making microbes more favorable. Furthermore, microorganisms reduce HMs and enhance soil fertility and plant growth [28].

Microbes can transform harmful substances into less harmful compounds through a process known as mineralization [25, 28]. Bacteria, fungi, algae, and

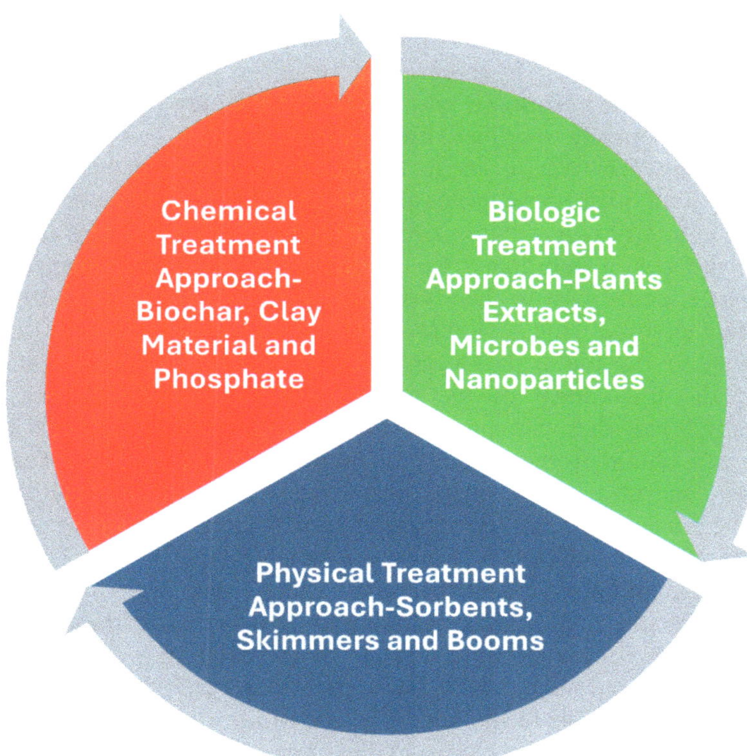

Figure 10.2. Types of bioremediation.

various other microorganisms can be used for bioremediation, as they flourish in peculiar settings where they can take in a broad range of contaminants [29]. Moreover, their capacity to resist harsh surroundings amplifies their efficacy; for example, acidophiles can withstand acidic environments, psychrophiles flourish in cold climes, and halophiles endure saltwater regions [30, 31]. This chapter examines the significance, scope, and utilization of magnetic sorbents in microbial bioremediation.

10.2 Bioremediation techniques and microbes

Microbial bioremediation is the employment of microorganisms as well as their byproducts to remove contaminants from the environment [32, 33]. It is important to keep in mind that because microorganisms are ubiquitous, pollutants in different environmental compartments frequently contact them [24, 34]. Pollutants are broken down by microbes through their intrinsic metabolic processes, sometimes with the help of small route modifications, which direct the pollutant into the typical microbial metabolic pathway for breakdown and biotransformation [35–37]. Microbes create enzymes that degrade compounds such as hydrocarbons and metals. Oil degradation requires a sufficient number of bacteria to break down the oil via metabolic pathways. Fortunately, nature has produced several bacteria that can do this [38]. According to Glazer *et al* [39], around 70 microbial genera worldwide can break down hydrocarbons. In order to support bioremediation, microorganisms can be introduced into a system if they are not already there. This could include genetically modified microbes or cultures cultivated from other polluted locations. However, even in the presence of these bacteria, degradation can only occur if all other prerequisites are satisfied [40]. Microbes are adaptable and may flourish in a variety of environments, including subzero temperatures, intense heat, desert climates, water, anaerobic conditions that contain dangerous substances, and waste streams. Due to their biological systems, bacteria may be utilized to decontaminate environmental pollutants [41].

Through these enzymatic pathways, microorganisms function as biocatalysts, accelerating the breakdown of the target pollutant. Microorganisms react to pollution only when they can produce energy and nutrients for cell expansion, which they can obtain from a variety of materials and chemicals. The chemical makeup and concentration of toxins, the physicochemical characteristics of the surrounding environment, and the accessibility of microorganisms are some of the elements that affect the efficiency of bioremediation [1, 28, 35–40]. The absence of contact between pollutants and bacteria affects the pace of deterioration. Furthermore, the distribution of contaminants and germs in the environment is nonuniform. The control and optimization of bioremediation processes are complex and require different parameters [42]. These parameters include the degree to which the pollutants are accessible to the microbial community, the availability of oxygen or other substitute electron acceptors in the environment, and the presence of a microbial population capable of decomposing the pollutants [34].

Today, the use of sorbents is crucial in preventing or minimizing pollution and the effects it has on the environment [43]. Different kinds of sorbents are utilized worldwide to clean up various contaminated sites. Sorbents are used in sorption processes, as the name suggests. Certain compounds can be separated from other liquids or gases using these materials [27, 44]. Their solid surface's ability to bind molecules to its porous nature is what makes them valuable. They are a desirable alternative that is used by the petroleum sector to remove spills from the environment because of their adaptable cleaning capabilities [42]. Theoretically, sorption technology can be applied to any type of surface, including roadways, coastlines, and even vapor pollution. In addition to being safe and easy to use, this method's cost-effectiveness adds to its attractiveness [42]. The sorbent method's ability to create a barrier that acts as a protective embankment to stop future spills is another advantage for the environment [45]. These materials can be recycled, disposed of, or rehabilitated when they are no longer needed.

10.3 Environmental factors affecting microbial bioremediation

The interactions that take place during the bioremediation process are determined by the target pollutants' physicochemical qualities and the metabolic traits of the microorganisms [46]. However, the environmental circumstances of the interaction site determine whether or not the two interact well [47, 48]. Although biodegradation can take place at a variety of pH values, it functions best at pH values between 6.5 and 8.5 within most aquatic and terrestrial systems. Moisture impacts the rate of pollutant metabolism in both terrestrial and aquatic systems because it modifies the osmotic pressure, pH, and the kinds and amounts of soluble elements that are accessible [49]. The majority of the environmental parameters are listed in the following sections.

10.3.1 The availability of nutrients

The rate and efficiency of biodegradation are both impacted by increasing the nutrient level, which also modifies the critical nutritional balance required for microbial growth and reproduction [46]. Microorganisms require a variety of nutrients, including phosphorus, carbon, and nitrogen in order to live and carry out their microbiological functions. The pace at which hydrocarbons break down also approaches a limit at low concentrations. In cold conditions, the addition of sufficient nutrients can speed up the rate of biodegradation by increasing the metabolic activity of microorganisms [50, 51]. Biodegradation in aquatic environments is restricted by the availability of nutrients. Oil-consuming bacteria, like other organisms, depend on resources for their best growth and maturation. While certain nutrients are present in nature, they are present only in small quantities [52, 53].

10.3.2 Temperature

Temperature is the primary physical element affecting the ability of microorganisms to survive [54]. In freezing climates such as the Arctic, natural mechanisms break down pollutants very slowly, making it more difficult for microbes to clean up spilled oil. In these regions, most oleophilic bacteria become metabolically dormant due to the subzero water temperature, which can obstruct transport channels within microbial cells or even cause the cytoplasm to freeze [53, 55].

Temperature-dependent biological enzymes participating in the breakdown process have different optimum temperatures and go through different metabolic turnovers [46]. Furthermore, the degradation process of a given molecule requires a specific temperature. Microorganisms' physiological traits are also influenced by temperature, and this might cause them to expand or contract throughout the bioremediation process. The rate of microbial activity rises with temperature, reaching its peak at the optimal temperature. It began to diminish abruptly when the temperature increased or decreased, and ultimately stopped after reaching a certain point [46].

10.3.3 Oxygen concentration

Certain organisms require more oxygen than others; those that require more oxygen have the ability to accelerate the biodegradation process. Although most living things require oxygen, biological degradation occurs in both aerobic and anaerobic settings. Hydrocarbon metabolism frequently benefits from oxygen presence [53].

10.3.4 pH

The (metabolic) activities of bacteria as well as the rate at which elimination occurs are both influenced by a substance's pH. The pH level in the soil may indicate that microbial growth is possible [56]. Because metabolic systems are extremely sensitive to even little pH fluctuations, adverse outcomes are induced by higher or lower pH levels [57]

10.3.5 Site characterization and selection

In order to effectively estimate the extent and severity of pollution, enough remedial investigation work must be finished before offering a bioremediation option. This endeavor ought to, at a minimum, comprise the following components: calculating the contamination's horizontal and vertical extent with precision, listing the characteristics and sample sites along with the rationale for each, and delineating the steps that must be taken to collect and analyze samples [46].

10.3.6 Toxic compounds and metal ions

Certain pollutants have toxic properties that can harm microorganisms in high quantities and impede the process of decontamination. The concentration of specific toxicants, the microorganisms exposed to them, and their degree of toxicity all influence these factors. Certain inorganic and organic substances are harmful to specific types of life [47]. The breakdown process is also influenced by the concentrations of metal ions. While bacteria and fungi require small levels of metals for survival, excessive concentrations of metals prevent the metabolisms of organisms from functioning. Metal compounds influence the rate of deterioration both directly and indirectly [47].

10.4 The advantages and limitations of bioremediation

Microbial bioremediation has several benefits, one of which is that it is generally accepted to be a natural process [37]. It is typically a less expensive technology than alternative methods for cleaning up hazardous materials [34, 37]. The lengthy treatment timescale of bioremediation; its potential ineffectiveness for certain contaminants; the possibility that some byproducts of biodegradation are more hazardous or persistent than the original compound; the fact that biological processes are specific to microbial populations, pollutants, and environmental limitations; and the need for specialized knowledge during design and implementation are some of its notable drawbacks [46].

The nature of the organisms is another significant limitation. The removal of contaminants from the environment by organisms is not philanthropic but rather an act of survival. Most bioremediation organisms work best in environments that are conducive to their requirements. Therefore, to enable the organisms to break down or absorb the pollution at a tolerable pace, some kind of environmental alteration is required [46]. Often, an organism needs to be gradually exposed to low concentrations of the contaminant. As a result, the organism is prompted to create the metabolic pathways required for the pollutant's digestion. In most cases, it is necessary to add fertilizer or oxygen to the contaminated substance when utilizing bacteria or fungi. When done *in situ*, this may cause disturbances to other organisms [46].

10.5 Features of the magnetic sorbents used in environmental remediation

The characteristics of magnetic materials are influenced by the external magnetic field applied to them. Determining the magnetic orientation of a material aids in categorizing different forms of magnetism observed in natural substances. Five primary types of magnetism can be identified: antiferromagnetism, diamagnetism, ferrimagnetism, ferromagnetism, and paramagnetism [58]. Diamagnetism is an inherent characteristic of a material, in which the magnetization is relatively minimal and takes place in the opposite direction to that of the induced magnetic field. Many materials exhibit paramagnetic properties, in which the orbital momentum increases from zero. Ferromagnetism pertains to materials that possess their own natural magnetic alignment and can develop spontaneous magnetization even without an external magnetic field. Ferrimagnetism is characterized by distinct atoms that have different magnetic moments, which distinguishes it from ferromagnetism. However, both exhibit an organized state below a specific critical temperature. Susceptibility to magnetic compounds permits the classification of diamagnets, paramagnets, and ferromagnets. Materials with high surface-to-volume ratios frequently have distinctive magnetic characteristics [59, 60].

Size-dependent fluctuations in saturation magnetization (M_s) occur until the magnetization reaches a critical threshold, after which it stabilizes and approaches the value of the bulk material. Below this essential threshold, researchers have found a linear relationship between M_s and particle size in a variety of sectors. Spherical nanoparticles have greater magnetic anisotropy than cubic nanoparticles of similar volume [61–64]. A material's composition is often considered the primary factor in determining its specific magnetic characteristics. These characteristics can be seen in magnetic materials regardless of whether metal ions or atoms have unpaired valence electrons [65, 66]. When every magnetic dipole of a magnetic sorbent is aligned under an external magnetic field, the maximum magnetization that occurs is known as the saturation magnetization. Characteristic locations on the magnetization curve for ferromagnetic materials are shown in figure 10.3. These locations correspond to the remanent magnetization (M_r), the coercivity (H), and the saturation magnetization (M_s) [67].

Superparamagnetic materials respond to an external field by following a sigmoidal curve, but they do not exhibit hysteresis (green line), in contrast to ferromagnetic materials (red loop), which are hysteretic [67]. The design also displays the responses of diamagnetic (black line) and paramagnetic (blue line) materials. The M_s in figure 10.3 is temperature dependent and reaches its maximum at 0 K where there is a decrease in thermal vibration and a consequent decrease in the randomization of aligned moments. The superparamagnetic behavior of both ferromagnetic and ferrimagnetic materials occurs above a temperature known as the blocking temperature (T_B). This activity is characterized by rapid random reversals of magnetization that result in a zero time-average magnetic moment [67].

Recently, magnetic materials such as biosorbents have demonstrated notable progress in the field of environmental bioremediation. To achieve bioremediation through sorption, the target material is isolated from the liquid solution through the utilization of an external magnetic field. This is done by introducing functionalized magnetic materials (FMMs) into the sample and allowing these

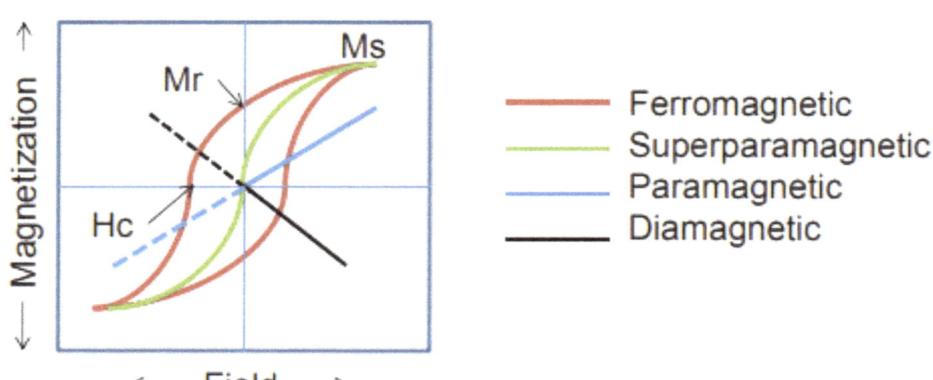

Figure 10.3. The magnetic behaviors of various types of materials subjected to an external field. Reproduced from [67]. CC BY 4.0.

particles to absorb the analyte onto their surface. The FMMs possess the ability to disperse again in a solution once the external magnetic field is no longer present [68, 69]. Moreover, FMMs have the potential to enhance mass transfer by expanding the contact surface area between the analyte and the solution [70]. Among FMMs, various iron oxides are commonly selected because of their advantageous qualities. Among these is their ease of large-scale preparation, which results from an easy-to-follow and effective procedure. Iron oxides also have many functional groups on their surface, which makes alterations easy. Their superparamagnetic characteristics make them extremely practical, since they do not require filtration or centrifugation to function. These materials also show good dispensability in aqueous solutions, which speeds up the attainment of extraction equilibrium and allows for easy recovery and reuse [70]. Because magnetic sorbents work well in a variety of fields, including science, technology, and engineering, their use has grown in popularity. In addition, they are increasingly being used for trace-metal analysis. Utilizing micro- or nanomagnetic particles, sophisticated techniques for determining and removing heavy metal ions from polluted samples have become increasingly prevalent in recent years [68, 71].

10.6 The application of magnetic sorbents in microbial bioremediation

This section and the rest of this chapter will provide an overview of recent research into the usage of magnetic substances such as zero-valent iron, iron oxides, and spinel ferrites in environmental remediation.

10.6.1 The elimination of organic substances

Once discharged into the aquatic environment, organic compounds can lead to a variety of environmental issues by negatively impacting aquatic organisms and various life forms. These pollutants encompass a range of substances such as hydrocarbons, oils, pesticides, phenols, dyes, biphenyls, fertilizers, detergents, plasticizers, and pharmaceuticals [72–74]. There is now compelling evidence indicating the presence of hazardous organic contaminants in drinking water sources, which poses significant challenges to the ecosystem. Various types of magnetic sorbents have been created to adsorb these contaminants from the environment. Chitosan is an abundant natural biopolymer derived from chitin hydrolysis; it has considerable sorption capacity due to its reactive amino and hydroxyl groups [75].

Indigotine blue dye (IBD) was adsorbed using a recently created chitosan complex with cobalt ferrite ($CoFe_2O_4$). The adsorbent was produced using coprecipitation to crosslink the chitosan polymer chains around $CoFe_2O_4$. The optimal parameters for the adsorption of IBD by the chitosan/cobalt ferrite complex were found to be a pH of 3 and an adsorbent dosage of 0.75 g l^{-1}. In less than 15 min, the process quickly reached equilibrium, and the pseudo first-order model adequately captured its characteristics. The maximal adsorption capacity of 380.88 mg g^{-1} at

328 K was effectively modeled using the Langmuir equations. Furthermore, it was discovered that the adsorption was endothermic, favorable, and spontaneous [76].

Singh *et al* [77] applied the *in situ* mineralization method to prepare chitosan–graphene oxide hydrogel nanocomposites with iron oxide. The preparation method was found to have a direct effect on the physical characteristics and organization of the complex, producing a porous structure with a greater surface area. This improved structure made it easier for the adsorbate to diffuse quickly. The ability of the produced composite materials to remove methylene blue, a cationic dye, from a water-based solution was tested. The effect of the solution's pH and ionic strength on the adsorption process suggested that an electrostatic interaction took place between the methylene blue molecules and the adsorbent. This electrostatic interaction seemed to be the main driver of the adsorption effectiveness, even though the nanocomposite had a comparatively small surface area. The Freundlich adsorption model provided a good representation of the equilibrium capacity, and the adsorption kinetics fitted a pseudo second-order model.

The biomass of *Astrocaryum aculeatum* was converted into magnetic activated carbon using a single-step pyrolysis method that included zinc and nickel chloride [78]. Nicotinamide and propranolol removal from a composite were examined. *A. aculeatum* seeds were combined with metal salts in a single step and then put in a quartz reactor under regulated conditions. Furthermore, the pore size increased, leading to a larger mesoporous area [78]. The (maximum) adsorption capacities of the activated carbon were found to be 199.3 for nicotinamide and 335.4 for propranolol. The adsorption was a good fit for the Liu isotherm model. The authors claimed that pore filling, which occurred when π-π interactions took place between the aromatic rings of the absorbed molecules propranolol and nicotinamide, was the primary mechanism responsible for the process's effectiveness [78].

Polycyclic aromatic hydrocarbons or PAHs are another group of organic pollutants that are associated with genetic, carcinogenic, and developmental effects [10]. It may take many years or even decades for their adverse effects on both land- and water-based creatures to completely heal. Thus, the potential for PAH cleanup is greatly increased by combining technology and microbial remediation while preventing the release of harmful byproducts such as carbon monoxide or inhibitory metabolites, which impair cellular processes and prevent other PAHs in the system from breaking down [79]

It has been shown that a biofunctionalized silica complex effectively adsorbed PAHs during the remediation of a benzo(a)pyrene-polluted soil. The sorbent consisted of lipid-coated silica materials and bacterial bilayers [80]. In addition, studies have shown that the presence of iron nanoparticles can lead to a significant increase in biodegradation efficiency to 79% within a span of 15 days for *Candida tropicalis* NN4 when dealing with indeno(1,2,3-cd) pyrene contamination—a complex characterized by six fused benzene rings [81].

Similarly, zinc oxide nanoparticles and biosurfactants were utilized to degrade some organic yeast substances, achieving a rapid degradation rate of 82.67% within six days [82]. In addition, following a six-day period, the introduction of zinc oxide nanoparticles enhanced the degradation of benzene rings in benzo(ghi)perylene by yeast by as much as 63.8% [79]. In comparison to earlier studies, strains of the yeast group with a zinc oxide complex have been recognized as a promising option for breaking down high concentrations of organic substances in just six days. In addition, it has been noted that magnetic materials not only enhance the remediation process but also aid in effectively isolating bacteria capable of degrading PAHs, such as *Pseudomonas* and *Sphingobium* sp., from polluted sites [83]. An electrokinetic system proved to be more effective in remediating PAH-contaminated soil when the electrode surface was enhanced by the addition of magnetite nanoparticles [84].

Significant progress has been made in the immobilization of bacterial and fungal enzymes in nanoscale materials, leading to effective solutions for environmental cleanup without causing harm to the environment [85]. This involves extracting specific enzymes from native microorganisms' cells and immobilizing them with nanoparticles, demonstrating the potential for the bioremediation of harmful PAHs through enzymatic conversion. Many enzymes, including lipase, laccase, and catalase, have been researched for immobilization in different metallic materials in an effort to successfully eliminate PAHs. Even in harsh environments with varying pH values and temperature ranges and long storage times, these enzymes show stability [86, 87].

There is evidence that nanoscale zero-valent iron (nZVI) can efficiently aid in the degradation of environmental contaminants [88, 89]. Zero-valent iron nanoparticles are frequently employed because of their large size and potent reactivity with pollutants such as PAHs. Bacterial strains have improved PAH accessibility, raised bacterial biomass, and accelerated the breakdown of several PAHs (pyrene and anthracene) by 85% [90]. Generally, magnetic sorbents are promising options for the treatment of PAHs in natural environments [91]. Zinc oxide- and iron-based magnetic sorbents are actively produced by microorganisms, especially bacteria such as *Geobacillus stearothermophilus*, *Pseudomonas aeruginosa*, and *Bacillus subtilis*, for the purpose of remediating persistent contaminants in the environment [92–94].

10.6.2 The elimination of inorganic substances

Several authors have explored the ability of magnetic sorbents to capture toxic HMs. These HMs are considered emerging pollutants due to their well-documented toxicity and high risk of human exposure [95]. A variety of magnetic biomaterials have been created for the efficient removal of lead ions from water under a variety of operating conditions. [95–99]. Extended exposure to lead ions can have a detrimental effect on human health because of increased oxidative stress, which can harm the kidneys, reproductive system, hematological system, and central nervous system [100].

A recent study discussed a chitosan-modified magnetic material with significantly improved sorption capacity that was used to remove lead ions and cadmium ions. Isotherm investigations revealed that both metals exhibited optimal performance within a pH range of 6–7, aligning well with the Langmuir isotherm [101]. The adsorption process was endothermic and thermodynamically advantageous.

The increased removal effectiveness was ascribed to the biopolymer coating of the nanomaterials, which produced more amino functional groups for improved absorption [96, 101].

Moreover, it has been demonstrated that copper and lead may be successfully removed from the environment by the application of magnetic hydrogel beads containing gum tragacanth. The beads' maximum adsorption occurred at pH 6. The presence of several chelating groups, including amino, hydroxyl, sulfonic, and carboxylic acids, was associated with a considerable removal capability [102]. Yuwei and Jianlong [103] developed a magnetic chitosan nanobiosorbent containing superparamagnetic particles ranging from 8 to 40 nm. Their study demonstrated that the biosorbent reached its maximum adsorption capacity for copper ions of 35.5 mg g^{-1} at 35 °C. Furthermore, it was found that the biosorbent could be effectively separated using a magnetic field while maintaining strong chelation of metal ions [103].

Fungal substances are a notable type of biosorbent that has proven to be very efficient and relevant in various research investigations. For instance, Contreras-Cortés *et al* [104] demonstrated the adsorption of copper ions using cross-linked composite beads containing a copper-tolerant biomass from *Aspergillus australensis*, achieving an effectiveness rate of 79% [105].

Due to their low cost, wide availability, ease of preparation, ease of manipulation, and great effectiveness, magnetic biosorbents derived from waste materials have piqued the interest of researchers [106]. In 2021, Franco *et al* [107] investigated the removal of gentian violet color by *Tipuana tipu* and *Inga marginata*. Each of these pods underwent separate cleaning and drying at 60 °C. Because carbonyl and hydroxyl groups were present on the surfaces of the *I. marginata* and *T. tipu* biosorbents, these materials absorbed cationic dye molecules at 77.65% and 68.71%, respectively [107]. Southichak *et al* [108] investigated the removal of nickel salts using cellulose and lignin. They added sodium hydroxide and adjusted the pH of the raw sorbent before drying it at 90 °C. Following these treatments, the sorbent's efficacy when exposed to a concentration of nickel rose dramatically from just 14% for the untreated material to 81% for the treated one. A minor rise in the biosorbent's negative zeta potential, which led to improved positively charged nickel ion adsorption, was suggested as the cause of this improvement [108].

Scientists have been using yeast for the production of biological sorbents over the years. Using yeast, do Nascimento *et al* [109] were able to extract 76% of copper ions in their investigation. The same yeast was used by De Rossi *et al* [110] to extract hexavalent chromium ions from actual wastewater. About 95% of the yeast proved to be efficient when enclosed in alginate beads. After rinsing the yeast with water, a biosorbent was created by mixing it with a sodium alginate solution in a phosphate buffer for ten minutes. This combination was then added to a calcium chloride solution for two hours, allowing the yeast to form capsules [110]. After that, the capsules were cleaned and baked at 50 °C for 24 h. The surface pores of the dried alginate beads were crucial for achieving significant removal efficiency [110].

Yeast was employed by Yu *et al* [111] to absorb copper and lead ions. There were two primary parts in the preparation process: first, the yeast was mixed with a glutaraldehyde solution and left to sit at room temperature for a whole day.

Once that was done, it was cleaned and freeze-dried. The following phase involved mixing ethylenediaminetetraacetic dianhydride with a solution of N, N-dimethylacetamide that contained yeast cross-linked biomass from the previous stage. After the combination was agitated for four hours at 60 °C, the unreacted reagent was removed by freeze-drying the mixture. By complexing ions with ethylenediaminetetraacetic acid, this procedure enhanced the modified biosorbent's active adsorption sites and resulted in a removal efficiency of more than 90% for both metals [111].

10.7 Conclusions

Magnetic sorbents have arisen as a viable substitute for conventional sorbents in the removal of environmental contaminants. These biosorbents offer two key benefits: magnetic retrieval and tailored attraction to various pollutants. It is essential to ensure that the surface structure of these materials meets the requirements for creating durable and high-absorption-capacity biosorbents that can be reused. Magnetic bio-based absorbents can be customized or created to possess specific properties that are advantageous for eliminating targeted pollutants. The chemical composition and functional groups of magnetic adsorbents enhance their adsorption capacity, which requires further exploration. Limited research has been conducted into the practical application of magnetic sorbents, indicating a need for additional investigation before they can be brought into commercial use. Future studies should focus on creating a more efficient operational process for the bioremediation of organic and inorganic pollutants. Close collaboration between environmental scientists and engineers is vital for the successful implementation of this approach.

References and further reading

[1] Ayilara M S and Babalola O O 2023 Bioremediation of environmental wastes: the role of microorganisms *Front. Agron.* **5** 1183691

[2] Oliomogbe T I, Emegha J O and Ukhurebor K E 2023 Microorganism derived biosorbent in the sequestration of contaminants from the soil *Adsorption Applications for Environmental Sustainability* (Bristol: IOP Publishing)

[3] Ukhurebor K E, Aigbe U O, Onyanche R B *et al* 2021 Developments, utilization and applications of nanobiosensors for environmental sustainability and safety *Bionanomaterials for Environmental and Agricultural Applications* (Bristol: IOP Publishing)

[4] Ukhurebor K E, Aigbe U O, Onyanche R B *et al* 2023 Introduction to the state of the art and relevant aspects of the applications of adsorption for environmental safety and sustainability *Adsorption Applications for Environmental Sustainability* (Bristol: IOP Publishing)

[5] Methneni N, Morales-Gonzalez J A, Jaziri A, Mansour H B and Fernandez-Serrano M 2021 Persistent organic and inorganic pollutants in the effluents from the textile dyeing industries: ecotoxicology appraisal via a battery of biotests *Environ. Res.* **196** 110956

[6] Egboduku W O, Egboduku T, Emegha J O and Imarhiagbe O 2023 The use of biosorbents derived from invasive plants for environmental remediation *Adsorption Applications for Environmental Sustainability* (Bristol: IOP Publishing) pp 9.1–9.15

[7] Ukhurebor K E, Aigbe U O, Olayinka A S, Nwankwo W and Emegha J O 2020 Climatic change and pesticides usage: a brief review of their implicative relationship *AU-eJIR* **5** 44–9

[8] Prabagar S, Dharmadasa R M, Lintha A, Thuraisingam S and Prabagar J 2021 Accumulation of heavy metals in grape fruit, leaves, soil and water: a study of influential factors and evaluating ecological risks in Jaffna, Sri Lanka *Environ. Sustain. Indic.* **12** 100147

[9] AL-Huqail A A, Kumar P, Eid E M, Adelodun B, Abou Fayssal S, Singh J *et al* 2022 Risk assessment of heavy metals contamination in soil and two rice (*Oryza sativa* L.) varieties irrigated with paper mill effluent *Agriculture* **12** 1864

[10] Emegha J O, Oliomogbe T I, Okpoghono J, Babalola A V, Ejelonu C A, Dennis Eyetan Elete D E and Ukhurebor K E 2023 Green biosorbents for the degradation of petroleum contaminants *Adsorption Applications for Environmental Sustainability* (Bristol: IOP Publishing)

[11] Ukhurebor K E, Aigbe U O, Onyanche R B *et al* 2023 The challenges of, and perspectives on adsorption applications for environmental sustainability *Adsorption Applications for Environmental Sustainability* (Bristol: IOP Publishing)

[12] Selvi A, Rajasekar A, Theerthagiri J, Ananthaselvam A, Sathishkumar K, Madhavan J and Rahman P K S M 2019 Integrated remediation processes toward heavy metal removal/recovery from various environments-a review *Front. Environ. Sci.* **7** 66

[13] Zaied B K, Rashid M, Nasrullah M *et al* 2020 A comprehensive review on contaminants removal from pharmaceutical wastewater by electrocoagulation process *Sci. Total Environ.* **726** 138095

[14] Rivera-Utrilla J, Sánchez-Polo M, Ferro-García M A, Prados-Joya G and Ocampo-Pérez R 2013 Pharmaceuticals as emerging contaminants and their removal from water: a review. *Chemosphere* **93** 1268–87

[15] Pham V H T, Kim J, Chang S and Chung W 2022 Bacterial biosorbents, an efficient heavy metals green clean-up strategy: prospects, challenges, and opportunities *Microorganisms* **10** 610

[16] Mukhopadhyay S and Maiti S 2010 Phytoremediation of metal mine waste *Appl. Ecol. Environ. Res.* **8** 207–22 https://www.aloki.hu/indvol08_3.htm

[17] Pansambal S, Roy A, Mohamed H E A *et al* 2022 Recent developments on magnetically separable ferrite-based nanomaterials for removal of environmental pollutants *Hindawi J. Nanomater.* **2022** 8560069

[18] Fashola M O, Ngole-Jeme V M and Babalola O O 2020 Heavy metal immobilization potential of indigenous bacteria isolated from gold mine tailings *Int. J. Environ. Res.* **14** 71–86

[19] Zwolak A, Sarzyńska M, Szpyrka E and Stawarczyk K 2019 Sources of soil pollution by heavy metals and their accumulation in vegetables: a review *Water Air Soil Pollut.* **230** 1–9

[20] Mehta D, Mazumdar S and Singh S K 2015 Magnetic adsorbents for the treatment of water/wastewater—a review *JWPE* **7** 244–65

[21] Akchiche Z, Abba A B and Saggai S 2021 Magnetic nanoparticles for the removal of heavy metals from industrial wastewater: review *Algerian J. Chem. Eng.* **01** 08–15

[22] Dua M, Singh A, Sethunathan N *et al* 2002 Biotechnology and bioremediation: successes and limitations *Appl. Microbiol. Biotechnol.* **59** 143–52

[23] Boopathy R 2000 Factors limiting bioremediation technologies *Bioresour. Technol.* **74** 63–7

[24] Kumar A, Bisht B S, Joshi V D and Dhewa T 2011 Review on bioremediation of polluted environment: a management tool *Int. J. Environ. Sci.* **1** 1079 https://www.indianjournals.com/ijor.aspx?target=ijor:ijes&volume=1&issue=6&article=004&type=pdf

[25] Azubuike C C, Chikere C B and Okpokwasili G C 2016 Bioremediation techniques—classification based on site of application: principles, advantages, limitations and prospects *World J. Microbiol. Biotechnol.* **32** 180
[26] Mandeep and Shukla P 2020 Microbial nanotechnology for bioremediation of industrial wastewater *Front. Microbiol.* **11** 590631
[27] Tripathi M, Singh P, Singh R, Bala S, Pathak N, Singh S, Chauhan R S and Singh P K 2023 Microbial biosorbent for remediation of dyes and heavy metals pollution: a green strategy for sustainable environment *Front. Microbiol.* **14** 1168954
[28] Liu S H, Zeng G M, Niu Q Y, Liu Y, Zhou L, Jiang L H *et al* 2017 Bioremediation mechanisms of combined pollution of PAHs and heavy metals by bacteria and fungi: a mini review *Bioresour. Technol.* **224** 25–33
[29] Gupta A, Joia J, Sood A, Sood R, Sidhu C *et al* 2016 Microbes as potential tool for remediation of heavy metals: a review *J. Microb. Biochem. Technol.* **8** 364–72
[30] Qiu, G, Y-l, Li, Zhao, K and Thiobacillus 2006 Thioparus immobilized by magnetic porous beads: preparation and characteristic *Enzyme Microb. Technol.* **39** 770–7
[31] Gómez-Pastora J, Bringas E and Ortiz I 2014 Recent progress and future challenges on the use of high performance magnetic nano-adsorbents in environmental applications *J. Chem. Eng* **256** 187–204
[32] Gargouri B, Karray F, Mhiri N, Aloui F and Sayadi S 2011 Application of a continuously stirred tank bioreactor (CSTR) for bioremediation of hydrocarbon-rich industrial wastewater effluents *J. Hazard. Mater.* **189** 427–34
[33] Srivastava J, Naraian R, Kalra S J and Chandra H 2014 Advances in microbial bioremediation and the factors influencing the process *Int. J. Eng. Sci. Technol.* **11** 1787–800
[34] Sharma S 2012 Bioremediation: features, strategies and applications *Asian J. Pharm. Life Sci.* **2** 202–13
[35] Quintero J C, Lu-Chau T A, Moreira M T, Feijoo G and Lema J M 2007 Bioremediation of HCH present in soil by the whiterot fungus *Bjerkandera adusta* in a slurry batch bioreactor *Int. Biodeterior. Biodegrad.* **60** 319–26
[36] Pino-Herrera D O *et al* 2017 Removal mechanisms in aerobic slurry bioreactors for remediation of soils and sediments polluted with hydrophobic organic compounds: an overview *J. Hazard. Mater.* **339** 427–49
[37] Chikere C B, Chikere B O and Okpokwasili G C 2012 Bioreactor-based bioremediation of hydrocarbonpolluted Niger Delta marine sediment, Nigeria *Biotech.* **2** 53–66
[38] Zeyaullah M, Atif M, Islam B, Abdelkafe A S *et al* 2009 Bioremediation: a tool for environmental cleaning *Afr. J. Microbiol. Res.* **3** 310–4
[39] Glazer A N and Nikaido H 1995 *Microbial Biotechnology.* (New York: W.H.Freeman)
[40] Validi M 2001 Bioremediation. An overview *Pure Appl. Chem.* **73** 1163–72
[41] Ruldolph F B and McIntire L V 1996 *Biotechnology: Science, Engineering and Ethical Challenges for the 21st Century* (Washington, DC: Joseph Henry Press)
[42] Emegha J O, Oliomogbe T I and Babalola A V 2023 Magnetic nanomaterials-based sensors for the detection and monitoring of toxic gases *Magnetic Nanomaterials. Engineering Materials* ed U O Aigbe, K E Ukhurebor and R B Onyancha (Cham: Springer)
[43] Singh N B, Nagpal G, Agrawal S and Rachna 2018 Water purification by using adsorbents: a review *Environ. Technol. Innov.* **11** 187–240
[44] Hyung Min Choi H M and Cloud R M 1992 Natural sorbents in oil spill cleanup *Environ. Sci. Technol.* **26** 772–6

[45] Haque E, Won Jun J and Jhung S H 2011 Adsorptive removal of methyl orange and methylene blue from aqueous solution with a metal-organic framework material, iron terephthalate (MOF-235) *J. Hazard. Mater.* **185** 507–11
[46] Abatenh E, Gizaw B, Tsegaye Z and Wassie M 2017 The role of microorganisms in bioremediation—a review *Open J Environ Biol* **2** 038–46
[47] Madhavi G N and Mohini D D 2012 Review paper on—parameters affecting bioremediation *Int. J. Life Sci. Pharma Res.* **2** 77–80
[48] Adams G O, Fufeyin P T, Okoro S E and Ehinomen I 2015 Bioremediation, biostimulation and bioaugmention: a review *Int. J. Environ. Bioremediat. Biodegrad* **3** 28–39
[49] Cases I and de Lorenzo V 2005 Genetically modified organisms for the environment: stories of success and failure and what we have learned from them *Int. J. Microbiol.* **8** 213–22
[50] Couto N, Fritt-Rasmussen J, Jensen P E, Højrup M, Rodrigo A P *et al* 2014 Suitability of oil bioremediation in an artic soil using surplus heating from an incineration facility *Environ. Sci. Pollut. Res.* **21** 6221–7
[51] Phulia V, Jamwal A, Saxena N, Chadha N K, Muralidhar *et al* 2013 Technologies in aquatic bioremediation *Freshwater Ecosystem and Xenobiotics* (New Delhi: Discovery Publishing House Pvt Ltd) pp 65–91
[52] Thavasi R, Jayalakshmi S and Banat I M 2011 Application of biosurfactant produced from peanut oil cake by *Lactobacillus delbrueckii* in biodegradation of crude oil *Bioresour. Technol.* **102** 3366–72
[53] Macaulay B M 2014 Understanding the behavior of oil-degrading microorganisms to enhance the microbial remediation of spilled petroleum *Appl. Ecol. Environ. Res.*. **13** 247–62
[54] Das N and Chandran P 2011 Microbial degradation of petroleum hydrocarbon contaminants: an overview *Biotechnol Res Int* **2011** 1–13
[55] Yang S Z, Jin H J, Wei Z, He R X, Ji Y J *et al* 2009 Bioremediation of oil spills in cold environments: a review *Pedosphere* **19** 371–81
[56] Asira E E 2013 Factors that determine bioremediation of organic compounds in the soil *Acad. J. Interdiscip. Stud.* **2** 125–8
[57] Wang Q, Zhang S, Li Y and Klassen W 2011 Potential approaches to improving biodegradation of hydrocarbons for bioremediation of crude oil pollution *Environ. Protect. J.* **2** 47–55
[58] Akbarzadeh A, Samiei M and Davaran S 2012 Magnetic nanoparticles: preparation, physical properties, and applications in biomedicine *Nanoscale Res. Lett.* **7** 1–13
[59] Frey N A, Peng S, Cheng K and Sun S 2009 Magnetic nanoparticles: synthesis, functionalization, and applications in bioimaging and magnetic energy storage *Chem. Soc. Rev.* **38** 2532–42
[60] Singamaneni S, Bliznyuk V N, Binek C and Tsymbal E Y 2011 Magnetic nanoparticles: recent advances in synthesis, self-assembly and applications *J. Mater. Chem.* **21** 16819–45
[61] Noh S H, Na W, Jang J T *et al* 2012 Nanoscale magnetism control via surface and exchange anisotropy for optimized ferrimagnetic hysteresis *Nano Lett.* **12** 3716–21
[62] Song Q and Zhang J Z 2004 Shape control and associated magnetic properties of spinel cobalt ferrite nanocrystals *J. Am. Chem. Soc.* **126** 6164–8
[63] Salgueiro A M, Daniel-da-Silva A L, Girão A V, Pinheiro P C and Trindade T 2013 Unusual dye adsorption behavior of κ-carrageenan coated superparamagnetic nanoparticles *Chem. Eng. J.* **229** 276–84

[64] Salazar-Alvarez G, Qin J, Šepelák V *et al* 2008 Cubic versus spherical magnetic nanoparticles: the role of surface anisotropy *J. Am. Chem. Soc.* **130** 13234–9

[65] Zhen G, Mui W B, Moffat B A *et al* 2011 Comparative study of the magnetic behavior of spherical and cubic superparamagnetic iron oxide nanoparticles *J. Phys. Chem.* C **115** 327–34

[66] Williams A, Moruzzi V L, Gelatt C D, Kübler J and Schwarz K 1982 Aspects of transition-metal magnetism *J. Appl. Phys.* **53** 2019–23

[67] Kolhatkar A G, Jamison A C, Litvinov D, Willson R C and Randall Lee T 2013 . Tuning the magnetic properties of nanoparticles *Int. J. Mol. Sci.* **14** 15977–6009

[68] Kanjilal T and Bhattacharjee C 2018 Green applications of magnetic sorbents for environmental remediation *Organic Pollutants in Wastewater I: Methods of Analysis, Removal and Treatment* (Millersville, PA: Materials Research Forum)

[69] Lakshmanan R 2013 Application of magnetic nanoparticles and reactive filter materials for wastewater treatment *PhD Thesis* Royal Institute of Technology, School of Biotechnology, Stockholm https://www.diva-portal.org/smash/get/diva2:665773/FULLTEXT01.pdf

[70] Li X S, Zhu G T, Luo Y B, Yuan B F and Feng Y Q 2013 Synthesis and applications of functionalized magnetic materials in sample preparation *Trends Anal. Chem.* **45** 233–47

[71] Cavallini M, Bystrenova E, Timko M, Koneracka M, Zavisova V and Kopcansky P 2008 Multiple-length-scale patterning of magnetic nanoparticles by stamp assisted deposition *J. Phys. Condens. Matter* **21** 204144

[72] Ali I, Asim M and Khan T A 2012 Low-cost adsorbents for the removal of organic pollutants from wastewater *J. Environ. Manage.* **113** 170–83

[73] Miranda-García N I, Maldonado M, Coronado J M and Malato S 2010 Degradation study of 15 emerging contaminants at low concentration by immobilized TiO_2 in a pilot plant *Catal. Today* **151** 107–13

[74] Pal A, Gin K Y, Lin A Y-C and Reinhard M 2010 Impacts of emerging organic contaminants on freshwater resources: review of recent occurrences, sources, fate and effects *Sci. Total Environ.* **408** 6062–9

[75] Vieira Y, Lima E C and Dotto G L 2022 Attraction to adsorption: preparation methods and performance of novel magnetic biochars for water and wastewater treatment *Nano-Biosorbents for Decontamination of Water, Air, and Soil Pollution* (Amsterdam: Elsevier)

[76] dos Santos J M N, Pereira C R, Pinto L A A, Frantz T *et al* 2019 Synthesis of a novel CoFe2O4/chitosan magnetic composite for fast adsorption of indigotine blue dye *Carbohydr. Polym.* **217** 6–14

[77] Singh N, Riyajuddin S, Ghosh K, Mehta S K and Dan A 2019 Chitosan-graphene oxide hydrogels with embedded magnetic iron oxide nanoparticles for dye removal *ACS Appl. Nano Mater.* **2019** 7379–92

[78] Thue P S, Umpierres C S, Lima E C *et al* 2020 Single-step pyrolysis for producing magnetic activated carbon from tucumã (*Astrocaryum aculeatum*) seed and nickel(II) chloride and zinc(II) chloride. Application for removal of nicotinamide and propranolol *J. Hazard. Mater* **398** 122903

[79] Rajput V D, Kumari S, Minkina T *et al* 2023 Nano-Enhanced Microbial Remediation of PAHs Contaminated Soil. *Air, Soil Water Res.* **vol 16** 1–7 Air, Soil and Water Research

[80] Wang H, Kim B and Wunder S L 2015 Nanoparticle-supported lipid bilayers as an *in situ* remediation strategy for hydrophobic organic contaminants in soils *Environ. Sci. Technol.* **49** 529–36

[81] Ojha N, Mandal S K and Das N 2019 Enhanced degradation of indeno(1,2,3-CD) pyrene using *Candida tropicalis* NN4 in presence of iron nanoparticles and produced biosurfactant: a statistical approach *3 Biotech.* **9** 1–13
[82] Mandal S K, Ojha N and Das N 2018 Optimization of process parameters for the yeast mediated degradation of benzo[a]pyrene in presence of ZnO nanoparticles and produced biosurfactant using 3-level Box–Behnken design *Ecol. Eng.* **120** 497–503
[83] Li J, Luo C, Zhang G and Zhang D 2018 Coupling magnetic-nanoparticle mediated isolation (MMI) and stable isotope probing (SIP) for identifying and isolating the active microbes involved in phenanthrene degradation in wastewater with higher resolution and accuracy *Water Res.* **144** 226–34
[84] Pourfadakari S, Ahmadi M, Jaafarzadeh N *et al* 2019 Remediation of pahs contaminated soil using a sequence of soil washing with biosurfactant produced by *Pseudomonas aeruginosa* strain PF2 and electrokinetic oxidation of desorbed solution, effect of electrode modification with Fe_3O_4 nanoparticles *J. Hazard. Mater.* **379** 120839
[85] Gao Y *et al* 2022 Immobilized fungal enzymes: innovations and potential applications in biodegradation and biosynthesis *Biotechnol. Adv.* **57** 107936
[86] Hou J, Dong G, Ye Y and Chen V 2014 Laccase immobilization on titania nanoparticles and titania-functionalized membranes *J. Membr. Sci.* **452** 229–40
[87] Acevedo F *et al* 2010 Degradation of polycyclic aromatic hydrocarbons by free and nanoclay-immobilized manganese peroxidase from *Anthracophyllum discolor Chemosphere* **80** 271–8
[88] Benjamin S R, Lima F D, Florean E T and Guedes M I F 2019 Current trends in nanotechnology for bioremediation *Int. J. Environ. Pollut.* **66** 19–40
[89] Galdames A, Mendoza A *et al* 2017 Development of new remediation technologies for contaminated soils based on the application of zero-valent iron nanoparticles and bioremediation with compost *Resource-Efficient Technol.* **3** 166–76
[90] Parthipan P, Cheng L, Dhandapani P, Elumalai P, Huang M and Rajasekar A 2022 Impact of biosurfactant and iron nanoparticles on biodegradation of polyaromatic hydrocarbons (PAHS) *Environ. Pollut.* **306** 119384
[91] Binh N D *et al* 2016 Sequential anaerobic-aerobic biodegradation of 2,3,7,8-TCDD contaminated soil in the presence of CMC-coated nZVI and surfactant *Environ. Pollut.* **37** 388–98
[92] Mukherjee P *et al* 2002 Extracellular synthesis of gold nanoparticles by the fungus *Fusarium oxysporum ChemBioChem.* **3** 461–3
[93] Singh B N, Rawat K, Khan W, Naqvi A H and Singh B R 2014 Biosynthesis of stable antioxidant ZnO nanoparticles by *Pseudomonas aeruginosa* rhamnolipids *PLoS One* **9** e106937
[94] Sundaram P A, Augustine R and Kannan M 2012 Extracellular biosynthesis of iron oxide nanoparticles by *Bacillus subtilis* strains isolated from rhizosphere soil *Biotechnol. Bioprocess. Eng.* **17** 835–40
[95] Bashir A *et al* 2019 Removal of heavy metal ions from aqueous system by ion-exchange and biosorption methods *Environ. Chem. Lett.* **17** 729–54
[96] Ghotekar S *et al* 2022 Magnetic nanomaterials-based biosorbents *Micro and Nano Technologies, Nano-Biosorbents for Decontamination of Water, Air, and Soil Pollution* (Amsterdam: Elsevier) pp 605–14

[97] Luo X *et al* 2016 Adsorptive removal of lead from water by the effective and reusable magnetic cellulose nanocomposite beads entrapping activated bentonite *Carbohydr. Polym.* **151** 640–8
[98] Wang Y, Li L, Luo C, Wang X and Duan H 2016 Removal of Pb^{2+} from water environment using a novel magnetic chitosan/graphene oxide imprinted Pb^{2+} *Int. J. Biol. Macromol.* **86** 505–11
[99] Luo X, Zeng J, Liu S and Zhang L 2015 An effective and recyclable adsorbent for the removal of heavy metal ions from aqueous system: magnetic chitosan/cellulose microspheres *Bioresource Technol.* **194** 403–6
[100] Flora G, Gupta D and Tiwari A 2012 Toxicity of lead: a review with recent updates *Interdiscip. Toxicol* **5** 47–58
[101] Li B *et al* 2017 Environmentally friendly chitosan/PEI-grafted magnetic gelatin for the highly effective removal of heavy metals from drinking water *Sci. Rep.* **7** 1–9
[102] Sahraei R, Pour Z S and Ghaemy M 2017 Novel magnetic bio-sorbent hydrogel beads based on modified gum tragacanth/graphene oxide: removal of heavy metals and dyes from water *J. Clean. Prod.* **142** 2973–84
[103] Yuwei C and Jianlong W 2011 Preparation and characterization of magnetic chitosan nanoparticles and their application for Cu(II) removal *Chem. Eng. J.* **168** 286–92
[104] Contreras-Cortés A G *et al* 2019 Toxicological assessment of crosslinked beads of chitosan-alginate and *Aspergillus australensis* biomass, with efficiency as biosorbent for copper removal *Polymers.* **11** 222
[105] Osman A L, Abd El-Monaem E M and Elgarahy A M 2023 Methods to prepare biosorbents and magnetic sorbents for water treatment: a review *Environ. Chem. Lett.* **21** 2337–98
[106] Gemici B T *et al* 2021 Removal of methylene blue onto forest wastes: adsorption isotherms, kinetics and thermodynamic analysis *Environ. Technol. Innov* **22** 101501
[107] Franco D S P *et al* 2021 Conversion of the forest species *Inga marginata* and *Tipuana tipu* wastes into biosorbents: dye biosorption study from isotherm to mass transfer *Environ. Technol. Innov.* **22** 101521
[108] Southichak B *et al* 2006 *Phragmites australis*: a novel biosorbent for the removal of heavy metals from aqueous solution *Water Res.* **40** 2295–302
[109] do Nascimento J M *et al* 2019 Biosorption Cu (II) by the yeast *Saccharomyces cerevisiae Biotechnol. Rep.* **21** e00315
[110] De Rossi A *et al* 2020 Synthesis, characterization, and application of *Saccharomyces cerevisiae*/alginate composites beads for adsorption of heavy metals *J. Environ. Chem. Eng.* **8** 104009
[111] Yu J *et al* 2008 Enhanced and selective adsorption of Pb^{2+} and Cu^{2+} by EDTAD-modified biomass of Baker's yeast *Bioresour. Technol.* **99** 2588–93
[112] El-Sheshtawy H S and Ahmed W 2017 Bioremediation of crude oil by *Bacillus licheniformis* in the presence of different concentration nanoparticles and produced biosurfactant *Int. J. Eng. Sci. Technol.* **14** 1603–14

www.ingramcontent.com/pod-product-compliance
Ingram Content Group UK Ltd.
Pitfield, Milton Keynes, MK11 3LW, UK
UKHW051857100926
8556IPUK00005B/661

9 780750 359078